Magnetic Nanomaterials

Applications in Catalysis and Life Sciences

Smart Materials

Series editors:
Hans-Jörg Schneider, *Saarland University, Germany*
Mohsen Shahinpoor, *University of Maine, USA*

Titles in this series:

1: Janus Particle Synthesis, Self-Assembly and Applications
2: Smart Materials for Drug Delivery: Volume 1
3: Smart Materials for Drug Delivery: Volume 2
4: Materials Design Inspired by Nature
5: Responsive Photonic Nanostructures: Smart Nanoscale Optical Materials
6: Magnetorheology: Advances and Applications
7: Functional Nanometer-Sized Clusters of Transition Metals: Synthesis, Properties and Applications
8: Mechanochromic Fluorescent Materials: Phenomena, Materials and Applications
9: Cell Surface Engineering: Fabrication of Functional Nanoshells
10: Biointerfaces: Where Material Meets Biology
11: Semiconductor Nanowires: From Next-Generation Electronics to Sustainable Energy
12: Supramolecular Materials for Opto-Electronics
13: Photocured Materials
14: Chemoresponsive Materials: Stimulation by Chemical and Biological Signals
15: Functional Metallosupramolecular Materials
16: Bio-Synthetic Hybrid Materials and Bionanoparticles: A Biological Chemical Approach Towards Material Science
17: Ionic Polymer Metal Composites (IPMCs): Smart Multi-Functional Materials and Artificial Muscles Volume 1
18: Ionic Polymer Metal Composites (IPMCs): Smart Multi-Functional Materials and Artificial Muscles Volume 2
19: Conducting Polymers: Bioinspired Intelligent Materials and Devices
20: Smart Materials for Advanced Environmental Applications
21: Self-cleaning Coatings: Structure, Fabrication and Application
22: Functional Polymer Composites with Nanoclays
23: Bioactive Glasses: Fundamentals, Technology and Applications

How to obtain future titles on publication:
A standing order plan is available for this series. A standing order will bring delivery of each new volume immediately on publication.

For further information please contact:
Book Sales Department, Royal Society of Chemistry, Thomas Graham House, Science Park, Milton Road, Cambridge, CB4 0WF, UK
Telephone: +44 (0)1223 420066, Fax: +44 (0)1223 420247
Email: booksales@rsc.org
Visit our website at www.rsc.org/books

Magnetic Nanomaterials
Applications in Catalysis and Life Sciences

Edited by

Stefan H. Bossmann
Kansas State University, Manhattan, USA
Email: sbossman@ksu.edu

and

Hongwang Wang
Kansas State University, Manhattan, USA
Email: hongwang@ksu.edu

ROYAL SOCIETY OF CHEMISTRY

THE QUEEN'S AWARDS
FOR ENTERPRISE:
INTERNATIONAL TRADE
2013

Smart Materials No. 26

Print ISBN: 978-1-78262-788-3
PDF eISBN: 978-1-78801-037-5
EPUB eISBN: 978-1-78801-168-6
ISSN: 2046-0066

A catalogue record for this book is available from the British Library

The Royal Society of Chemistry is a charity, registered in England and Wales, Number 207890, and a company incorporated in England by Royal Charter (Registered No. RC000524), registered office: Burlington House, Piccadilly, London W1J 0BA, UK, Telephone: +44 (0) 207 4378 6556.

For further information see our web site at www.rsc.org

Printed in the United Kingdom by CPI Group (UK) Ltd, Croydon, CR0 4YY, UK

Preface

Magnetic nanomaterials have undergone a significant evolution during the last decade, enabling physical performances that could not have been envisioned at the beginning of this millennium. It is the conviction of all authors of this edited book that nanomagnetic research is a truly international endeavor that has the potential of growing into a vibrant international community able to bridge cultural and political differences.

This edited book covers the underlying physical principles and nine cutting edge applications of magnetic nanomaterials. The reader will have the opportunity to learn about the frontier of magnetic nanotechnology from scientists, engineers and physicians that have shaped this unique and highly collaborative field of research. Magnetic nanomaterials have reached very high levels of performance and versatility. We are at the beginning of the "age of nanomagnetism", which is the next phase of the "nanorevolution" and will truly reshape numerous fields. Examples of the superior performance of nanomagnetic materials comprise chemical catalysis, pathways towards sustainable chemical industry, theranostic applications and cell-based nanomedicine. The book chapters are concerned with the mechanisms of magnetism in nanomaterials, the use of magnetic nanoparticles in (pseudo-) homogeneous and heterogeneous catalysis, recyclable nanomaterials for biomass conversion, magnetic materials for NMR- and MRI-enhancement and nanoparticles designed for magnetic hyperthermia, as well as image-guided/theranostic cancer therapy, and cell-based magnetic cancer therapy. The authors are currently performing research in the UK, Hungary, China, Russia, and the USA.

Smart Materials No. 26
Magnetic Nanomaterials: Applications in Catalysis and Life Sciences
Edited by Stefan H. Bossmann and Hongwang Wang

Published by the Royal Society of Chemistry, www.rsc.org

This book will provide the academic reader with a deep insight into the underlying paradigms that have shaped these highly collaborative endeavors. At the same time, it provides detailed perspectives of the anticipated transformations through the use of nanomagnetic materials, that will be important for policy makers and teachers alike.

Hongwang Wang and Stefan H. Bossmann
Manhattan, Kansas

Contents

Smart Materials No. 26
Magnetic Nanomaterials: Applications in Catalysis and Life Sciences
Edited by Stefan H. Bossmann and Hongwang Wang

Published by the Royal Society of Chemistry, www.rsc.org

CHAPTER 1

Magnetism in Nanomaterials: Heat and Force from Colloidal Magnetic Particles

GEORGE PODARU[a] AND VIKTOR CHIKAN*[a]

[a]Kansas State University, Department of Chemistry, 213 CBC Building, Manhattan, KS 66506-0401, USA
*E-mail: vchikan@ksu.edu

1.1 Introduction

Magnetic materials have fascinated people ever since the development of the first compass in 200–300 BC. In a compass, the earth magnetic field interacts with the lodestone providing information about the direction of the magnetic north pole. This piece of information was crucial to establish a universal navigation system leading to close interaction of various cultures and economies throughout this planet. The magnetic compass has gone through several development stages, but the basic idea remains in today's navigational devices. In nature one finds examples of animals utilizing magnetic nanomaterials for navigation even in the absence of awareness of magnetic fields and magnetic materials. The sensing of the Earth's magnetic field, called magnetoreception, is present in bacteria, arthropods, mollusks, and even in vertebrates. For instance, Magnetotactic bacteria[1] contain several small magnetite crystals that act together as a compass needle to orient these bacteria along the geomagnetic field lines. The small magnetite crystals are

Smart Materials No. 26
Magnetic Nanomaterials: Applications in Catalysis and Life Sciences
Edited by Stefan H. Bossmann and Hongwang Wang

Published by the Royal Society of Chemistry, www.rsc.org

created *via* very well controlled biomineralization processes[2] to produce single domain magnetite crystals analogous to the artificial colloidal synthesis of magnetic nanomaterials. Humans are thought to be unaware of the Earth's magnetic field, however there are molecules such as cryptochrome that, principally, could serve this purpose. From a thermodynamic perspective, magnetic materials can convert the external energy of a magnetic field into heat or energy/work. The type of interaction depends on the nature of the magnetic field, such as homogeneous, inhomogeneous, static or time dependent (dynamic) magnetic fields. These possible interactions between various types of magnetic fields and materials are listed in Table 1.1 adapted from ref. 3. As one can see there is a wide range of effects on materials in general from the different types of magnetic fields. In this chapter, we will review some basic information about magnetic materials with a focus on colloidal magnetic nanoparticles. We will also focus on how colloidal magnetic nanostructures can produce heat when placed in alternating magnetic fields and how these particles can be used to produce mechanical action, *e.g.* translational motion in the presence of inhomogeneous magnetic fields, specifically the generation of sound waves and their potential applications.

1.2 Magnetism in Nanoparticles

Magnetic nanoparticles are a group of engineered particles (typically smaller than 150 nm) that can be controlled under the influence of an external magnetic field.[4] In these particles, the magnetic field $\vec{B}$ is produced by magnetic materials or by free electric currents:

$$\vec{B} = \mu_0(\vec{H} + \vec{M})$$

μ_0 represents the magnetic permeability of a vacuum and the contribution of $\vec{H}$ arises from the generation of a free current while the magnetization $\vec{M}$ can be defined as the induced magnetic dipole moments in a magnetic material. The factors that influence the magnetic properties of nanoparticles include their chemical composition, crystal lattice, particle shape and size, morphology and finally the manner in which the magnetic nanoparticles interact with adjacent particles or with the surrounding matrix.[5] When the size of the nanoparticles decreases, the surface-to-volume ratio increases. Exploring the properties of the large surface-to-volume ratio of the nanoparticles is of critical importance and has led to novel physical, chemical, and mechanical property discoveries compared to those of the corresponding bulk material.[6] For example, Kenneth *et al.*[7] have shown that reducing the terminal sugar in a dextran coating of iron oxide nanoparticles has had an important impact on the stability of the particles while retaining the same magnetic properties. The magnetic behavior and size dependence of magnetic nanoparticles is controlled by the domain structure of the nanoparticles.[8] The factors that influence the critical size of the single domain consist of the shape of the particles, strength of the crystal anisotropy, value of the magnetic saturation,

Table 1.1 The impact of various magnetic fields on materials.

	Static magnetic fields		Dynamic magnetic fields	
	Homogeneous	Inhomogeneous	Rotating	Oscillating
Ferromagnetic materials	Alignment and steering	Alignment and propulsion	Rotational movement	Heat and motion
Paramagnetic materials	Alignment and steering	Alignment and propulsion, magnetic trapping	Rotational movement of chains of paramagnetic particles	Induce beating pattern in paramagnetic bead chains Manipulate self-assembly of paramagnetic beads
Diamagnetic materials	Alignment and deformation	Magnetic levitation and separation, magnetic trapping	Alignment along multiple axis	No known examples so far

and domain wall energy. With increasing values of the radius, magnetic nanoparticles can be classified as superparamagnetic, single domain and multi-domain. The domain wall in magnetic particles is important, because it has a major impact on the magnetization of the particles. In general, the magnetic moments of the particles will scale with the volume of the particle, but because of the presence of domain walls, this scaling is abruptly halted when multiple magnetic domains are formed in a single particle. The critical radius (r_c) represents the point where it is energetically favored for the magnetic particle to exist without a domain wall:[9]

$$r_c \sim f\left(\frac{\sigma}{M_s^2}\right)$$

where σ represents the magnetic moment per unit mass and $M_s{}^2$ is the saturation magnetization. In Figure 1.1 it is shown that the coercivity varies dramatically with the particle diameter, unlike the magnetization. In the multi-domain region, it can be observed that the coercivity decreases with the increase of the particle diameter (due to subdivision into domains), while in the single domain region it increases with the radius of the particle. The calculated critical radii for some common magnetic material are: Fe_3O_4 (4 nm), Co (8 nm), Fe (1 nm) and Ni (35 nm). In other words, a typical colloidal synthetic method is able to produce magnetic nanoparticles that fall into the single domain region. The single domain region can be further divided into two subregions: the superparamagnetic region, where the coercivity is 0 due to the randomizing effects of the thermal energy, and the ferromagnetic region, where the coercivity increases dramatically with the particle diameter. In the absence of a magnetic field, the *net magnetic moment* is 0 for superparamagnetic nanoparticles, while in an applied magnetic field there will be a *net alignment* of the magnetic moments. This phenomenon is like paramagnetism with the exception that the magnetic moment originates from a single domain particle (10^5 atoms) and not from the magnetic moment of a

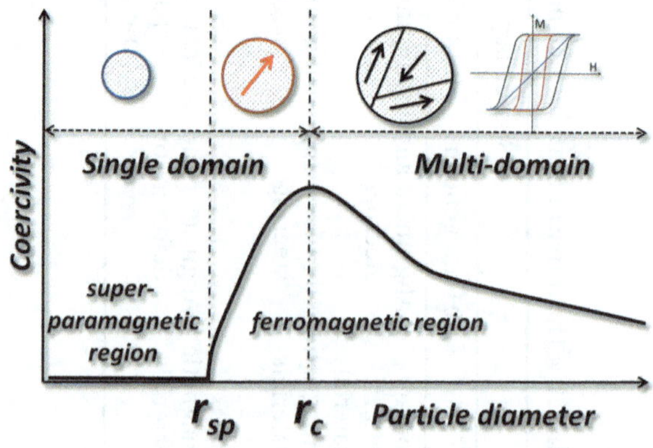

Figure 1.1 The change in coercivity of magnetic particles as a function of size.

single atom, as observed in paramagnetism. This also means that the saturation magnetization is typically larger in those materials exhibiting paramagnetic behavior.

There are multiple ways in which magnetic nanoparticles can be classified based on their properties. In this chapter, we will concentrate on biological applications. First, magnetic nanoparticles must be biocompatible and non-toxic. Second, the size related properties become important to yield reasonable colloidal and chemical stability. Choosing a versatile magnetic nanoparticle can be very helpful in those cases where the nanoparticle suspension is delivered intravenously, because aggregation (*e.g.* in blood vessels) must be avoided. For these reasons, in recent years, magnetic nanoparticles have been researched intensively. One solution to this problem has been developed by tailoring their surface chemistry.[10] Recently, Li *et al.* reported a new synthesis of sodium-citrate-modified iron oxide nanoparticles where the 3T3 cell line maintained a cell viability higher than 70% at medically relevant concentrations.[11]

Regarding diamagnetic shell-enhanced magnetic properties, Ye *et al.* have shown that mesoporous silica coated superparamagnetic iron oxide nanoparticles exhibited a 21-fold enhanced MRI efficiency when compared to a commercial T_2 contrast agent.[12] A high magnetization value is critical for magnetic nanoparticles, because it will permit control over the movement of particles in blood or lymphatic fluid. Magnetic nanoparticles can be moved closer to a targeted pathological tissue aided by strong magnetic fields.[13] Practically, the desired high magnetization can be obtained by synthesizing nanoparticles that belong to the group of transition metals and their oxides. A list of commonly used magnetic nanoparticles in biomedical applications and their magnetic properties is summarized in Table 1.2. From Table 1.2 one can see the reason why iron oxide is widely used in cancer research. For example, 17 nm iron oxide magnetic nanoparticles have a high saturation magnetization (82 emu g^{-1}), and, at the same time, a high coercivity value (364 G).

1.3 Impact of Static and Dynamic Magnetic Fields on Biological Systems

Natural and human-made magnetic fields permeate our environment. The Earth's magnetic fields typically do not exceed 100 mT, but human-made magnetic fields in medical devices or industrial processes, such as electrolysis or welding machines, do exceed this level. In research facilities, magnetic fields as high as 100 Tesla have become a possibility (the strongest pulsed non-destructive magnetic field produced in a laboratory was at the Pulsed Field Facility at the National High Magnetic Field Laboratory's Los Alamos National Laboratory). Magnetic fields will impact not only magnetic materials, but materials that do not show intrinsic magnetism, *e.g.* water that is a diamagnetic material opposes the magnetic field. The interaction of diamagnetic

Table 1.2 Variation of coercivity in typical magnetic nanoparticles as a function of size.

Magnetic nanoparticle	Size (nm)	Saturation magnetization(emu/g)	Coercivity (G)	References
Fe_3O_4	4.2	75	318	Guardia *et al.*[14]
	7.4	70	270	
	8.1	65	70	
	17	82	364	
	45	92	340	
Fe_3O_4	4.9	60.4		Pereira *et al.*[15]
	6.3	64.8		
	8.6	58		
$CoFe_2O_4$	4.2	30.6		
	4.8	46		
	18.6	48.8		
$MnFe_2O_4$	9.3	57.1		
	11.7	54.6		
	59.5	35.2		
Ni	24	25.3	120	He *et al.*[16]
	50	32.3	79	
	96	40.6	18	
	165	46.7	146	
	200	52	158	

materials with magnetic fields is typically much less than that of magnetic materials. The magnetic susceptibility of water is small, -0.9×10^{-8} m^3 kg^{-1}, in comparison with the susceptibility of typical ferromagnetic materials (1000–10 000 $\times$ 10^{-8} m^3 kg^{-1}). A typical example of the interactions of diamagnetic materials with static magnetic fields is magnetic levitation. For water, the levitation condition is when the magnetic field gradient is around 1400 T^2 m^{-1}. For a small "object", such as a frog[17] or a strawberry, this condition is met in a static magnetic field with about 10–20 T. Dynamical magnetic fields can have similar effects on metals. High frequency (10–30 kHz) induction heating is used to magnetically heat and levitate molten metal. Pulsed magnetic fields are known to have an impact on the nerve cells of animals. For instance, Yamaguchi *et al.* have shown that ventricular defibrillation cannot be attained by a magnetic stimulus with a flux density of 9.2 T or below.[18] Strong magnetic pulses can stimulate biological processes not only at a macroscopic scale, but also at the cellular level. An interesting effect of pulsed magnetic fields (7 T, 2 s) on algae-like fungi (*Achlya americana*) and protozoans (*Saprolegnia diclina*) is to open up pores in the lipid bilayers, which can be performed either reversibly or irreversibly.[19] At the molecular scale, magnetic fields can interact with the presence of unpaired electrons in radicals and transition metal complexes, as well as the charges present at the surfaces of larger molecular structures, such as proteins or lipids. It has been shown that the torque on proteins from magnetic fields or magnetic field gradients can positively influence protein crystallization.[20] Magnetic fields can also facilitate the alignment of lipid bilayers. In biological systems, the presence of static and dynamic magnetic fields can contribute to the change in the apparent rates of chemical reactions. The impact of typical dynamic and static magnetic fields are minimal on single reactions. However, in biological systems where there is significant feedback in complex reaction systems, the small impact of the magnetic fields can cause large effects. For instance, small 50 Hz alternating magnetic fields are capable of influencing the free O_2 radical concentration in rat lymphocytes *in vitro*.[21,22] The European Union[23] has recently surveyed the literature data on the human health effects of RF (low frequency and static electromagnetic radiations). The conclusion of their work is that there is some data available especially at frequencies important for human communication devices. However, there are areas where there is limited data available. Clearly, there is a critical need to further investigate the impact of magnetic fields on biologically important materials at a fundamental level.

1.4 Heating of Magnetic Particles Under the Influence of an External AC Field

It has been observed that when colloidal magnetic nanoparticles are placed in external alternating magnetic fields, the solution can be rapidly heated. The heating takes place by power absorption by the magnetic particles due to the presence of a high frequency alternating magnetic field. The source of power absorption varies with (nano)particle size, as discussed later in this section. In multi-domain particles, the dominant heating is the hysteresis loss due to the movement of domain walls. Although multi-domain magnetic particles were prepared early on, the heating efficacies of these particles are relatively low.[24] The two main contributing mechanisms of SAR (Specific Absorption Rates) in single domain magnetic nanoparticles are Brownian motion[25] (rotation of the entire nanoparticle) and Néel relaxation (random flipping of the spin without rotation of the particle).[26] The transition occurs between 5–40 nm, depending on the materials, but it also varies with frequency and the viscosity of the system. In Table 1.3, calculated transition sizes are shown for typical magnetic NPs in water and in lipids. The latter represent the response in biological media. At low frequencies and larger nanoparticles, Brownian relaxation is

Table 1.3 Crossover size from Néel to Brownian relaxation of various spherical magnetic NPs in water and in lipids (cell membranes).

NP material	Crossover size in H_2O	Crossover size in viscous lipids
Co	5 nm (12.3 MHz)	6.3 nm(47 kHz)
FePt	6.7 nm(5.7 MHz)	8.1 nm(25 kHz)
$CoFe_2O_4$	8.2 nm(3.5 MHz)	10.1 nm(15 kHz)
Fe	11.9 nm(1.4 MHz)	13.8 nm(7 KHz)
Magnetite	20.8 nm(316 kHz)	24.7 nm(1.4 kHz)
γ-Fe_2O_3	27.0 nm(153 kHz)	30.9 nm(757 Hz)
Fe_3Si	31.7 nm(98 kHz)	37.0 nm(454 Hz)
FeCo	40.0 nm(50 KHz)	45.5 nm(>300 Hz)

the dominant process, while at high frequencies and small nanoparticle sizes Néel relaxation is the main contributor to the heat dissipation. When Brownian relaxation is dominant, the energy from the magnetic field is deposited in a coherent manner (mechanical movement) into the nanoparticles and their surroundings. The important factor for magnetic heating experiments is the specific absorption rate or SAR, which is determined by SAR = $C^*\delta T\,\delta t^{-1}$, where C is the specific heat capacity of the sample and T and t are the temperature and time, respectively. SAR is very sensitive to the materials' properties, which are also dependent on the type and frequency of magnetic field that is applied. Figure 1.2 shows the SAR values of typical magnetic NPs at 100 kHz magnetic field with an amplitude of 5000 A m^{-1}.

The heating capability of magnetic nanoparticles (characterized by their SAR values as shown in Figure 1.2) can be further increased by combining hard and soft magnetic materials at the nanoscale. Magnetic materials in close proximity interact *via* exchange coupling thus allowing the fine tuning of the magnetocrystalline anisotropy of the combined material. These so-called exchanged coupled magnetic materials provide an increase in SAR values of the NPs by a factor of 5–10. As an example Lee *et al.* have developed several new exchange coupled systems including $CoFe_2O_4$@$MnFe_2O_4$, $CoFe_2O_4$@Fe_3O_4, $MnFe_2O_4$@$CoFe_2O_4$, Fe_3O_4@$CoFe_2O_4$ and $Zn_{0.4}Co_{0.6}Fe_2O_4$@$Zn_{0.4}Mn_{0.6}Fe_2O_4$. All of these exchange coupled particles exhibit 1000–4000 W g^{-1} specific loss power compared to the non-exchange coupled constituent of nanoparticles, which are in the range of 100–400 W g^{-1}.[27,28] The higher heating efficiencies of these materials mean shorter treatment times and smaller required nanoparticle concentrations (fewer potential side effects) in magnetic hyperthermia treatments. The incremental increase of SAR values *via* various mechanisms could

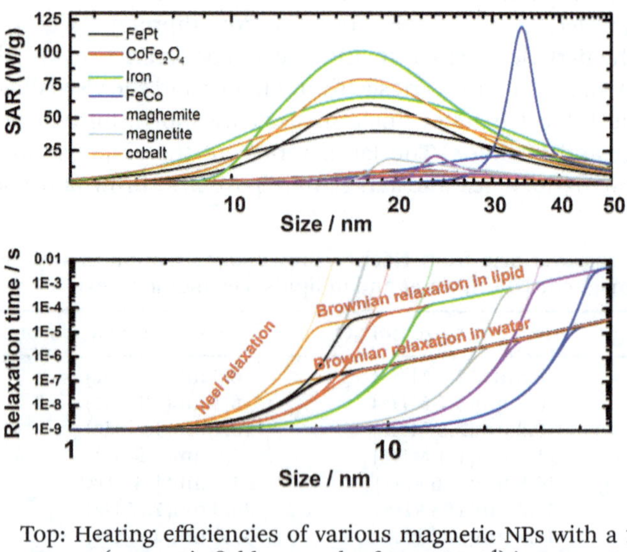

Figure 1.2 Top: Heating efficiencies of various magnetic NPs with a 100 kHz AC magnet (magnetic field strength of 5000 A m^{-1}) in water; Bottom: Magnetic Relaxation of magnetic NPs in water and in lipids.

potentially lead to much more viable physical treatment methods that could be used for treating a variety of conditions. In the next paragraph, the impact of the magnetic field types on the heating process will be reviewed.

According to the literature,[29,30] the ranges of frequency (f) and amplitude (H) that can be tolerated by humans are considered to be $f < 1.2$ MHz and $H < 15$ kA m^{-1}, respectively, with their product $f \times H$ not exceeding ~5×10^9 A m^{-1}. The colloidal magnetic nanomaterials will heat better at higher frequencies, but the contrast between tissue heating and nanoparticle-mediated heating will decrease, due to changes in tissue conductivity. Therefore, in typical treatments the frequencies remain below 100 kHz. There is little room to further increase the magnetic field amplitude or the frequency for more efficient treatment. However, the heating efficiencies of magnetic NPs can be further increased by using rotating magnetic fields as opposed to alternating magnetic fields. As shown in Figure 1.3, in the case of rotating magnetic fields the magnetic field is always "on", but changes direction, which produces heat more efficiently from magnetic nanoparticles than alternating magnetic fields, which are "on" only a fraction of the time. There are examples in literature demonstrating that rotating magnetic fields can be more efficient than AC magnetic fields. Sharapova *et al.*[31] have demonstrated that superparamagnetic nanoparticles produce 2–3 times more heat in a rotating magnetic field than in alternating magnetic fields. Raikher *et al.*[32,33] have developed a simple model that shows that the observed heating is consistent with the expectations based on the underlying physics of the particles. The implementation of rotating magnetic field magnetic hyperthermia machines is challenging. The small magnetic field amplitudes from such machines are able to produce a heating effect, but not the sustained rotation of magnetic particles. In order to utilize the small particles as "tiny drill bits",[34] a few important requirements

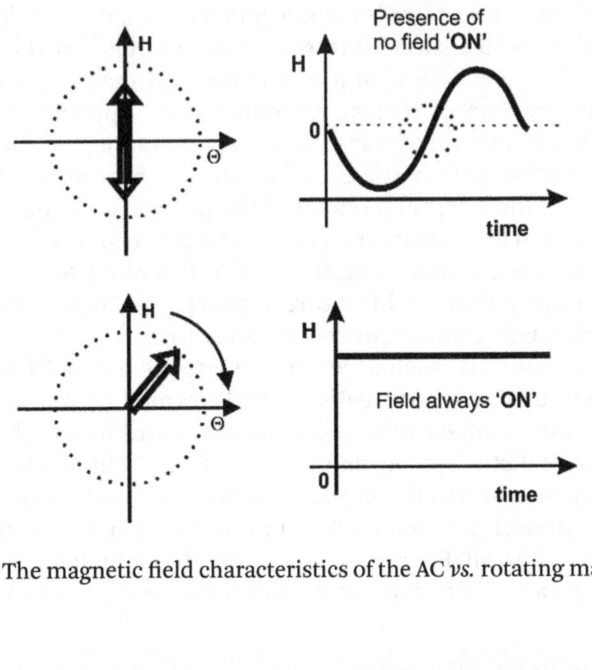

Figure 1.3 The magnetic field characteristics of the AC *vs.* rotating magnetic fields.

have to be met that will be discussed in Section 3.5. The magnetic field needs to be increased further to be able to maintain a hold on the small magnetic particles against the drag forces of the medium and the thermal fluctuation of the magnetic moment. One needs to point out that the increase of magnetic fields for rotating magnetic particles does not fall under the restriction ($f \times H$ not exceeding ~5×10^9 A m^{-1}) because the large magnetic fields required are most suitable for low duty cycle-pulsed magnets. In addition, the therapeutic effect of the magnetic particles is not the result of heating the medium, but of the mechanical action of the particles. These ideas will be discussed later in the chapter, but the first questions that will be dealt with are the requirements for rotating tiny magnetic particles in solution.

1.5 Mechanical Rotation of Magnetic Particles in Colloidal Solutions Due to External Rotating Magnetic Fields

The forces acting on colloidal magnetic particles are governed by the volume (V) of the particles, the difference in magnetic susceptibilities ($\Delta\chi$) of the medium, such as water and the magnetic particle, as well as the magnetic field strength (***B***) and its gradient ($\nabla \cdot B$).

$$F = \frac{V\Delta\chi}{\mu_0}(B \cdot \nabla)B$$

In inhomogeneous magnetic fields, the particles will be moving while in homogeneous magnetic fields the particles merely align along the external magnetic field. An important question is how fast different sized magnetic nanoparticles can be rotated in homogeneous magnetic fields and what size of magnetic field is needed to overcome the random thermal motion of the particles in colloidal solution. The internal magnetic field of single domain magnetic nanoparticles responds to the application of external magnetic fields in two main ways: Néel (superparamagnetic particles) and Brownian (ferromagnetic particles) relaxation.[35] Both mechanisms involve the alignment of the internal moment of the particle with the external field, but only the Brownian relaxation results in the physical rotation of the particles. Calculations of the total relaxation times (1/total = 1/Néel + 1/Brownian) in Figure 1.2 are based on literature reports.[36,37] These data show that most common single domain magnetic nanoparticles are available for these experiments in solvents, such as water and viscous media (lipids). In order to make the simulations more realistic, the Brownian relaxation component also includes the diameter of a typical ligand (1 nm in length) on the surface to mimic realistic hydrodynamic radii of MNPs in solution. This simple calculation also shows the impact of solvent viscosity. The viscosity will change the crossover between Néel and Brownian relaxation. In addition, it will restrict how fast MNPs can respond to rapidly changing magnetic fields. Increasing the viscosity of the solvent slows the rotational dynamics of the particles. For this example calculation, the viscosity of a typical lipid is used because of the potential application of small rotating magnetic nanoparticles in cancer therapy (drilling cells) for destroying cancerous cells.[38,39] Liu *et al.* have shown that relatively weak magnetic fields (40 and 75 mT in magnitude) can be used to enhance cell membrane poration (magnetoporation) with the help of polymer-coated multi-walled carbon nanotubes, which form rotating bundles on exposure to magnetic fields. Although this is a macroscopic phenomenon, one would expect that similar effects can be shown with very tiny nanometer-sized individual magnetic particles as well.

The viscosity has a significant impact on the rotation and movement of particles in liquids. Fluorescence[40] anisotropy measurement of a fluorescent dye suggests that the lipid viscosity in its fluid state is on the order of 0.1 Pa s.[41] Significant slowing of the rotation frequencies of the particles in viscous biological media is anticipated. Experimental confirmation of the rotation frequency of sub-20 nm magnetic nanoparticles has already been reported in the literature. For example, experiments by Feyen *et al.*[42] have shown that sub-10 nm Co nanoparticles can be mechanically rotated by using relatively weak magnetic fields. These results indicate that Co particles, with increasing ligand length, can be rotated at a relatively slow rate; *e.g.* the highest possible rotation frequency is decreased from 19 000 Hz to about 170 Hz for particles with hydrodynamic radii from 17–38 nm. The AC susceptometry in their experiment only uses a 0.25 mT AC (non-rotating) magnetic field.

The relaxation time is one of the many important considerations in relation to how fast magnetic particles can be aligned in solution in the presence of an external magnetic field. Rikken *et al.* have pointed out that smaller particles are more easily disturbed by the Brownian motion.[3] The torque on the particles is defined as $\tau = B \times m$ where B corresponds to the magnetic field and m is the magnetic moment on the particle.[43,44] Figure 1.4 shows a calculation

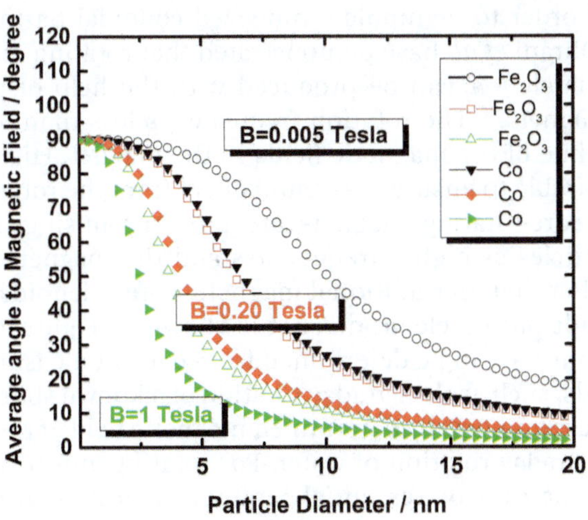

Figure 1.4 The calculated magnetic alignment of spherical cobalt and Fe_2O_3 nanoparticles in water at room temperature in various magnetic fields.

on the average angle of selected nanoparticles in the presence of 0.005, 0.2 and 1 T magnetic fields at room temperature. It is important to point out that while the relaxation times are controlled by the anisotropy of the crystal and the saturation magnetization, the alignment primarily depends on the latter. Figure 1.4 shows the calculated alignment of the Co and Fe_2O_3 MNPs in 0.2 and 1 T magnetic fields. The data also indicates that reasonable alignment can be achieved even with relatively small magnetic fields. The 0.005 T magnetic field can be easily produced in laboratories where continuous electromagnets are used. At higher magnetic fields, it is apparent that almost complete alignment can be achieved for particles with sizes of 20 nm with sub-T magnetic fields. This is also supported by the observation that the typical saturation magnetization of magnetic nanoparticles takes place on the order of a 1 T magnetic field.[45] In order to achieve better alignment with smaller particles, larger magnetic fields are needed. Conventional continuously operated water-cooled electromagnets are unable to produce fields large enough to rotate sub-20 nm magnetic particles. For these rotating magnetic fields, electromagnets are operated using pulse power systems as described in the next section.

1.6 Pulsed Electromagnets to Produce Homogeneous Rotating Magnetic Fields

There is already evidence in the literature that small magnetic particles can be controlled by strong magnetic fields in colloidal solutions.[3] An important factor to successfully achieve rotation in small magnetic colloids is use of a device that is capable of producing strong rotating magnetic pulses. Because of their simplicity and the reduced cooling needs, low duty cycle pulse magnets are convenient experimental tools to generate large magnetic fields in order to manipulate nm-sized colloidal particles.[3] In their recent work, Ohtani *et al.* have demonstrated that rotating magnetic fields with approximately 6 T can be produced with the help of pulsed superconducting magnets.[46] The rotation frequency is less than 500 Hz due to the slow rise time of the magnetic fields in the magnet. Higher frequency rotation is desirable to ensure maximum effect from the rotating magnetic particles. Air core-rotating magnets are more promising, because they can rotate particles at higher frequencies and the magnetic field can be increased further than in traditional magnetic core solenoids. As an example, a homebuilt pulsed electromagnet is shown in Figure 1.5. The magnetic field amplitude can be determined from current measurements or by optical methods, such as the Faraday rotation of a known substance. Figure 1.5 shows the experimentally determined magnetic field of the pulsed magnet from the Faraday rotation of water. For rotating anisotropic magnetic nanoparticles, there are two essential parameters for the construction of an electromagnet. First, the magnetic fields must be homogeneous to ensure

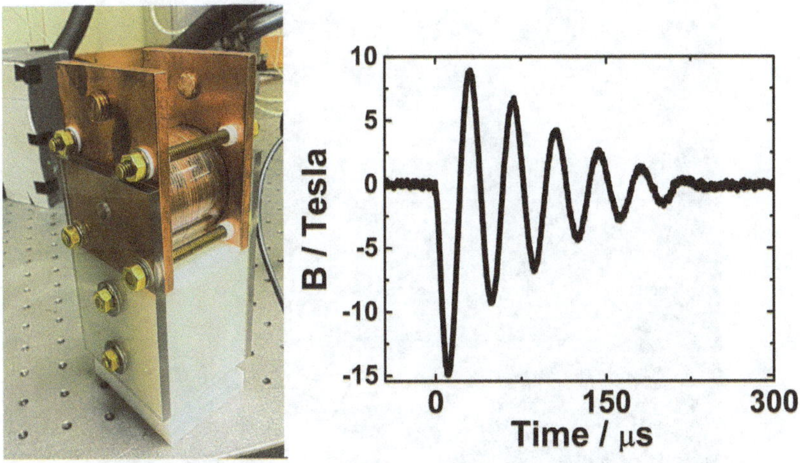

Figure 1.5 Left: Picture of a pulsed magnet; Right: Measured magnetic field of the pulsed magnet. Reprinted with permission from ref. 57. Copyright 2016 American Chemical Society.

that no translation of the particles can take place. Second, the magnetic field has to change rapidly, which brings challenges associated with high frequencies. Rotating magnetic fields are routinely produced in electrical motors, but the magnetic fields are spatially inhomogeneous, relatively weak, and insufficient to rotate small magnetic nanoparticles. Chikan *et al.* have recently developed a prototype of a rotating magnetic field generator consisting of a pair of high voltage Helmholtz coils (Figure 1.6).[47] The device combines high voltage technology with a so-called Bitter coil design.[48] This technology is useful for the production of large currents for the strong rotating magnetic fields. The nested Helmholtz coils consist of copper/beryllium[49] alloy disks that are electro-coated with silver, which ensures the highest electrical conductivity. The conductors are separated by means of a mica insulator. The two nested coils are rotated 90 degrees (orthogonally) with respect to each other. This arrangement makes it possible to produce linear or rotating magnetic fields in the 2D plane by controlling the currents independently in each coil. The coil driver utilizes two discharge capacitors as a power source triggered by a pair of homebuilt high-voltage spark gap switches. The calculated peak current *via* the triggered spark gaps can reach several thousands of amperes. In this device, the magnitude of the magnetic field is directly proportional to the magnitude of the current and can be as high as 40 000 A. The coil is not only capable of producing rotating, but also elliptical and linear magnetic pulses. As mentioned above, the magnetic field of the coil is monitored either optically by Faraday rotation measurements[50] of a known substance (water) or electrically by Rogowski coils,[51] which measure the current derivative.

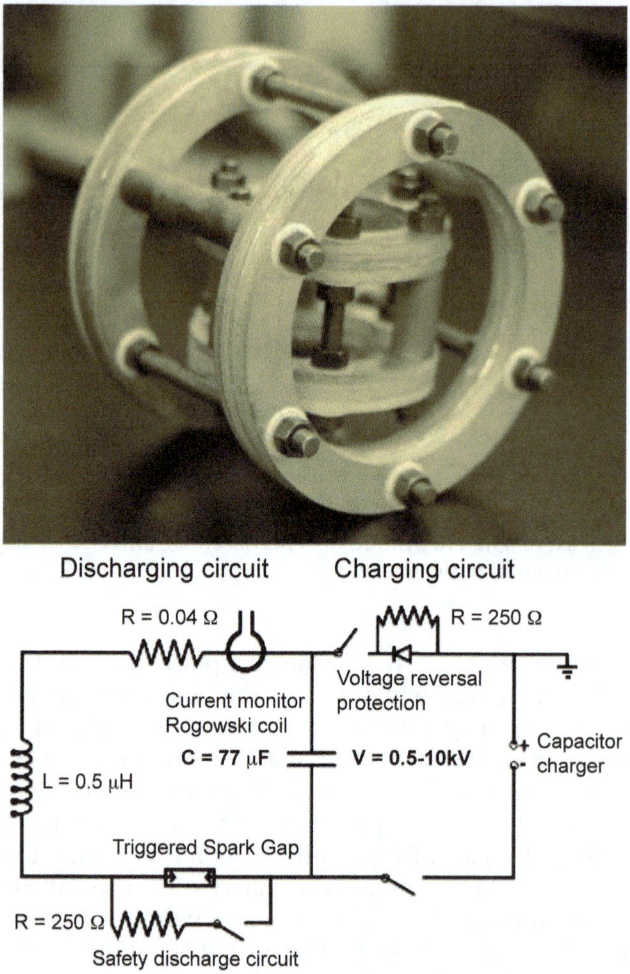

Figure 1.6 Top: Picture of the high voltage 2D Helmholtz magnet to produce homogeneous rotating magnetic pulses; Bottom: Sketch of the electrical circuit used to make a magnetic pulse.

The data in Figure 1.7 show a rotating magnetic pulse of approximately 200 μs duration at 20 KHz rotation (time axis not shown) and 200 mT amplitude. The magnetic field amplitude is not constant due to the operational mode of the electromagnet. The current and the magnetic field amplitudes are controlled by the initial voltage of the discharge capacitors. The rotation frequency is controlled by both the capacitance of the capacitor and the inductance of the coil, which change the resonance frequency of the RLC circuit. Pulsed magnets producing rotating homogeneous or pulsed inhomogeneous magnetic fields can be effectively used to manipulate tiny magnetic particles, as will be shown.

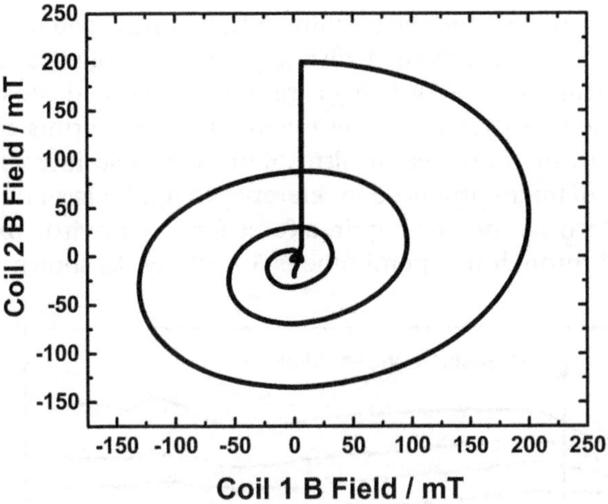

Figure 1.7 The measured magnetic field strength of the rotating magnetic pulse from the 2D Helmholtz coil in Figure 1.6.

1.7 Sound from Magnetic Particles

The energy from the magnetic field can be converted not only into rotational motion, but also into translational motion. Colloidal magnetic particles in the presence of external homogeneous magnetic fields will align according to the field direction. However, they will exhibit no translation motion. This challenge can be circumvented in various ways. For instance, it is possible to convert the rotational motion into linear motion as Fischer *et al.* have demonstrated. They have used DNA fragments as propellers on magnetic particles to turn the rotation of the particles into linear motion.[52,53] With their approach, as they have pointed out, potential micro surgical tools can be developed. In order to achieve translational motion without the need for molecules attached to the particles, the magnetic field has to be inhomogeneous in which the particle will move from a region of low magnetic fields to the high magnetic fields. The linear motion of the particles can lead to induced sound waves in the colloidal solution or ultrasound if the frequency of the applied inhomogeneous magnetic field is larger than 20 kHz. Figure 1.8 shows the sound power as a function frequency from increasing concentrations of colloidal iron oxide nanoparticles. With the increase in concentration of the particles, the sound power increases. The ultrasound power is roughly constant in the 20–500 kHz frequency range. Table 1.4 summarizes the actual pressures and forces from iron oxide nanoparticles in various magnetic fields. The magnitude of the forces and pressures from these particles is not as dramatic as other types of sources of ultrasound, but it has its origin in magnetic fields. This is rather important since the electromagnetic radiation does not suffer the same attenuation in the human body as ultrasound does, therefore in applications where the ultrasound needs to be generated locally well within the human body, it is anticipated that the above technique may have a very distinct advantage. The role of ultrasound in magnetic hyperthermia is currently being carefully investigated. Most recent data suggest that the effectiveness of the magnetic hyperthermia treatments is increased due to the presence of ultrasound. It is also interesting to note the magnitude of forces produced in reference to lipid bilayer permeability.[54] According to Barnes *et al.*, the required force for a 10 nm iron oxide (SPION) particle to pull through the membrane is 5×10^{-6} N. As shown in Table 1.4,

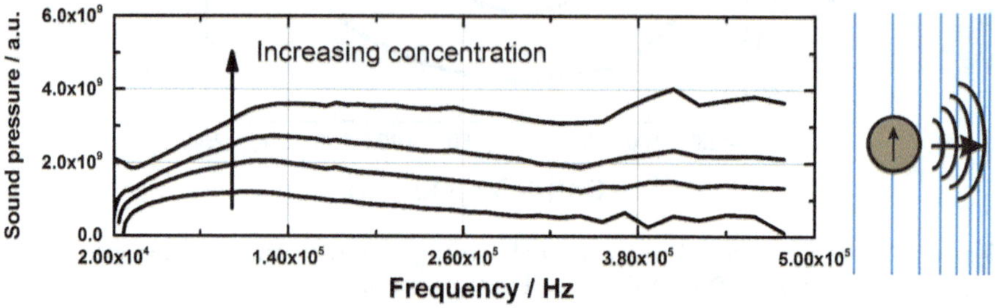

Figure 1.8 The ultrasound power from colloidal magnetic iron oxide in inhomogeneous magnetic fields as a function of particle concentrations.

Table 1.4 The measured and calculated pressures and forces from colloidal iron oxide in homogeneous and inhomogeneous magnetic fields. Note, that coils designed to provide homogeneous magnetic fields will have contributions from inhomogeneous magnetic fields in these experiments. The calculated and measured forces from the particles are in very good agreement. The data reproduced with permission from ref. 57. Copyright (2016) American Chemical Society.

	Pulsed coils		AC coils	
	Helmholtz	Anti-Helmholtz	Helmholtz	Anti-Helmholtz
Peak magnetic field	15 Tesla	5 Tesla	2.7 mTesla	1.9 mTesla
Average magnetic field	4.97 Tesla	1.57 Tesla	1.1 mTesla	0.7 mTesla
Peak magnetic field gradient	321	475	0.138	0.221
Average magnetic field gradient	0.173	0.168	0.049	0.064
Estimated peak force on 10 nm particle	1.46×10^{-18} N	2.17×10^{-18} N	6.31×10^{-22} N	1.01×10^{-21} N
Estimated average force on 10 nm particle	7.92×10^{-22} N	7.69×10^{-22} N	2.24×10^{-22} N	2.92×10^{-22} N
Measured max. pressure	334 Pa	294 Pa	10 Pa (@200 KHz)	18 Pa (@200 KHz)
Particle concentration	5.8×10^{15} NP ml^{-1}	5.8×10^{15} NP ml^{-1}	3.8×10^{16} NP ml^{-1}	3.8×10^{16} NP ml^{-1}
Measured force/ particle	1.18×10^{-18} N	1.04×10^{-18} N	6.9×10^{-21} N	1.2×10^{-20} N

the forces are much smaller, which suggests that particles in typical magnetic fields will impact the membrane permeability in a different way. The literature[55–58] shows that drug release can be induced from liposomes loaded with magnetic nanoparticles. These results indicate that ultrasound has a critical role in increasing the membrane permeability, which will be important in future biomedical applications (Figure. 1.9).

Ultrasound from magnetic particles can have many potentially important applications. A specific example of such a practical application is to use the ultrasound to trigger drug release from liposomes commonly used to carry drugs or other substances. Liposomes were first described in the literature in 1961.[59] Liposomes are quite stable in blood and able to keep their content intact under physiological conditions.[60–63] When incubated with plasma constituents, they retain a spherical shape.[64,65] Liposomes made from L,α-dipalmitoylphosphatidylcholine (DPPC) are widely used for the intravenous delivery of drugs, because they are relatively inexpensive and feature suitable biophysical properties. The higher phase transition temperature (T_m) of DPPC is 310 K. When the temperature is higher than the transition temperature, liposomes can be filtered through porous membrane filters. In this process the multi-lamellar liposomes become spherical unilamellar liposomes with a very small polydispersity (PD < 1.15).[66] In order to slow the removal of the liposomes from circulation by the macrophages and monocytes of the reticuloendothelial system, a polyethylene glycol coating is attached to their exterior.[67] Polyethylene glycol serves as a steric block at the exterior of the liposome. Polyethylene glycol molecules have only limited interaction with recognition molecules. Therefore, it is possible to prevent their adsorption at the liposomes' exterior surfaces. Consequently, they are (mostly) ignored by the reticuloendothelial system. The liposomes coated with polyethylene glycol are known as stealth liposomes.[67] To date, several liposomal drug delivery systems have been developed (*e.g.* NICODERM and others)[68,69] that rely on the slow release (several minutes to hours) of their payload. However, for medical treatment applications, fast delivery of liposomal payloads is desirable once the target has been reached. Therefore, a suitable release trigger has to be designed. It should be noted that the concept of using magnetic fields and nanoparticles for drug delivery is not new. AC-magnetic hyperthermia is commonly used to trigger the release of payloads from magnetoliposomes

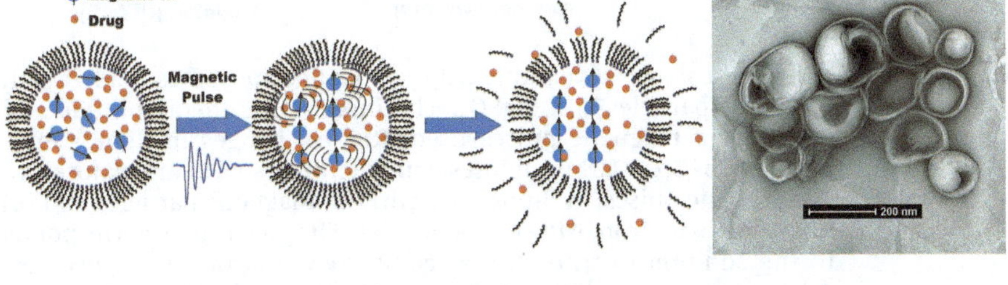

Figure 1.9 Drug release from magnetoliposomes as a result of sound from dynamical inhomogeneous magnetic fields. Reprinted with permission from ref. 55. Copyright (2014) American Chemical Society.

by heating magnetic nanoparticles and their surroundings. As a result of heating the supramolecular nanostructures of liposomes either burst or (partially) dissolve in the surrounding aqueous medium.[70] One concern with this approach is that the heat may damage the payload of the liposomes. For instance, anticancer drugs like SN-38 and its prodrug irinotecan[71] and si-RNA[72] that can degenerate when heated above $T = 330$ K.

Liposomes loaded with magnetic particles can be prepared by using various types of nanoparticles, such as iron, cobalt, or iron oxide. The nanoparticles induce efficient drug release from these so-called magnetoliposomes. Figure 1.10a shows the drug release from magnetoliposomes that are loaded with $MgSO_4$ (model drug) and PtFe magnetic particles. The $MgSO_4$ release can be monitored *via* AC impedance measurements. This analytical technique provides a sensitive probe of the effect of the pulsed magnetic fields on the magnetoliposomes. In the example provided in Figure 1.10, several short 50 μs magnetic pulses (the actual magnetic pulses are shown in the inset of Figure 1.10) are applied to the liposomes in solution. The AC resistance (real part of impedance) of the solution decreases as $MgSO_4$ is released from the liposomes. Figure 1.10b shows the release of another model drug (5(6)-Carboxyfluorescein) from magnetoliposomes. This model drug molecule is highly fluorescent when it is at low concentrations in aqueous solution, but will generate very low emissions when the molecule is at high concentrations inside the liposomes. This property of the 5(6)-Carboxyfluorescein is used in permeability assays[73] to assess the release of larger molecules from liposomes. When 5(6)-Carboxyfluorescein is released, the fluorescence increases. This change of fluorescence intensity permits the quantification of the model drug release. In Figure 1.10b, the actual drug release is only a few percent, which shows that there is more room for improvement of

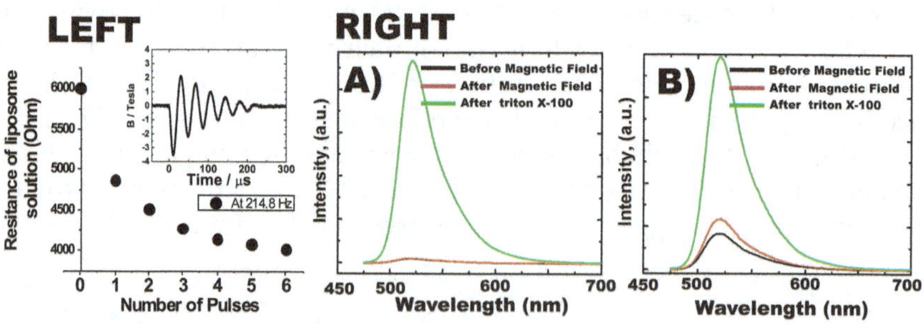

Figure 1.10 Left: Decrease of AC conductivity of magnet liposomes loaded with PtFe nanoparticles and $MgSO_4$. The inset shows the amplitude and duration of magnetic pulses used; Right: Changes in fluorescence intensity of 5(6)-Carboxyfluorescein when liposomes are exposed to 10 magnetic pulses, (a) liposomes without magnetic particles and (b) liposomes with iron oxide nanoparticles. The green curve corresponds to the addition of triton-X that completely destroys the liposomes. Adapted with permission from ref. 55. Copyright (2014) American Chemical Society.

ultrasound-induced drug release from magnetic liposomes. The following general conclusions can be drawn from the experimental work on rapid drug release from magnetic liposomes from pulsed magnetic fields: (1) It is possible to achieve drug release up to 40% with the application of few 3–4 T magnetic pulses and a total length of the pulse train of approx. 200 μs. (2) A load of magnetic nanoparticles that is too high results in a decrease in sensitivity of detecting drug release due to the competition between the magnetic nanoparticles and 5(6)-Carboxyfluorescein to fill the same volume inside the liposome. (3) The solubility of nanoparticles and therefore the stability of ligands on the surface of the nanoparticles have a major impact on how efficiently nanoparticles can be loaded into liposomes. (4) The separation of magnetoliposomes from nanoparticle aggregates is challenging. (5) Using amphiphilic, hydrophilic or hydrophobic ligands on the surface makes little difference for drug release efficiency. (6) Due to the integration of nanoparticles into the lipid bilayer, the use of hydrophilic ligands may cause passive drug leakage, which reduces the shelf life of the magnetoliposomes. (7) Magnetoliposomes exhibit increased coalescence when exposed to centrifuging because of the increased magnetic interaction between them. (8) The magnetoliposomes when triggered with inhomogeneous pulsed magnetic fields release drug molecules.[55] (9) The mechanism of drug release is the result of the ultrasound generation from small magnetic nanoparticles.[56] (10) The ultrasound is generated from the movement of particles in inhomogeneous magnetic fields as opposed to the much weaker magnetostriction effect due to homogeneous magnetic fields.[74] In parallel with the work of Chikan *et al.*, research groups in Europe (Piero Baglioni[56] at University of Florence, Stefania Petralito[58] at University of Rome and Erik Reimhult[74] at the University of Natural Resources and Life Sciences, Vienna) have shown that low frequency magnetic fields (a few mT) can trigger drug release from magnetoliposomes on a time scale of minutes/hours. Their work also suggests that the release of drug molecules is not simply the result of heating liposomes.

1.7.1 Potential Applications of Ultrasound from Colloidal Magnetic Particles

There are several potential applications of ultrasound from magnetic nanoparticles in biological media (see Figure 1.11). (1) In a recent study, Kinoshita *et al.*[75] have shown that the blood–brain barrier can be made penetrable for drug molecules *via* application of focused ultrasound. In their study, they have successfully delivered HERCEPTIN (trastuzumab), a humanized antihuman epidermal growth factor receptor 2 (HER2/c-erbB2) monoclonal antibody, to treat brain metastasis of cancer. A crucial barrier for more effective use of the blood–brain barrier opening is the relatively low penetration of ultrasound through the skull.[76] The successful transfer of the ultrasound-induced technology for opening the blood–brain barrier to medical practice will require the use of alternative means of generating the ultrasound. Current clinical studies of magnetic hyperthermia on human

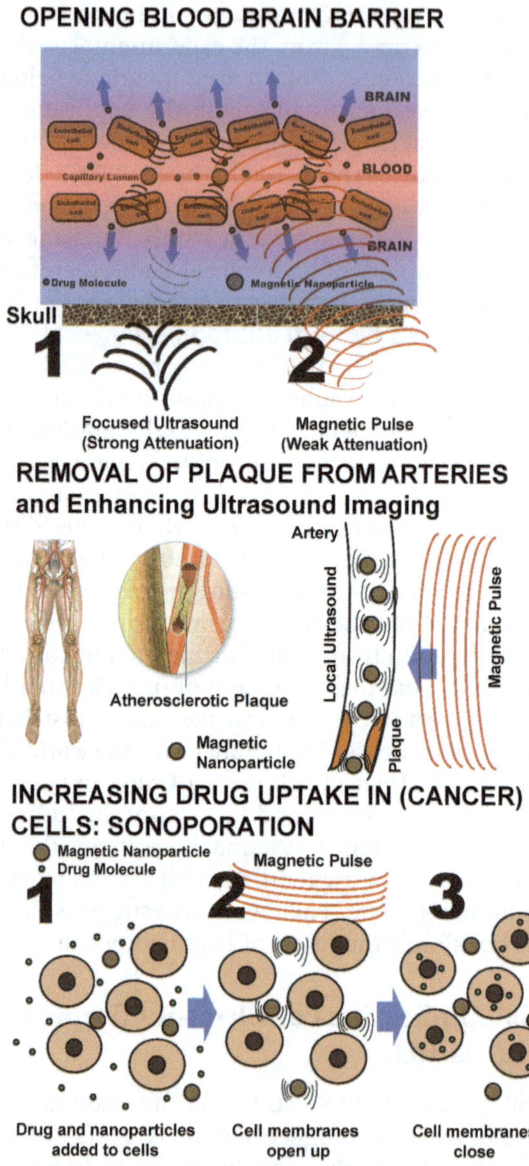

Figure 1.11 Potential applications of ultrasound from colloidal magnetic particles.

subjects with brain tumors utilize the direct injection method, which could be detrimental to the patients due to excessive bleeding. (2) A similar application is the delivery of drug molecules to drug resistant cancer cells. It is anticipated that the application of ultrasound from these magnetic structures can potentially open up lipid bilayers of cancerous cells for drug delivery. (3) In addition to the above-mentioned applications, ultrasound may be used in cleaning and imaging of clogged arteries in human patients.

First, the arteries can be isolated and filled with a medium containing magnetic nanoparticles. Second, the nanoparticles can be imaged either by MRI or ultrasonic imaging where the ultrasound is locally generated in the arteries. Third, the nanoparticles are exposed to an inhomogeneous magnetic field to create ultrasound for cleaning the arteries. Fourth, the arteries are imaged again for assessing the results. Finally, the magnetic nanoparticles can be drained from the arteries.

References

1. A. Komeili, Z. Li, D. K. Newman and G. J. Jensen, *Science*, 2006, **311**, 242–245.
2. D. A. Bazylinski, R. B. Frankel, B. R. Heywood, S. Mann, J. W. King, P. L. Donaghay and A. K. Hanson, *Appl. Environ. Microbiol.*, 1995, **61**, 3232–3239.
3. R. S. M. Rikken, R. J. M. Nolte, J. C. Maan, J. C. M. van Hest, D. A. Wilson and P. C. M. Christianen, *Soft Matter*, 2014, **10**, 1295–1308.
4. T. Indira and P. Lakshmi, *Int. J. Pharm. Sci. Nanotechnol.*, 2010, **3**, 1035–1042.
5. Y. A. Koksharov, *Magnetic Nanoparticles*, Wiley-VCH Verlag GmbH & Co. KGaA, 2009, pp. 197–254.
6. B. Issa, I. M. Obaidat, B. A. Albiss and Y. Haik, *Int. J. Mol. Sci.*, 2013, **14**, 21266–21305.
7. K. G. Paul, T. B. Frigo, J. Y. Groman and E. V. Groman, *Bioconjugate Chem.*, 2004, **15**, 394–401.
8. A. Akbarzadeh, M. Samiei and S. Davaran, *Nanoscale Res. Lett.*, 2012, **7**, 1–13.
9. A. G. Kolhatkar, A. C. Jamison, D. Litvinov, R. C. Willson and T. R. Lee, *Int. J. Mol. Sci.*, 2013, **14**, 15977–16009.
10. G. Orive, O. A. Ali, E. Anitua, J. L. Pedraz and D. F. Emerich, *Biochim. Biophys. Acta, Rev. Cancer*, 2010, **1806**, 96–107.
11. J. J. Li, Y. Liu, R. T. Cha, B. Ran, K. W. Mou, H. S. Wang, Q. Xie, J. S. Sun and X. Y. Jiang, *RSC Adv.*, 2016, **6**, 14329–14334.
12. F. Ye, S. Laurent, A. Fornara, L. Astolfi, J. Qin, A. Roch, A. Martini, M. S. Toprak, R. N. Muller and M. Muhammed, *Contrast Media Mol. Imaging*, 2012, 7, 460–468.
13. X. Batlle, N. Perez, P. Guardia, O. Iglesias, A. Labarta, F. Bartolome, L. M. Garcia, J. Bartolome, A. G. Roca, M. P. Morales and C. J. Serna, *J. Appl. Phys.*, 2011, **109**, 07B524.
14. P. Guardia, A. Labarta and X. Batlle, *J. Phys. Chem. C*, 2011, **115**, 390–396.
15. C. Pereira, A. M. Pereira, C. Fernandes, M. Rocha, R. Mendes, M. P. Fernandez-Garcia, A. Guedes, P. B. Tavares, J. M. Greneche, J. P. Araujo and C. Freire, *Chem. Mater.*, 2012, **24**, 1496–1504.
16. X. M. He and H. G. Shi, *Particuology*, 2012, **10**, 497–502.
17. M. D. Simon and A. K. Geim, *J. Appl. Phys.*, 2000, **87**, 6200–6204.

18. M. Yamaguchi, T. Andoh, T. Goto, A. Hosono, T. Kawakami, F. Okumura, T. Takenaka and I. Yamamoto, *IEEE Trans. Biomed. Eng.*, 1994, **41**, 1188–1191.
19. V. Novickij, A. Grainys, J. Novickij and S. Markovskaja, *IET Nanobiotechnol.*, 2014, **8**, 157–162.
20. D. C. Yin, *Prog. Cryst. Growth Charact. Mater.*, 2015, **61**, 1–26.
21. J. M. Zheng, W. C. Chin, E. Khijniak, E. Khijniak and G. H. Pollack, *Adv. Colloid Interface Sci.*, 2006, **127**, 19–27.
22. M. Zmyslony, E. Rajkowska, P. Mamrot, P. Politanski and J. Jajte, *Bioelectromagnetics*, 2004, **25**, 607–612.
23. A. Ahlbom, J. Bridges, R. de Seze, L. Hillert, J. Juutilainen, M. O. Mattsson, G. Neubauer, J. Schuz, M. Simko and K. Bromen, *Toxicology*, 2008, **246**, 248–250.
24. A. Jordan, *Int. J. Hyperthermia*, 2009, **25**, 512–516.
25. A. B. Pakhomov, Y. Bao and K. M. Krishnan, *J. Appl. Phys.*, 2005, **97**, 10Q305.
26. R. Kötitz, P. C. Fannin and L. Trahms, *J. Magn. Magn. Mater.*, 1995, **149**, 42–46.
27. R. Hergt, R. Hiergeist, M. Zeisberger, G. Glockl, W. Weitschies, P. Ramirez, I. Hilger and W. A. Kaiser, *J. Magn. Magn. Mater.*, 2004, **280**, 358–368.
28. A. K. Gupta and M. Gupta, *Biomaterials*, 2005, **26**, 3995–4021.
29. Q. A. Pankhurst, J. Connolly, S. K. Jones and J. Dobson, *J. Phys. D: Appl. Phys.*, 2003, **36**, R167–R181.
30. P. Guardia, R. Di Corato, L. Lartigue, C. Wilhelm, A. Espinosa, M. Garcia-Hernandez, F. Gazeau, L. Manna and T. Pellegrino, *ACS Nano*, 2012, **6**, 3080–3091.
31. V. A. Sharapova, M. A. Uimin, A. A. Mysik and A. E. Ermakov, *Phys. Met. Metallogr.*, 2010, **110**, 5–12.
32. Y. L. Raikher and V. I. Stepanov, *J. Exp. Theor. Phys.*, 2011, **112**, 173–177.
33. Y. L. Raikher and V. I. Stepanov, *Phys. Rev. E*, 2011, **83**, 021401.
34. J. G. Gibbs and P. Fischer, *Chem. Commun.*, 2015, **51**, 4192–4195.
35. R. Kotitz, P. C. Fannin and L. Trahms, *J. Magn. Magn. Mater.*, 1995, **149**, 42–46.
36. A. H. Habib, C. L. Ondeck, P. Chaudhary, M. R. Bockstaller and M. E. McHenry, *J. Appl. Phys.*, 2008, **103**, 07A307.
37. J. P. Fortin, C. Wilhelm, J. Servais, C. Menager, J. C. Bacri and F. Gazeau, *J. Am. Chem. Soc.*, 2007, **129**, 2628–2635.
38. S. Balivada, R. S. Rachakatla, H. Wang, T. Samarakoon, R. K. Dani, M. Pyle, F. Kroh, B. Walker, X. Leaym, O. Koper, M. Tamura, V. Chikan, S. Bossmann and D. Troyer, *BMC Cancer*, 2010, **10**, 119.
39. M. T. Basel, S. Balivada, H. Wang, T. B. Shrestha, G. M. Seo, M. Pyle, G. Abayaweera, R. Dani, O. B. Koper, M. Tamura, V. Chikan, S. H. Bossmann and D. L. Troyer, *Int. J. Nanomed.*, 2012, **7**, 297–306.
40. D. Liu, L. J. Wang, Z. G. Wang and A. Cuschieri, *Nano Lett.*, 2012, **12**, 5117–5121.

41. M. M. G. Krishna and N. Periasamy, *J. Fluoresc.*, 1998, **8**, 81–91.
42. M. Feyen, E. Heim, F. Ludwig and A. M. Schmidt, *Chem. Mater.*, 2008, **20**, 2942–2948.
43. R. Martin and E. Andreas, *Appl. Organomet. Chem.*, 2004, **18**, 536–541.
44. J. A. Bain and W. F. Egelhoff, *Appl. Phys. Lett.*, 2006, **88**, 242508.
45. Y. W. Jun, J. W. Seo and A. Cheon, *Acc. Chem. Res.*, 2008, **41**, 179–189.
46. I. Ohtani, H. Matsuzaki, Y. Kimura, E. Morita, H. Ogata, T. Ida, M. Izumi, M. Miki and M. Kitano, *Supercond. Sci. Technol.*, 2006, **19**, S521–S524.
47. G. Podaru, J. Moore, R. K. Dani, P. Prakash and V. Chikan, *Rev. Sci. Instrum.*, 2015, **86**, 034701.
48. R. Kratz and P. Wyder, *Principles of Pulsed Magnet Design*, Springer-Verlag, Berlin Heidelberg, 2002.
49. E. A. Chaniotakis, L. Bromberg and D. R. Cohn, *J. Fusion Energy*, 1994, **13**, 291–294.
50. M. A. Machado Gama, *Opt. Quantum Electron.*, 1975, **7**, 334.
51. D. G. Pellinen, M. S. Dicapua, S. E. Sampayan, H. Gerbracht and M. Wang, *Rev. Sci. Instrum.*, 1980, **51**, 1535–1540.
52. A. Ghosh and P. Fischer, *Nano Lett.*, 2009, **9**, 2243–2245.
53. P. Fischer and A. Ghosh, *Nanoscale*, 2011, **3**, 557–563.
54. A. L. Barnes, R. A. Wassel, F. Mondalek, K. Chen, K. J. Dormer and R. D. Kopke, *Biomagn. Res. Technol.*, 2007, **5**, 1.
55. G. Podaru, S. Ogden, A. Baxter, T. Shrestha, S. Q. Ren, P. Thapa, R. K. Dani, H. W. Wang, M. T. Basel, P. Prakash, S. H. Bossmann and V. Chikan, *J. Phys. Chem. B*, 2014, **118**, 11715–11722.
56. S. Nappini, F. B. Bombelli, M. Bonini, B. Norden and P. Baglioni, *Soft Matter*, 2010, **6**, 154–162.
57. G. V. Podaru and V. Chikan, *J. Phys. Chem. C*, 2016, **120**, 2386–2391.
58. R. Spera, F. Apollonio, M. Liberti, A. Paffi, C. Merla, R. Pinto and S. Petralito, *Colloids Surf., B*, 2015, **131**, 136–140.
59. A. D. Bangham and R. W. Horne, *J. Mol. Biol.*, 1964, **8**, 660–668.
60. G. Gregoriadis and B. E. Ryman, *Eur. J. Biochem.*, 1972, **24**, 485–491.
61. G. Gregoriadis, D. Putman, L. Louis and D. Neerunjun, *Biochem. J.*, 1974, **140**, 323–330.
62. G. Gregoriadis, in *Enzyme Replacement Therapy of Lysosomal Storage Diseases*, ed. J. M. Tager, G. J. M. Hooghwinkel and W. T. Daems, North Holland Publishing Company, 1974, pp. 131–148.
63. A. A. Gabizon, *Nanotechnol. Cancer Ther.*, 2007, 595–611.
64. D. B. Fenske and P. R. Cullis, *Methods Enzymol.*, 2005, **391**, 7–40.
65. A. D. Bangham, M. W. Hill and N. G. A. Miller, in *Methods in Membrane Biology*, ed. E. P. E. D. Korn, Plenium Press, New York, 1974, pp. 1–68.
66. A. Gueven, M. Ortiz, M. Constanti and C. K. O'Sullivan, *J. Liposome Res.*, 2009, **19**, 148–154.
67. D. D. Lasic, F. J. Martin, A. Gabizon, S. K. Huang and D. Papahadjopoulos, *Biochim. Biophys. Acta, Biomembr.*, 1991, **1070**, 187–192.
68. M. A. Moses, H. Brem and R. Langer, *Cancer Cell*, 2003, **4**, 337–341.

69. T. M. Allen and P. R. Cullis, *Science*, 2004, **303**, 1818–1822.
70. M. T. Basel, R. K. Dani, M. Kang, M. Pavlenok, V. Chikan, P. E. Smith, M. Niederweis and S. H. Bossmann, *ACS Nano*, 2009, **3**, 462–466.
71. A. Sparreboom, D. F. S. Kehrer, R. H. J. Mathijssen, R. Xie, M. J. A. de Jonge, P. de Bruijn, A. S. T. Planting, F. A. L. M. Eskens, C. Verheij, G. de Heus, A. Klaren, S. Zhang, T. Verhaeghe, P. A. Palmer and J. Verweij, *Br. J. Cancer*, 2004, **90**, 1508–1515.
72. E. Rozners, *Latv. Kim. Z.*, 2006, **1**, 28–40.
73. G. R. Anyarambhatla and D. Needham, *J. Liposome Res.*, 1999, **9**, 491–506.
74. B. S. Shaghasemi and E. Reimhult, *250th American Chemical Society National Meeting & Exposition*, Boston, 2015.
75. M. Kinoshita, N. McDannold, F. A. Jolesz and K. Hynynen, *Proc. Natl. Acad. Sci. U. S. A.*, 2006, **103**, 11719–11723.
76. F. Xie, M. D. Boska, J. Lof, M. G. Uberti, J. M. Tsutsui and T. R. Porter, *Ultrasound Med. Biol.*, 2008, **34**, 2028–2034.

CHAPTER 2

Magnetic Nanoparticle Design and Application in Magnetic Hyperthermia

ANDREW KONECNY[a], JOSE COVARRUBIAS[a] AND HONGWANG WANG*[a]

[a]Department of Chemistry, Kansas State University, Manhattan, KS, 66506, USA
*E-mail: hongwang@ksu.edu

2.1 Introduction

Hyperthermia is a novel cancer treatment modality which uses heat to destroy cancer cells.[1] Owing to their rapid metabolic rate, cancer cells are more sensitive than healthy cells to elevated temperatures. For example, in a temperature range of 41–46 °C, heat stress occurs in cancer cells, leading to different types of intra- and extracellular degradations, such as protein denaturation, protein folding, aggregation, DNA cross linking, and eventually triggered cell death.[2] In addition, hyperthermia renders cancer cells more susceptible to other treatment modalities, such as chemotherapy and radiation therapy. Hyperthermia-induced thermal sensitization causes a change of vascular permeability, resulting in increased tumor oxygenation, which in turn enhances the cytotoxicity of chemo-/radiotherapy. Therefore, hyperthermia is usually administered clinically in combination with chemo-/radiotherapy.[3]

Smart Materials No. 26
Magnetic Nanomaterials: Applications in Catalysis and Life Sciences
Edited by Stefan H. Bossmann and Hongwang Wang

Published by the Royal Society of Chemistry, www.rsc.org

Heat delivery, heat distribution, and thermal dosage are crucial issues in hyperthermia. In ideal situations, heat should be localized only in the tumor area in order to minimize the damage to the surrounding normal tissue; in addition, heat should be distributed evenly inside the tumor to achieve total elimination of cancer cells to prevent relapses. Furthermore, a reliable thermal dosage should be applied. Conventional hyperthermia uses an external heat source such as ultrasound, radiofrequency, microwave, infrared radiation or tubes with hot water, to transfer heat from outside the cells toward the inside. During the thermal energy transfer, a temperature gradient is generated along the path of radiation, dissipating a large amount of energy in the healthy tissue. This non-thermal discrimination of tissue heating can lead to serious side effects. In this type of hyperthermia, it is also impossible to achieve even distribution of thermal energy inside the tumor. Due to the heterogenic energy distribution, proper thermal dosage is hard to evaluate.[4]

Magnetic material has the capability of converting magnetic energy to thermal energy. Magnetic hyperthermia for cancer treatment was first demonstrated in 1957 by Gilchrist *et al.*[5]; the development of magnetic nanomaterials was then pursued for 60 years.[6] Magnetic nanomaterials in the lens of life sciences house great potential for therapeutic and diagnostic devices in the growing field of personalized medicine.[7] Due to their size, nanoparticles, defined as particles of less than 100 nm in all dimensions, are able to interact with larger biomolecules such as DNA, RNA, and protein. The size range of 10–100 nm also presents optimal tumoral retention with a slow barrier diffusion.[7a] Intrinsically, magnetic nanomaterials contain novel physical attributes, which include: a large surface-to-volume ratio, multifunctionality with respect to a diverse pool of coatings, and diversity in architecture in terms of shape, size, and structure. More importantly, magnetic nanoparticles have demonstrated far better capability for heat generation than their micro sized counterparts in an alternating magnetic field (AMF).[8] With the development of stable and biocompatible MNP colloids, MNP-based magnetic hyperthermia has demonstrated many advantages compared to conventional hyperthermia. First, due to the nanoscale size, the heat probe can be administered through a number of delivery pathways, and can be accumulated at the tumor site either passively by the Enhanced Permeation and Retention (EPR) effect,[9] or actively by the cancer-specific binding moieties on the nanoparticles.[10] Therefore, upon applying AMF, more specific and effective hyperthermia can be achieved. Second, after being enriched at the tumor site, MNPs can be internalized by cancer cells through different pathways depending on the size of the particles and where intracellular hyperthermia occurs with heat localized within cancer cells.[11] Third, due to the multi-functionality of MNPs, magnetic hyperthermia-based controlled drug delivery systems have been developed.[12] The success of magnetic hyperthermia research has gradually transformed into clinical applications. In 2004, Charité-Medical University of Berlin developed the first clinical magnetic hyperthermia treatment system,[13] and in 2010, two products based on magnetic Fe_3O_4 nanoparticles, NANOTHERM® and NANO-ACTIVATOR®, developed by MagForce, were approved for magnetic hyperthermia treatment of brain tumors in Europe. In 2015, MagForce installed the first clinical NANOACTIVATOR® in USA, which extended its capability to treating prostate cancer in addition to the treatment of brain tumors.[14]

Clinically, MNP-based magnetic hyperthermia involves multi-step operations and mechanisms including: the administration of MNP dispersions, preferable accumulation of MNP at the tumor site, and distribution of MNPs in tumor cells, followed by exposure to an external alternating magnetic field.[15] Acting as heat mediator, magnetic nanoparticles play a crucial role in the success of magnetic hyperthermia. Several criteria have to be met when designing magnetic nanoparticles to achieve widespread use of MNP-based magnetic hyperthermia in clinical treatment of different types of cancers. First, MNPs should have the capability of generating heat sufficiently in a safe applied-AC magnetic field;[16] second, MNPs should have sufficient aqueous dispersibility, biological stability, and biocompatibility in order to be used in biomedical applications;[17] third, MNPs should act as a centerpiece for constructing a multi-functional nanoplatform, which is capable of targeting tumors and releasing preloaded anticancer drugs, in addition to the hyperthermia treatment.[18]

This chapter will cover three sections, including: the design and synthesis of MNPs, which have optimal heating ability in a biomedically safe AC magnetic field; surface coating/decorating of the MNPs to render solubility, biocompatibility, and multi-functionality to the MNPs; and clinical MNP-based magnetic hyperthermia applications.

2.2 Design and Synthesis of MNPs for Magnetic Hyperthermia

2.2.1 Heating Mechanisms

MNP-mediated conversion of electromagnetic energy to thermal energy is mainly according to two mechanisms: relaxation losses (Néel and Brownian relaxation) occur in superparamagnetic or single-domain magnetic particles, and hysteresis losses occur in multi-domain magnetic particles. Néel relaxation involves the movement of the magnetic moments relative to the crystal lattice of the single-domain MNP without the rotation of the particle, and in contrast, the Brownian relaxation pertains to the rotation of the MNP relative to the surrounding environment.[19]

After the external magnetic field is temporarily removed, the previously allied magnetic moments of the MNPs relax back to random distribution in a typical relaxation time. For Néel relaxation, the characteristic relaxation time τ_N can be expressed in the following equation:

$$\tau_N = \left(\frac{\sqrt{\pi}}{2}\right)\tau_0 \frac{\exp\left(\frac{KV}{k_B T}\right)}{\sqrt{\frac{KV}{k_B T}}}$$

where K is the magnetic anisotropy, and V is the volume of the particle.

And for Brownian relaxation, the characteristic relaxation time τ_B is expressed in the following equation:

$$\tau_B = \frac{3\eta V_H}{k_B T}$$

where η is the dynamic viscosity of the media, and V_H is the hydrodynamic volume of the particle.

The overall effective relaxation time is determined by a collective contribution of both Néel and Brownian relaxation processes, and can be expressed as:

$$\frac{1}{\tau} = \frac{1}{\tau_B} + \frac{1}{\tau_N}$$

The power dissipation (P) of magnetic nanoparticles inside an alternating magnetic field can be expressed by Rosensweig's linear response theory (LRT), in the following equation:[20]

$$P = \mu_0 \pi \chi_0 f H^2 \left(\frac{2\pi f \tau}{1 + [2\pi f \tau]^2} \right)$$

where μ_0 is the permeability of free space, χ_0 is the actual susceptibility, f and H are the frequency and amplitude of the AC magnetic field, τ is the effective relaxation time.

The heating efficiency of MNPs is evaluated by the specific absorption rate (SAR), which is defined as the ratio of the dissipated heat power over the mass of the magnetic elements:

$$\text{SAR} = \frac{P}{m_{\text{mag}}}$$

The SAR also can be calculated by the following equation:

$$\text{SAR} = C \frac{\Delta T}{\Delta t} \frac{1}{m_{\text{mag}}}$$

where C is the specific heat of solvent ($C_{\text{water}} = 4.18$ J g^{-1} °C), $\Delta T/\Delta t$ is the initial slope of the time-dependent temperature curve, and m_{mag} is the weight fraction of the magnetic element in the MNP sample.

For given MNPs, higher SAR values are usually observed at a higher frequency and amplitude of external magnetic field, however, for the safe application of hyperthermia on patients, the product of the magnetic field frequency and field amplitude should not exceed the threshold of 5×10^9 A m^{-1}s^{-1}.[21]

2.2.2 Design of MNPs for Magnetic Hyperthermia

To this end, we ask the question: what parameters need to be considered when designing MNPs for magnetic hyperthermia applications?

First, the materials used to prepare the MNPs have to be biocompatible, or have virtually no cytotoxicity in order to be used in a biomedical application. Owing to their well-known biocompatibility, biodegradability, and abundant supply, iron oxide based MNPs become the preferred candidates.[7a,17]

Second, the size and the polydispersity of the MNPs determines their performance. For superparamagnetic nanoparticles, their Néel and Brownian relaxation losses are size dependent. The magnetization of MNPs decreases with reduced particle size.[22] According to the linear response theory (LRT), the power dissipation (P) of magnetic nanoparticles inside an alternating magnetic field is governed by the relaxation time and magnetization of MNPs. Therefore, there is an optimum size that will lead to enhanced hyperthermia for a given magnetic material. It is also of vital importance that the polydispersity of the MNPs be controlled as narrowly as possible in order to maintain the proportion of the heat-generating particles.[16b]

Third, the shape of MNPs is an important factor. MNPs of different shapes possess different shape anisotropy, which strongly influences the MNP's heating efficiency in AMF.[23] Bauer *et al.* have demonstrated that with a similar magnetic volume, the iron oxide nanocubes heat more efficiently than iron oxide nanospheres.[24] In addition, nanoscale iron oxide particles with different morphologies, such as nano-octapods, nanorods, nanoflowers, were reported to show higher SAR values than their spherical counterparts due to their enhanced shape anisotropy.[25]

Fourth, the magnetization of MNPs is a decisive factor. The SAR of MNPs in a given AC magnetic field is proportional to their saturation magnetization. For example, by changing the chemical composition, the Jacobsite ($MnFe_2O_4$), and Zn^{2+}-ion-doped spinel ferrite ($(Zn_{0.4}Fe_{0.6})Fe_2O_4$) nanoparticles demonstrated higher magnetization than their Fe_3O_4 analog.[24,26] Both of these types of particles showed higher SAR values, compared to Fe_3O_4-based particles of a similar size.

2.3 Synthesis Strategies

2.3.1 Mechanism for the Formation of Monodisperse Nanoparticles

In 1950, LaMer and Dinegar proposed a mechanism for the formation of monodisperse nanoparticles: the classical model of burst nucleation and growth.[27] Three consecutive steps are involved as demonstrated in Figure 2.1. In Step I, the concentration of the "monomers" increases with the progress of reaction. In Step II, when the concentration of the "monomers" reaches the nucleation threshold, nuclei form rapidly, and this process is termed a "burst nucleation". A large number of nuclei form in a short period of time, and as a consequence the monomer concentration drops below the nucleation threshold. In Step III, driven by the free energy, the monomers in the reaction solution attach to the preformed nuclei instead of forming new nuclei, and finally result in the formation of monodisperse nanoparticles.

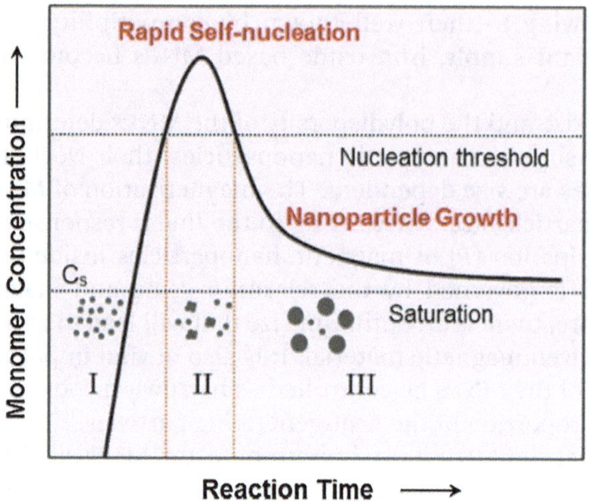

Figure 2.1 LaMer model of the nucleation and growth process of monodisperse NPs. Reprinted with permission from ref. 41. Copyright (2016) American Chemical Society.

2.3.2 Co-Precipitation

Co-precipitation is a convenient method of preparing iron oxide nanoparticles (Fe_3O_4, γFe_2O_3) by reacting a 1 : 2 molar ratio of ferric and ferrous ions in a basic aqueous medium at an ambient or elevated temperature.[28] The reaction can be described in the following two equations. First, magnetite (Fe_3O_4) particles are formed in inert atmosphere according to eqn (2.1):

$$Fe^{2+} + 2Fe^{3+} + 8OH^{-} \rightarrow Fe_3O_4 + 4H_2O \quad (2.1)$$

Second, since magnetite (Fe_3O_4) particles are not stable, they are easily oxidized to hematite (Fe_2O_3) or maghemite based on eqn (2.2):

$$Fe_3O_4 + 2H^{+} \rightarrow Fe_2O_3 + Fe^{2+} + H_2O \quad (2.2)$$

Co-precipitation gains popularity because of the mild reaction conditions, short reaction times, and easy scalability. However, this method has very limited control over particle size and shape, and usually produces magnetic nanoparticles with a broad size distribution. In addition, the saturation magnetization values of the MNPs prepared through co-precipitation are usually much lower than the bulk value. Therefore, the MNPs prepared *via* co-precipitation found limited success in the magnetic hyperthermia application.

2.3.3 Microemulsions

Microemulsion is a thermodynamically stable isotropic dispersion formed by two immiscible liquids with the help of surfactant.[29] In water-in-oil microemulsions, water disperses as 1 to 50 nm microdroplets, which are surrounded by a monolayer of surfactant molecules in the hydrophobic phase. The size of the water microdroplets can be tuned by varying the volume ratio of the two phases and the concentration of the surfactant.[30] Microemulsion can act as a nanoreactor for the formation of magnetic nanoparticles. In a typical synthesis, reactants are first dissolved in the aqueous phase. Then the aqueous solution is mixed with a hydrophobic phase in the presence of a surfactant and a co-surfactant to form water-in-oil microemulsions. Upon mixing these microemulsions, their microdroplets containing different reactants will collide, break, and coalesce. The previously isolated reactants will be mixed and react in a confined environment, leading to the formation of controllably sized nanoparticles.[31] At the end of the reaction, nanoparticles can be collected by first breaking the microdroplets with acetone or ethanol, and then followed by centrifugation and washing steps.

Liu *et al.* reported the synthesis of single crystalline spinel ferrite $MnFe_2O_4$ in reverse micelles formed by water-in-toluene microemulsion in the presence of sodium dodecylbenzenesulfonate (NaDBS) as surfactant.[32] The size of the nanoparticles can be controlled from 4 to 15 nm by changing the volume ratio of water and toluene. The size distribution is usually about 15%, and can be narrowed down to 9% in some optimized cases. Neutron diffraction and magnetic measurements showed that 7.7 ± 0.7 nm particles are superparamagnetic. Gupta *et al.* synthesized superparamagnetic iron oxide nanoparticles in aerosol-OT (AOT)/n-hexane reverse micelles.[33] Fe^{3+} and Fe^{2+} salts with molar ratio of 2 : 1 were dissolved in the aqueous core of the reverse micelles formed by AOT in *n*-hexane. Co-precipitation was induced by using a deoxygenated solution of sodium hydroxide. The size of the nanoparticles was controlled under 15 nm with narrow size distribution, and the nanoparticles showed high magnetization values.[34] Vidal-Vidal *et al.* reported a one-pot emulsion method for the synthesis of both coated and uncoated monodisperse maghemite nanoparticles. Using Brij97A as a non-ionic surfactant, a stable water-in-oil emulsion was created by mixing water into cyclohexane phase. Maghemite nanoparticles were formed by co-precipitation of ferrous and ferric salts in the water-in-oil emulsion with cyclohexylamine and oleylamine as bases. Spherical nanoparticles with narrow size distribution were obtained by this method, and despite the small size, a high saturation magnetization of 76.3 A m^2 kg^{-1} was reported for the 3.5 nm bare nanoparticles.[34]

One of the major advantages of the microemulsion method is that the size of the microdroplets can be fine-tuned by modulating the ratio of the two phases and the concentration of surfactant. The formation of microemulsions is a simple procedure because of the well-defined thermodynamic stability of microemulsion systems. The main drawback of this system is the relatively low yield of nanoparticles compared to other methods; in addition, large amounts of organic solvents have to be used in this system, which limits the large-scale production.

2.3.4 Hydrothermal Synthesis

Hydrothermal synthesis of nanoparticles is based on a wet-chemical synthesis in sealed reactors or autoclaves at high vapor pressure (from 0.3 to 4 MPa) and elevated temperature (130 to 250 °C). High reaction temperature leads to rapid nucleation and fast growth of the particles. Hydrolysis/oxidation and neutralization of mixed metal hydroxides are two major routes to achieve ferrite nanoparticle formation under hydrothermal conditions.[35] This technique has also been used to grow dislocation-free single crystal particles, and the particles formed *via* this method usually demonstrate a better crystallinity than those from other processes. In hydrothermal synthesis, the geometry of the product is affected by the reaction time, temperature, solvents, and metal precursors. For example, at given reaction conditions, a higher reaction temperature results in smaller particles, because of the faster nucleation rate at higher temperatures. On the contrary, longer reaction times lead to the formation of larger particles.[36]

Ge *et al.* reported a facile one-step synthesis of Fe_3O_4 nanoparticle under hydrothermal conditions.[37] The synthesis is based on the oxidation of $FeCl_2{\cdot}4H_2O$ in basic solution at elevated temperature and pressure. The size of the nanoparticles can be controlled from 15 to 31 nm with narrow size distribution through modulation of reaction parameters. The obtained nanoparticles are highly crystalline, and exhibit high saturation magnetization values (53.3–97.4 emu g^{-1}). 20 nm magnetite nanoparticles were synthesized by mixing of $FeCl_2{\cdot}4H_2O$ and $FeCl_3{\cdot}6H_2O$ at 473 K for 24 hours in a Teflon-lined vessel in the presence of oleate or PAA solutions. The nanoparticles showed high saturation magnetization (84 emu g^{-1}).[38] Larger sized magnetite nanoparticles were synthesized through a two-step reaction: first, ferrous Fe^{2+} and ferric Fe^{3+} ions were co-precipitated at 70 °C using $N(CH_3)_4OH$ as a base, second, the precipitation underwent thermal treatment at 250 °C for 24 hours. A significant increase in the saturation magnetization (59.8–82.5 emu g^{-1}) of the nanoparticles was observed after the thermal treatment.[39]

Fe_3O_4 nanorods of tunable aspect ratio were synthesized using the hydrothermal method with iron pentacarbonyl as a precursor in the presence of oleic acid and hexadecylamine (HAD) (see Figure 2.2).[40] Due to the larger saturation magnetization and effective anisotropy, the Fe_3O_4 nanorods demonstrated far better heating efficiency (862 W g^{-1}) compared to cubic Fe_3O_4 (314 W g^{-1}) and spherical Fe_3O_4 (140 W g^{-1}) with similar volumes.

2.3.5 Thermal Decomposition

Thermal decomposition of iron precursors in high boiling-point organic solvents in the presence of surfactants produces monodisperse magnetic nanoparticles. In addition, this method has extreme control over the particle size, shape, and morphology. At high reaction temperatures, iron precursors decompose and undergo burst nucleation rapidly, which is followed by steady growth to form monodisperse nanoparticles. The most

Figure 2.2 Scheme for the synthesis of tunable aspect ratio Fe_3O_4 nanorods. All the reactions were performed in an autoclave with a Teflon liner at 200 °C, varying the oleic acid (OA) to hexadecylamine (HDA) ratio. Shown in samples S1–S3 are TEM images of the as-synthesized Fe_3O_4 nanorods. Scale bar is 200 nm. Photographs on the top left reveal the highly monodisperse nature of Fe_3O_4 nanorods (sample S1), which were dispersed in hexane in (A) the absence and (B) the presence of a permanent magnet. Reprinted with permission from ref. 40. Copyright (2016) American Chemical Society.

commonly used iron precursors are [Fe(acac)$_3$] (acac = acetylacetonate), [Fe(Cup)$_3$] (Cup = *N*-nitrosophenylhydroxylamine), [Fe(CO)$_5$], and Fe[N-(SiMe$_3$)]$_2$. The high boiling-point solvents of choice are phenyl ether, benzyl ether, octyl ether, 1-hexadecane, 1-octadecene, 1-eicosene hexadecanodiol, and trioctylamine. Fatty acids, oleic acid, and long chain amines are commonly used surfactants.[41]

Fe_3O_4 nanoparticles are produced through the thermal decomposition of [Fe(acac)$_3$] in the presence of a reducing agent with oleic acid or oleylamine as surfactant. 4–6 nm Fe_3O_4 nanoparticles are obtained in this reaction. In addition, larger Fe_3O_4 NPs can be synthesized through a seed-mediated-growth, by further reductive decomposition of [Fe(acac)$_3$] with the small Fe_3O_4 NPs serving as seeds. The reaction results in magnetic nanoparticles with controllable size ranging from 5–20 nm. The particles exhibit high crystallinity as well as uniform size distribution, as clearly illustrated in Figure 2.3.[42]

Kim *et al.* reported the synthesis of Fe_3O_4 nanocubes by thermal decomposition of [Fe(acac)$_3$] in benzyl ether using oleic acid as surfactant. The size of the particles can be tuned from 20–160 nm with narrow distribution, as seen in Figure 2.4.[43]

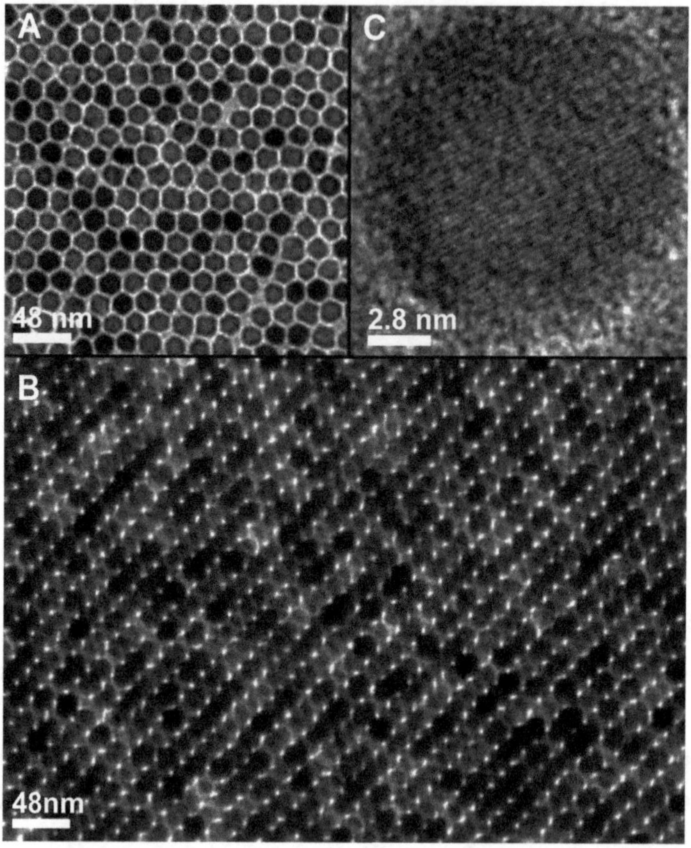

Figure 2.3 TEM bright field image of 16 nm Fe_3O_4 nanoparticles deposited from their dodecane dispersion on amorphous carbon surface and dried at 60 °C for 30 min: (A) a monolayer assembly, (B) a multilayer assembly, (C) HRTEM image of a single Fe_3O_4 nanoparticle. Reprinted with permission from ref. 42. Copyright 2002 American Chemical Society.

Similarly, using decanoic acid as surfactant, and by simply varying the degassing time and temperature, Guardia *et al.* synthesized Fe_3O_4 nanocubes of 12, 19, 25, and 38 nm.[44] Further exploring this system, Wang *et al.* reported the synthesis of hexagonal nanoplatelets using oleic acid/stearic acid as dual surfactants, as illustrated in Figure 2.5.[45] The obtained nanoplatelets are highly crystalline, and have demonstrated excellent heating ability in AC magnetic fields.

Zinc-doped $(Zn_{0.4}Fe_{0.6})Fe_2O_4$ nanoparticles were prepared by thermal decomposition of $Fe(acac)_3$ in the presence of $ZnCl_2$ and $FeCl_2$ in octyl ether at 300 °C, with oleic acid and oleylamine as surfactants. This type of MNPs demonstrated higher magnetization and higher SAR value than their un-doped Fe_3O_4 counterparts.[46] Furthermore, cubic $(Zn_{0.4}Fe_{0.6})Fe_2O_4$ nanoparticles were prepared by fine-tuning the reaction conditions. The SAR value of cubic $(Zn_{0.4}Fe_{0.6})Fe_2O_4$ nanoparticles was found to be two times higher than that of the spherical $(Zn_{0.4}Fe_{0.6})Fe_2O_4$ nanoparticles (see Figure 2.6).[24]

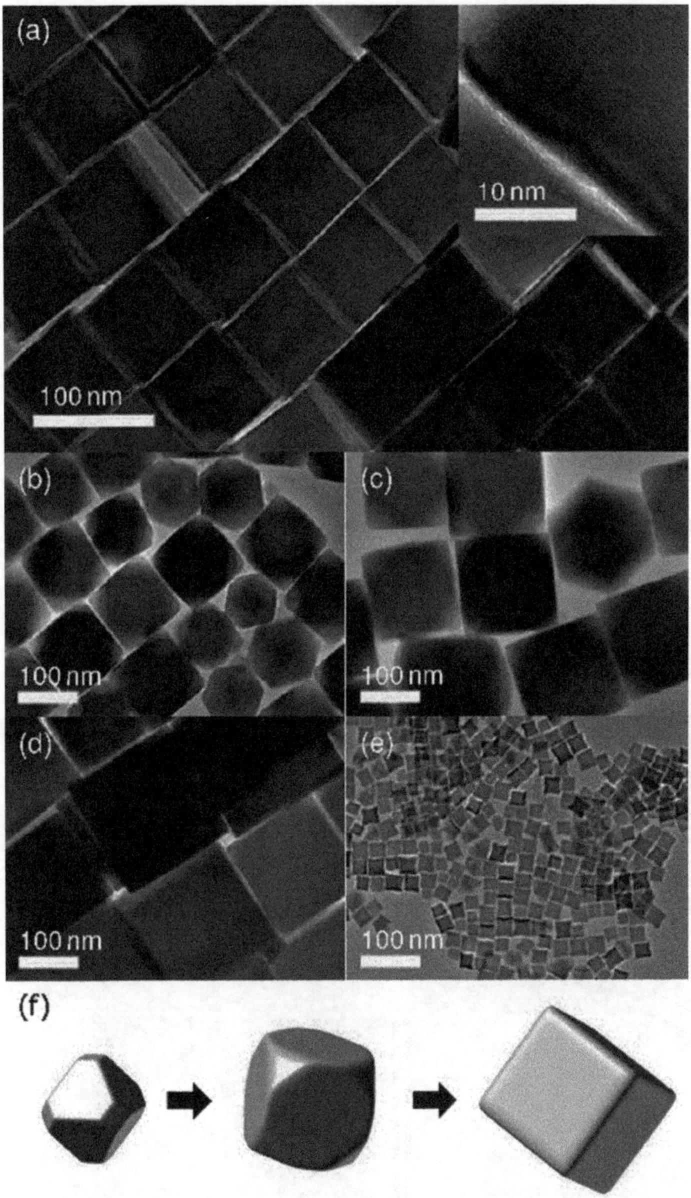

Figure 2.4 TEM images of (a) 79 nm-sized Fe_3O_4 nanocubes (inset: HRTEM image); (b) mixture of truncated cubic and truncated octahedral nanoparticles with an average dimension of 110 nm; (c) 150 nm-sized truncated nanocubes; (d) 160 nm-sized nanocubes; (e) 22 nm-sized nanocubes; (f) schematics showing the overall shape evolution of the Fe_3O_4 nanoparticles. Reproduced with permission from ref. 43. Copyright (2009) American Chemical Society.

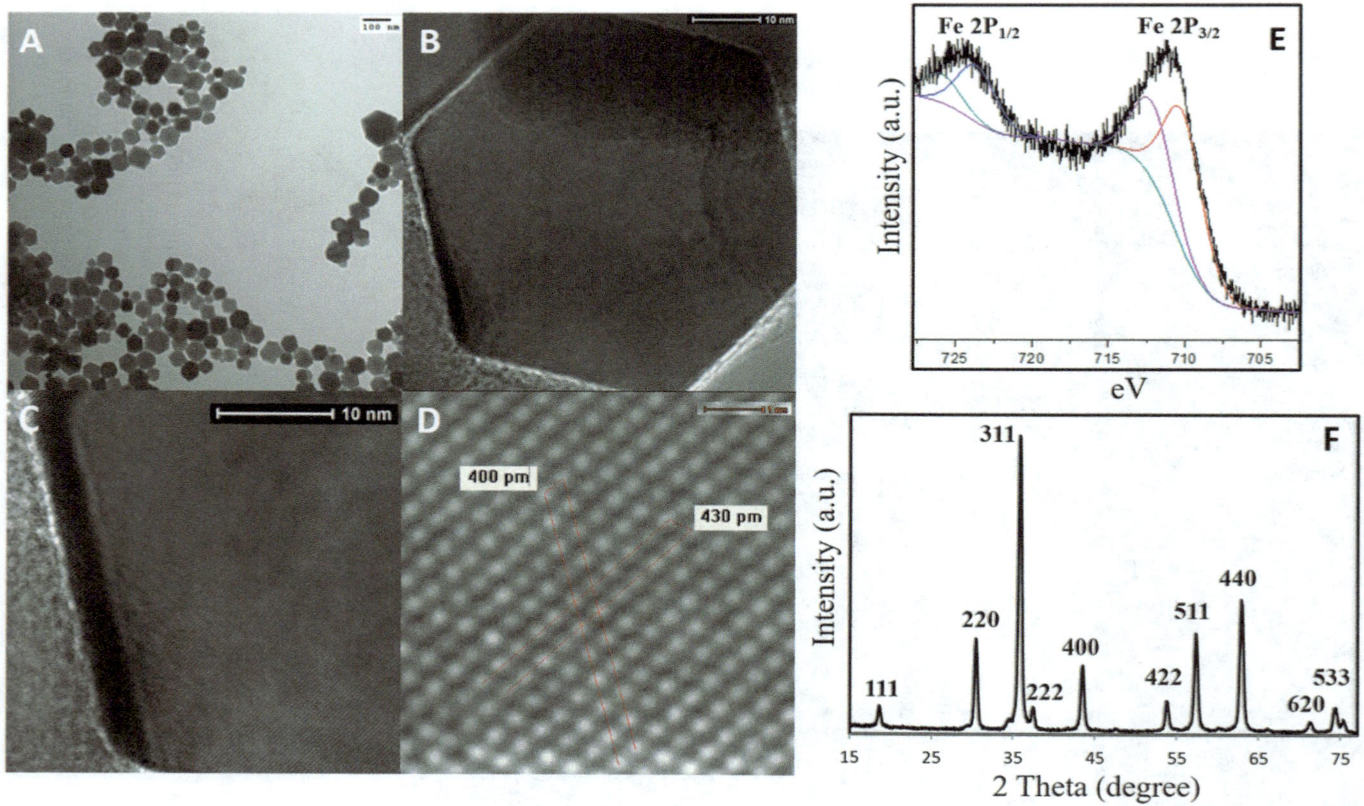

Figure 2.5 TEM and HRTEM of hexagonal Fe_3O_4 nanoplatelets: (A) TEM of nanoplatelets; (B) HRTEM of a representative hexagonal nanoplatelet; (C) HRTEM side-view of the depth of the nanoplatelet; (D) HRTEM of monocrystalline hexagonal nanoplatelet. The (111) planes of the iron bcc crystal structure are highlighted; (E) XPS spectrum of iron oxide (Fe_3O_4) nanoplatelets; (F) XRD patterns of iron oxide (Fe_3O_4) nanoplatelets. Reproduced from ref. 45 with permission from The Royal Society of Chemistry.

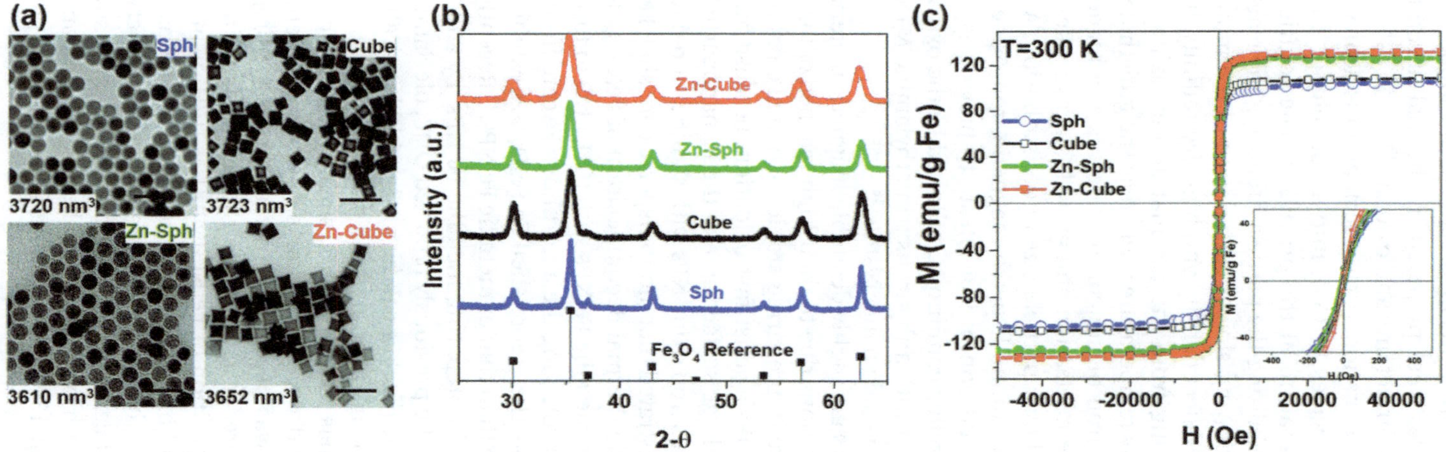

Figure 2.6 (a) Transmission electron microscopy images of magnetite spherical NPs (Fe_3O_4,Sph), cubic NPs (Fe_3O_4,Cube), and the corresponding zinc-doped magnetite spherical NPs ($Zn_{0.4}Fe_{2.6}O_4$,Zn–Sph), and cubic NPs ($Zn_{0.4}Fe_{2.6}O_4$,Zn–Cube) of equivalent volumes. The scale bars represent 50 nm; (b) the powder X-ray diffraction patterns; (c) the field-dependent magnetization curves of the IONP samples measured at 300 K. The inset displays the zoom-in portion of the field-dependent magnetization curves showing superparamagnetic behavior. Reproduced from ref. 24 with permission from The Royal Society of Chemistry.

Another commonly used iron precursor is $Fe(CO)_5$. Monodisperse crystalline γ-Fe_2O_3 nanoparticles were prepared by thermal decomposition of $Fe(CO)_5$ in the presence of surfactants and oxidant.[47]

The nucleation step of the nanoparticles' formation is achieved by rapid injection of precursor molecules $Fe(CO)_5$ into a hot solution heated above the decomposition temperature of the precursor. The solution contains coordinating ligands for stabilizing the particles, reducing the reactivity of the monomers and thereby allowing slow incorporation of the monomers into the particles with reduced defect density and high crystallinity. It was found that the nucleation resulting from the thermal decomposition of $Fe(CO)_5$ takes place at relatively low temperatures, whereas the growth derived from the decomposition of the iron oleate complex occurs at a higher temperature. The separation of the nucleation and growth stages makes it possible to achieve extreme control over particle size and distribution. A 1 nm-scale size-controlled synthesis of monodisperse magnetic iron oxide nanoparticles was achieved by the addition of iron oleate to previously synthesized iron nanoparticles under reflux conditions, followed by the oxidation of the obtained iron nanoparticles with anhydrous trimethylamine *N*-oxide to produce highly crystalline iron oxide nanocrystals.[48]

Metallic Fe nanoparticles are highly desirable magnetic materials in magnetic hyperthermia applications due to their high magnetizations (Ms = 218 emu g^{-1} Fe). One challenge in preparing stable metallic Fe nanoparticles is in how to stabilize these NPs against fast oxidation, because metallic Fe is chemically active, and the large surface area of the Fe nanoparticles enables efficient oxidation to various iron oxide NPs with significant reduced magnetizations. Sun *et al.* demonstrated that monodisperse core/shell Fe/Fe_3O_4 NPs can be prepared first by the thermal decomposition of $Fe(CO)_5$, followed by controlled oxidation of the metallic Fe NPs with $(CH_3)_3NO$ to form a crystalline iron oxide shell, thus protecting the amorphous Fe core from fast oxidation. Even with this method, these core/shell Fe/Fe_3O_4 NPs exhibit a relatively lower magnetization value than that of metallic Fe NPs.[49] Recently, Sun *et al.* achieved a major advance in preparing metallic Fe NPs. By adding hexadecylammonium chloride to the previously reported ODE, oleylamine, $Fe(CO)_5$ system, body-centered cubic (bcc) Fe can be made with drastically increased stability and magnetization, as illustrated in Figure 2.7.[50]

$Fe(CO)_5$ as a precursor has its limitations, such as high toxicity, low boiling-point, and complicated decomposition pathways. To overcome these drawbacks, $Fe[N(SiMe_3)]_2$ was used as the precursor for the preparation of iron nanoparticles. Chaudret *et al.* reported the synthesis of iron cubes by reductive thermal decomposition of $Fe[N(SiMe_3)]_2$ with H_2 using hexadecylamine (HAD) and oleic acid or hexadecylammonium chloride as surfactants (see Figure 2.8). The size of the nanocubes can be controlled between 7 and 8.3 nm by varying the concentration of the amine and acid surfactants.[51] Replacing the oleic acid with palmitic acid, the size of the nanocubes can be further controlled from 13–30 nm by changing the HAD : palmitic acid ratio.[52]

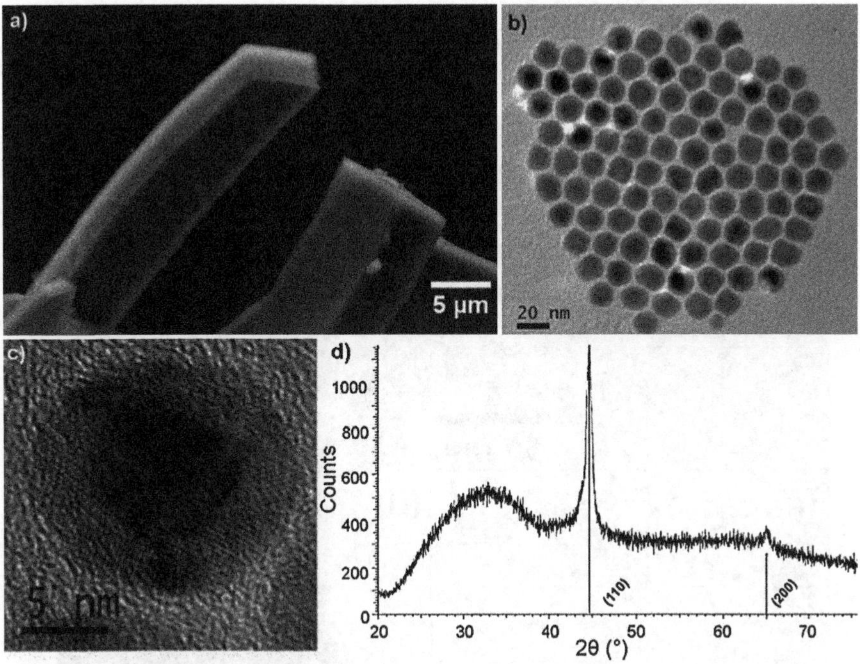

Figure 2.7 Morphological and structural characterization of bcc-Fe/Fe_3O_4 NPs: (a) SEM image of the platelike Fe NP assembly obtained directly from the synthesis solution; (b) TEM image of the 15 nm NPs obtained from the redispersion of the plate assembly in hexanes; (c) HRTEM image of a single NP revealing the metallic bcc-Fe core and Fe_3O_4 shell with Fe (110) and Fe_3O_4 (222) planes indicated; (d) XRD pattern of the bcc-Fe NPs. Reproduced with permission from ref. 50. Copyright (2011) American Chemical Society.

2.4 Functionalization of Magnetic Nanoparticles

Being able to use magnetic nanoparticles (MNPs) for hyperthermia depends not only on biocompatibility of the nanoparticles, but also on selecting the best functionalization groups that will optimize and allow MNPs to maintain the properties required for magnetic hyperthermia. Making magnetic particles biocompatible plays a crucial role if the goal is to use these particles for nano-bio applications. There are different ways and functional groups that can be used to fulfill the biocompatibility requirement. In this section, we will discuss different methods of functionalizing MNPs for hyperthermia, the benefits and different materials used for functionalization, and the bioconjugation of MNPs.

2.4.1 Functionalization Strategies of MNPs for Hyperthermia

The surface functionalization of MNPs is one of the most important steps in creating the ideal particles to be used for the selected nano-bio application. There are a few properties that MNPs must successfully have to be used in

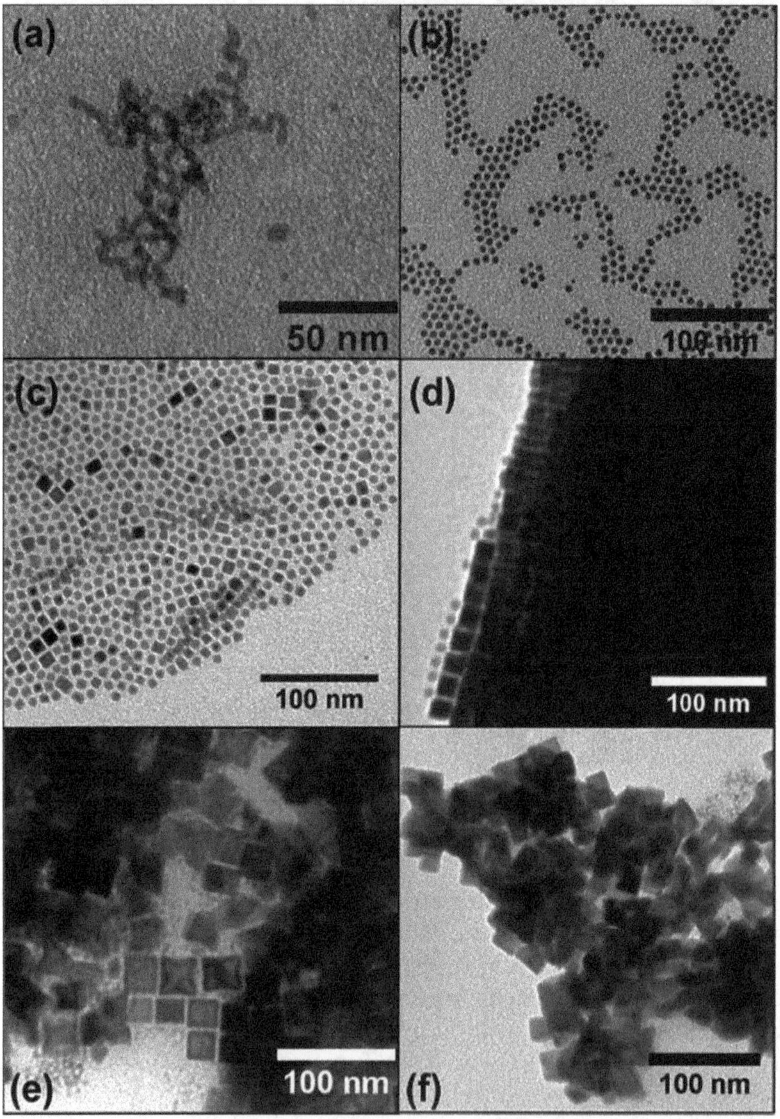

Figure 2.8 TEM pictures of NPs obtained by decomposition at 150 °C for 48 h in the presence of 2 equiv. of HDA but different acid concentrations: (a) 0, (b) 1, (c) 1.2, (d) 1.4, (e) 1.6, and (f) 1.8 equiv. Reproduced with permission from ref. 52. Copyright (2009) American Chemical Society.

different bio-applications. These properties include solubility, biocompatibility, protection and stabilization of MNPs. These methods and mechanisms of surface functionalization are discussed in more detail in the following sections.

2.4.2 Desired Properties of MNPs for Bio-Applications

The five most important properties of MNPs for successful utilization not only for hyperthermia but for other bio-applications are the following: prevention of aggregation, water compatibilization, magnetic controllability, protection and surface-stability, and biocompatibility. Prevention of aggregation and improvement of colloidal stability is a very important property required in order to obtain stable MNPs in both biological media and a magnetic field.[54] The colloidal stability can be improved by controlling the agglomeration of MNPs, which can be achieved by controlling the strength of repulsive forces. There are a couple of ways of achieving steric stabilization which will help improve colloidal stability. Steric stabilization can be improved by coating MNPs with a ligand shell, or with an inorganic, or polymeric matrix. Electrostatic stability also depends on the ionic strength and pH of the solution, and can be improved by studying the stability between the ratio of the number of collisions between particles and the number of collisions resulting in aggregation.[53] Compatibility of MNPs with water is another property that has to be taken into consideration when synthesizing MNPs for bio-applications since most biological reactions happen in an aqueous environment. That is why surface functionalization should help overcome this problem. Another important property is magnetic controllability. The magnetic property is very important not only because it allows MNPs to be magnetically separated, but also because it allows for easier transportation of MNPs toward specific locations. Also, the protection and stability of the surface of MNPs is very important in preventing side reactions with other biological components. Different coatings can be used to improve protection and stability of MNPs, for example the use of materials that can form oxygen-impermeable coatings.[54] Finally, the biocompatibility of MNPs needs to be non-toxic for biological applications. Biocompatibility can be achieved using various coatings. Negatively charged coatings exhibit good biocompatibility, whereas positively charged coatings show enhanced interaction with numerous biosystems.[54] All the mentioned properties can be achieved by surface functionalization of MNPs, an important step to create useful MNPs not only for hyperthermia but also for any intended biological application.

2.4.3 Methods and Mechanisms for MNP-Functionalization

There are two different methods and three mechanisms behind surface functionalization. The two methods are *in situ* surface functionalization and post synthesis surface functionalization, and the three mechanisms are ligand addition, ligand exchange, and encapsulation. *In situ* surface functionalization is a single step that helps achieve both synthesis and surface functionalization of MNPs, in which a reduced particle size and a narrow particle size distribution are achieved, but a limitation to this method is the incompatibility of functional groups with the preparation process.[1] The post synthesis surface functionalization is divided into two steps: synthesis and surface functionalization. The main advantage with post synthesis surface functionalization is the availability of a large number of coupling agents, while the main concern is that the functional groups must have a high affinity for the surface of MNPs.[1] The surface functionalization of MNPs is achieved by three different mechanisms as previously mentioned. Ligand addition is the addition of a ligand to the surface without removing any other functional groups. Ligand exchange is the replacement of a hydrophobic ligand with a hydrophilic ligand, thus achieving the water dispersibility of MNPs. Finally, the encapsulation mechanism is achieved by coating the MNPs with amphiphilic materials where the hydrophobic part is inserted with an initial hydrophobic ligand leaving the hydrophilic portion facing toward the solution, which results in the formation of water compatible MNPs.[41,55,56] Figure 2.9 illustrates the three different mechanisms for functionalization.

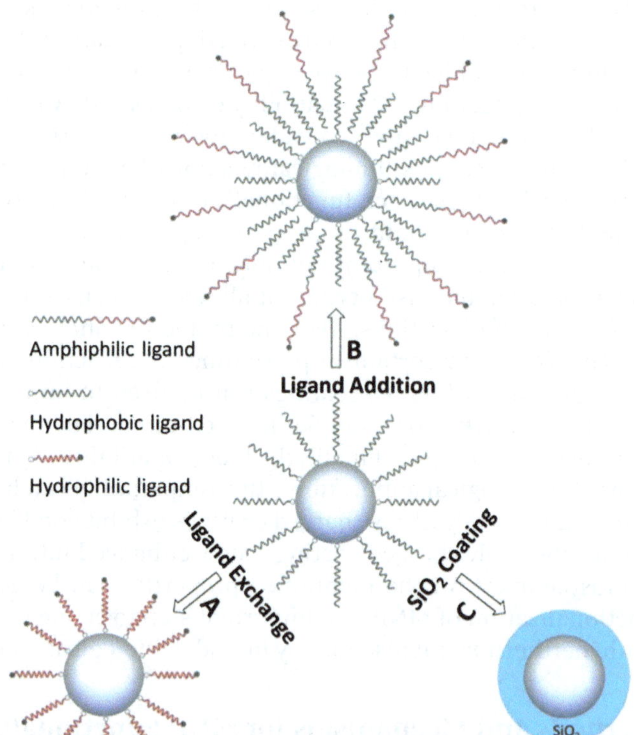

Figure 2.9 Schematic illustration of surface modification of hydrophobic magnetic NPs *via*: (A) ligand exchange; (B) ligand addition, and (C) silica coating. Reproduced with permission from ref. 41. Copyright American Chemical Society, 2016.

2.4.4 Benefits and Materials Used for the Functionalization of MNPs

There are a variety of possible materials that can be used for the surface functionalization of MNPs. The different materials used for functionalization provide different and tailored benefits. The list of materials will be divided into organic and inorganic categories, and will be discussed in brief as follows.

2.4.4.1 Organic Materials

Organic compounds are ideal to functionalize MNPs because they help conserve the magnetic properties, and offer good biocompatibility and biodegradability. Organic compounds are also of great use because they can introduce extra functional groups that can be used for further linkage of other substances. One possibility for the functionalization of MNPs is the use of monomers such as citrates or phosphates. The use of citrates helps with stabilization of MNPs in aqueous dispersion, as well as making MNPs hydrophilic by adsorbing acid and functional groups and leaving at least one carboxylic group exposed to the solvent.[7a,57] Phosphates are also a great material for the functionalization of stable MNPs. Mohapatra *et al.* demonstrated that Fe_3O_4 particles showing superparamagnetic properties could be prepared at room temperature by controlling co-precipitation of Fe^{2+} and Fe^{3+} in the presence of hydrophilic polyvinylalcohol phosphate (PVAP).[58] They also studied how the size of particles, particle size distribution, colloidal stability, and magnetic properties were impacted by changing the concentration of PVAP. Their findings show that using 1% of PVAP solution produced an aqueous suspension of stable MNPs for four weeks at pH 5–8.

Another type of organic compound useful for surface functionalization is the group of small molecules that includes amines, aminosilanes, and thiols. Small molecules provide the benefit of synthesizing MNPs with a smaller hydrodynamic radius and adding functional groups on the surface. The use of amines and thiols helps synthesize stable MNPs with high water solubility. But the difference is that thiols are used for functionalization intended for targeted drug delivery. Maurizi *et al.* developed a method for stabilizing free thiols on the surface of MNPs under physiological pH, but they also observed that the stability of thiols could be increased by protecting meso-2,3-dimercaptosuccinic acid (DMSA) with polyethylene glycol on the surface of MNPs.[59]

Other organic materials with ideal properties for surface functionalization are macromolecules, which could include either polymers or liposomes. The benefit of liposomes is that these are biocompatible, and amphipathic liposomes can be used to encapsulate MNPs. Since liposomes have a large encapsulation range, they can be used to encapsulate different biomolecules. On the other hand, polymers can be used to coat MNPs, and these can be either natural or synthetic. The benefit is that polymer coatings can help improve colloidal stability and increase repulsive forces, thus balancing the magnetic and

Van der Waals forces. Polymers can be functionalized to MNPs *via* either ligand exchange or ligand adsorption, but these coatings usually need the help of active groups present on the surface of MNPs.[60] Natural polymers include the use of dextran and chitosan. The main benefit of dextran is its biocompatibility, which has demonstrated that it can limit the particle size. Another advantage of dextran is that it can be animated by ammonia, thus allowing for easier conjugation of other biomolecules.[61] Pardoe *et al.* studied the synthesis of MNPs with dextran and concluded that the polymer limits the particle size.[62] Likewise, chitosan is a biocompatible, biodegradable, hydrophilic, and non-toxic polymer, which is an excellent carrier for nano-bio applications. In contrast, synthetic polymers include the use of polyethylene glycol (PEG) and polyvinyl alcohol (PVA). PEG is mainly used because it is biocompatible, flexible, and hydrophilic, and helps improve water compatibility as well as stabilize MNPs in physiological media. Fang *et al.* developed a new method for a nanoparticle platform with a superparamagnetic iron oxide core feathering surface functionalized PEG molecules with amine functional groups.[63] On the contrary, PVA is also a biocompatible, hydrophilic polymer which can help prevent self-agglomeration. Chastellain *et al.* synthesized MNPs coated with PVA and established that colloidal stability and particle size distribution were different for different iron : polymer ratios.[64]

2.4.4.2 Organic Materials Used for Hyperthermia

There have been numerous experiments performed targeting hyperthermia using MNPs functionalized with organic materials. Prasad *et al.* prepared an aqueous solution of $La_{1-x}Sr_xMnO_3$ nanoparticles using a polymer, ACRYPOL 934, and they examined the biocompatibility with HeLa cells. They found that nanoparticles of size range 20–100 nm had a saturation magnetization of around 38 emu g^{-1} at 20 kOe and a Curie temperature of 45 °C. They also found that a good heating ability could be obtained in the presence of an AC magnetic field, and that it could be increased with the amplitude of the field. Finally, they observed that these nanoparticles were biocompatible even after 96 h of treatment.[65] Thorat *et al.* prepared $La_{0.7}Sr_{0.3}MnO_3$ MNPs coated with a double layer of oleic acid (OA) to produce a water-based magnetic nanofluid. They studied the effect of the bilayer coating on magnetic properties, colloidal behavior in an aqueous medium, and the effects of pH and ionic strength on the colloidal stability. They found that the aggregation of MNPs did not occur even in high ionic strengths and at physiological pH. They also found that under induction heating, it was possible to obtain hyperthermia temperature (42–43 °C) at a magnetic field of 168–335 Oe and a frequency of 267 kHz. They concluded that the bilayer OA prevented agglomeration of MNPs and produced a stable suspension with improved hyperthermia properties.[66] Hardiansyah *et al.* developed a novel cancer treatment by combining chemotherapy and hyperthermia in synthesizing doxorubicin-loaded magnetic liposomes. The MNPs were coated with citric acid and encapsulated in liposomes along with doxorubicin. They concluded that these MNPs have an excellent biocompatibility and increased ability to kill cancer cells.[67] Meenach *et al.* prepared hydrogel nanocomposites based on poly(ethylene glycol) methyl ether methacrylate and dimethacrylate with iron oxide biomaterials for thermal cancer therapy. They found that both iron oxide nanoparticles and hydrogel nanocomposites had high cell viability for murine fibroblasts. They also found that the heating response of the hydrogels was dependent on both iron oxide loading and the strength of the magnetic field. They demonstrated the system had the ability to kill M059K glioblastoma cells when the hydrogels were exposed to an alternating magnetic field.[68] Kim *et al.* synthesized $CoFe_2O_4$ nanoparticles, dispersed in water, and investigated the use of these nanoparticles as heating agents for drug delivery and hyperthermia. They found that each nanoparticle could reach a temperature of 42.9 °C when a magnetic field of 127–634 Oe was applied at 231 or 266 kHz. They also found the heat generated was dependent on the intensity of the AC magnetic field and frequency applied. They concluded that the ideal heating for triggering drug delivery or hyperthermia could be achieved by adjusting the AC magnetic field and frequency.[69]

2.4.4.3 Inorganic Materials

A benefit of using inorganic compounds for surface coating is that MNPs can exhibit great enhanced antioxidant properties. In addition, the use of inorganic compounds to coat MNPs is promising for different nano-bio applications. As of today, there are three strategies that have been extensively used as coating materials. These include silica, metal element coatings, and metal oxides/sulfides. Coating MNPs with silica produces nanoparticles that are colloidally stable, photostable, robust, and water soluble. Other advantages of silica coating are that it can help prevent aggregation in liquid, improves chemical stability, and has better protection against toxicity. There are three possible methods of preparing magnetic silica nanospheres. The first method involves the *in situ* formation of silica through hydrolysis and condensation of a precursor, such as tetraethyl orthosilicate (TEOS). Im *et al.* used this method to prepare silica colloids with superparamagnetic iron oxide nanoparticles, and they discovered that the final size of silica colloids is dependent upon concentration of iron oxide nanoparticles and type of solvent.[70] The second method consists of the deposition of silica from silicic acid solution. This method is easy to perform, and can help control the particle size by changing the ratio of SiO_2 : MNPs or by repeating the coating procedure. The third method is an emulsion method in which micelles or inverse micelles are used to control silica coating. Yang *et al.* used this method to synthesize monodispersed silica coated iron oxide superparamagnetic nanoparticles; they were able to successfully synthesize the MNPs and demonstrate that biological macromolecules could be trapped in the pores of the nanoparticles.[71] The main advantage of using silica is that it leaves silanol groups expressed on the surface that can easily attach to other specific ligands.[72] On the contrary, the use of metal element coatings tends to act as a protective layer. Lyon *et al.* synthesized gold-coated iron oxide nanoparticles by reducing the precursors on iron oxide nanoparticles, and Wang *et al.* synthesized them by hydroxyl amine seeding.[73,74] Both were able to show that the magnetic properties remain independent of the degree of gold addition. Even though the study of silver and gold coatings have similar effects, silver coatings produce MNPs with antibacterial properties due to silver's strong sterilization ability.[75] On the other hand, metal oxides/sulfide coatings have unique physical and chemical properties. It has been observed that coating MNPs with magnetic materials produces a dramatic influence on the magnetic properties. Hong *et al.* synthesize ZnO-coated iron oxide MNPs by coating them by direct precipitation using ammonium carbonate and zinc acetate, and they found that ZnO-coated MNPs had a better antioxidant ability than bare iron oxide nanoparticles.[76]

2.4.4.4 Inorganic Materials Used for Hyperthermia

Even though the use of inorganic compounds as coating materials for MNPs is promising for different nano-bio applications, the use of these compounds for hyperthermia is very limited. Of the three different inorganic materials, silica is the most useful for hyperthermia. This is because when using MNPs and silica nanoparticles, a unique drug delivery system can be developed where MNPs can be used for hyperthermia and silica nanoparticles can be used for drug delivery. Thomas *et al.* incorporated zinc-doped iron oxide nanocrystals within a mesoporous silica nanoparticle with surface-modified pseudorotaxanes. They demonstrated that breast cancer cells were killed when these were treated with doxorubicin-loaded particles and exposed to an AC field. The application of an AC magnetic field generated local internal heating in the nanoparticles, which cause the molecular machines to disassemble and allowed the drug to be released.[77] In another case, Martín-Saavedra *et al.* studied the biocompatibility of maghemite nanoparticles embedded in a mesoporous silica matrix to form magnetic microspheres (MMS), and their ability to conduct magnetic hyperthermia upon exposure to a low-frequency alternating magnetic field (AMF). They concluded that exposing MMS to low-frequency AMF had the ability to induce magnetic hyperthermia. Their cell culture experiments showed that adjusting the amount of MMS and the time of exposure to AMF achieved different heat treatment intensities.[78]

2.4.5 Bioconjugation Strategies

We already discussed various materials that can be used for surface functionalization of MNPs, which play an important role in improving water compatibility, colloidal stability, and biocompatibility, but MNPs are also required to bind with biomolecules for the specific targeted biomedical application. There have been different methods used for bioconjugation which can be divided into the two categories of physical interactions and covalent chemistry, but the choice of bioconjugation chemistry depends on the chemical properties, the biomolecules linked, and the functional groups found on the MNP surface. The advantages of physical interactions are rapid binding, high efficiency, and no need for intermediate modification. Physical interactions include electrostatic, hydrophobic/hydrophilic, and affinity interactions. Electrostatic interactions use the positive charge on the MNP surface to bind negative charge biomolecules, but the pH and ionic strength need to be optimized. Hydrophobic/hydrophilic interactions have been useful for drug delivery. For this interaction, MNPs are synthesized and functionalized with a hydrophobic layer that is used to adsorb hydrophobic drugs, which can later be released when the coating gets degraded. On the other hand, affinity interactions have been useful for bioconjugation of targeted ligands to MNP surface, which are modified with streptavidin to bind biotinylated molecules. The advantage here is that this interaction is not sensitive to environmental conditions, such as pH, salinity, and hydrophilicity.[53]

Covalent chemistry produces strong and stable linkages with biomolecules. These linkages are formed either by a direct method or by using various strategies based on functional groups expressed on the MNPs' surface. Covalent chemistry includes direct MNP conjugation, carbodiimide coupling, maleimide coupling, and click chemistry. The direct MNP method conjugation is formed by the functional group present on the MNP surface with a ligand. For this method, the surfaces of MNPs are functionalized with different functional groups that bind with different biomolecules. The second method is the carbodiimide coupling reaction, which permits the chemical linkage between carboxylic acids and amines. This method has been used to link small molecules with amine groups. A major advantage of this method is that it does not require any lengthy linker species. The third method is maleimide coupling, which is used to link amines to thiol groups expressed on proteins, and mainly to couple DNA and proteins. The fourth method is click chemistry, which is a new method for direct conjugation developed by Kolb *et al.*[79] This new method involves the coupling of an alkyne to an azide under catalysis by copper. This reaction is fast, efficient, requires mild reaction conditions, an aqueous environment, and it creates water soluble and biocompatible linkages at neutral pH. The advantage is that the bonds formed are stable and rigid, which additionally helps prevent cross interactions from the MNPs. However, one major disadvantage of this method is that the preparation of azide- or alkyne-functionalized species is lengthy and low-yielding. Furthermore, the copper involved in driving the reaction can create toxicity problems due to improper purification of the nanoparticles before their use.

2.5 Magnetic Hyperthermia

With respect to the applications of magnetic nanomaterials, their potential as therapeutic and diagnostic devices for the growing field of personalized medicine is enormous. Magnetic nanomaterials contain unique physical properties, which make them ideal candidates. These properties include their ability to be manipulated by an external magnetic field, a large surface-to-volume ratio, and diversity in architecture in terms of shape, size, and structure. With respect to their size distribution, nanoparticles, defined as particles less than 100 nanometers in all dimensions (IUPAC), are optimal to interact with larger biomolecules, such as DNA, RNA, and proteins.[80] This size range also permits optimal tumoral retention.[7a] Finally, as already discussed, magnetic nanomaterials can be chemically modified by means of different coatings and ligand attachments to increase their solubility, biocompatibility, chemical protection, and stabilization.

A wealth of research has been devoted to characterizing iron oxide nanoparticles (IONPs) as a prospective material encompassing the characteristics previously mentioned. Additionally, IONPs are affordable to produce, biocompatible,[81] able to be metabolized into useful body stores of iron,[82] maintain low to no toxicity,[83,84] and are chemically stable in core–shell models or coatings.[85] To date, bio-applications that have utilized the magnetic property of IONPs include use as contrast agents for magnetic resonance imaging,[86–88] biomolecule separation,[89,90] bio probes,[91,92] and hyperthermic therapies. Of these, magnetic hyperthermia has the greatest clinical use for tumor reduction,[93–95] controlled drug delivery,[96] and stimulated tissue repair.[97] The following section will report the current progress of hyperthermia therapy for tumor reduction.

Cancer is a major public health concern and a leading cause of morbidity and mortality. Globally, cancer accounts for one out of every seven deaths.[98] The current standard of care methods implemented to treat cancer include chemotherapy, radiation, and surgery.[99] Although these techniques are widely used and have been expanded on, there exists a great need for the continual development of modern methods for preventative measures, early detection, cases with sensitive location of tumors, and progressive cases. Research in this area has been focused on gene therapy, immunotherapy, photodynamic therapy, hormone therapy, and hyperthermia.[100]

Hyperthermia for cancer therapy may be described in terms of its temperature range and location. Relatively low temperatures, ($T < 41$ °C), is commonly used in physiotherapies to treat rheumatic diseases through diathermia. Temperatures above ($T > 46$ °C) are used in thermal ablation, and when applied to a target tissue will cause direct necrosis from coagulation and carbonization. A moderate hyperthermia model aims at a temperature range from (41 °C $< T <$ 44 °C).[2,12a] This temperature range causes intra- and extracellular degradations, such as protein denaturation, protein folding, aggregation, and DNA cross linking. Ramifications from these modifications cause permanent damage to cellular functions resulting in induced apoptosis or increased susceptibility to other methods.[101] Cancer cells additionally show a higher susceptibility to damage by heat due to their differing physiology; a higher metabolic process, regions of hypoxia, and low pH.[2] Therefore, studies have shown moderate hyperthermia to successfully eliminate or shrink tumors while causing minimal damage to the surrounding tissue.[102] However, the success of hyperthermia as a treatment in any capacity will rely on the duration of exposure, the temperature generated at the target site, and morphology of the cancer.[103]

For cancer therapies, hyperthermia may be applied locally, regionally, or to the whole body. Regional hyperthermia applies a target temperature of 106–108 °F to large areas, limbs, or organs while whole body hyperthermia applies a target temperature of 107–108 °F throughout the body to treat metastatic cancers. In both of these models an external heat source, such as ultrasound, radiofrequency, microwave, infrared radiation and tubes with hot water, is used to transfer heat from the outside toward the inside.[104] During the thermal energy transfer, a temperature gradient is generated along the path of radiation, dissipating a large amount of energy in the healthy tissue. Due to the non-discrimination of the heat transfer and wide location applied, healthy tissue may be negatively affected during the application of either of these methods.[100] In local hyperthermia, heat is applied only to a small area, such as the tumor site. The specific heat subjection to only the intercellular target site makes local hyperthermia an ideal candidate for tumor reduction. More specifically, hyperthermia mediated through superparamagnetic particles in an AMF generates heat from the inside out eliminating thermal gradients that would harm healthy tissue.[105]

A modularity of hyperthermia for cancer reduction should be as follows. First, a removal of metals from the target tumor tissue area. Second, evaluation of the tumor size using MRI, PET, and CT individually or in combination to determine dosage. Once a dosage is determined, magnetic nanoparticles may be administered surgically or intravenously. Ideally, the distribution would be homogenized within the tumor to prevent hot spots and sufficient heating of all affected areas. Third, an alternating magnetic field may be applied to transform energy from magnetic to thermal energy. Heat generated intratumorally would then destroy tumor cells by induced apoptosis or thermal ablation, or make them more susceptible to an accompanying chemo-/radiotherapy.[94,95] During the process, it is important that the temperature be monitored or controlled. This has been accomplished by surgically implanted thermometers, software predicting programs,[106] and utilization of the Curie temperature of magnetic materials.[107] Finally, MNPs should act as a center piece for constructing a multi-functional nanoplatform, which is capable of targeting tumors and releasing preloaded anticancer drugs, in addition to the hyperthermia treatment.[12a] Each of these steps presents its own problems. Several have made their way to clinical trials, while others are still in the preclinical development of *in vitro* and *in vivo* research.

The use of hyperthermia for cancer reduction was first proposed by Gilchrist *et al.* in 1957 for use in removing metastases from lymph tissue that had been missed by surgery. During the period, heating of a material in an RF field was known to be caused by dielectric loss, eddy current loss, and hysteresis. Heat originating from the dielectric loss is due to the material not being a good conductor. The dielectric loss increases proportionally with frequency of the field, and in terms of a clinical approach, results in unspecific heating of tissue. Eddy current loss is present in materials that are good conductors when applied in an electromagnetic field, where heat is caused by the electrical resistance. Power absorbed from eddy current loss is directly related to the particle diameter as shown in eqn (2.3):

$$P_E = Ef^2 B_{\text{max}}{}^2 d^2 \quad (2.3)$$

where E = eddy current coefficient, f = frequency incycle, d = particle diameter and B_{max} = max flux density.

Unfortunately, power absorption originating from the eddy current loss portion of non-magnetic conductive material is negligible when reduced to a size suitable for injection to a tumor site. Unlike eddy current loss, hysteresis loss is independent of the size of material. Rather, hysteresis loss relies on the saturation of magnetization of the material.[5]

Therefore, the Gilchrist (1957) study utilized a ferromagnetic material Fe_2O_3 with a size distribution of 0.02 µm to 0.1 µm. The authors were successful in applying a specific hyperthermic treatment to lymph nodes containing carcinoma with the use of an external alternating magnetic field.[5] The studies that followed, worked on optimizing the SAR of magnetic materials for the application of hyperthermic tumor reduction through the use of differing materials with a higher magnetic saturation by incorporating cobalt alloys or through the increase in material using larger implantable thermo seeds.[108] Obstacles that still needed to be overcome or had yet to be addressed for this developing technology were found in optimizing the SAR, dose of magnetic material, biocompatibility, toxicity, specific loading of target tissue, the ability to monitor the tissue loaded, and ability to measure the temperature change during application.

The next milestone in the development of a hyperthermic treatment to eradicate cancer was completed in the classic review of magnetic material by Jordan *et al.* in 1993. Their findings concluded that "magnetic fluids", superparamagnetic single-domain suspensions of 3–10 nm, contained superior heating abilities to their thousand times larger counterparts, multi-domain ferrite particles, due to the intrinsically different mechanisms of producing heat, the Brownian and Nèel mechanisms.[109,110] This knowledge provided the field with a platform that was less invasive which may be injected to the cancer site at a lower dosage.

The present day success of hyperthermia has reached clinical use as either a monotherapy or adjunctive therapy in the presence of chemo/radiation. The first clinical treatment system was developed in 2004 by the Charité-Medical University of Berlin. Following in 2010, two products based on magnetic Fe_3O_4 nanoparticles, NANOTHERM® and NANOACTIVATOR® were developed by MagForce and approved for magnetic hyperthermia treatment of brain tumors in Europe. In 2015, MagForce installed the first clinical NANOACTIVATOR® in USA, which extended its capability of treating prostate cancer in addition to the treatment of brain tumors.[14] Since the development of these technologies, studies have been performed on patients with glioblastoma multiforme, prostate, esophagus, and liver cancers.[100]

The first clinical application of interstitial hyperthermia utilizing magnetic nanoparticles was conducted by Johannsen *et al.* in 2005 with a pilot study for pancreatic cancer. Magnetic nanoparticles with aminosilan-type shells and an average diameter of 15 nm were injected at ferrite concentrations of 120 mg ml^{-1} in aqueous solution. The magnetic field applicator utilized an alternating field at 100 kHz and variable field strength from 0–18 kA m^{-1}. For the study, the maximum field strength of 5 kA m^{-1} was not exceeded, but lowered to 4 kA m^{-1} when the patient complained of discomfort. Six treatments took place weekly after the initial injections. Magnetic nanoparticles were shown to be retained in the tumor tissue on Week Six, concluding that a significant number of the magnetic nanoparticles were not cleared by macrophages in the tumor and that sequential treatments could take place following the first injection of magnetic nanomaterial. Temperature was measured using invasive thermometry on the first and sixth treatments in addition to a software estimation that computed the temperature from a 3D CT reconstruction of the tumor and dispersed magnetic nanoparticles. A maximum temperature of 48.5 °C was reached on the first treatment and 42.5 °C on the sixth. The initial SAR of the nanoparticle suspensions in tumor tissue was calculated to be 288 W kg^{-1} assuming homogeneity. Per the CT scan homogeneity was not achieved and therefore a reliable SAR could not be determined. The treatment was well tolerated by the patients.[93]

Maier-Hauff *et al.* explored the combinational effects of magnetic fluid hyperthermia with radiotherapy in Phase I and Phase II trials of recurrent glioblastoma consisting of 14 and 66 patients respectively. An average of six treatments of thermotherapy (4–10 sessions) and 30 Gy of radiotherapy dose (2 Gy per fraction) were administered in both trials. In the Phase I trial, a medium maximum intratumoral temperature of 44.6 °C (42.4–49.5 °C) was recorded. No long term side effects were observed due to the hyperthermia treatment except for worsening of existing hemiparesis. Side effects of sweating, general sensation of warmth, increased body temperature, tachycardia, short term fluctuations of blood pressure, headaches, convulsions, and motor disturbances were observed in more than one patient. With regard to safety and because of its generally mild side effects, hyperthermia in conjunction with reduced radio therapy is a promising alternative to conventional methods. For both trials, an increase in overall survival was seen when compared to conventional methods.[94,95]

2.6 Conclusion

This chapter discusses the design and synthesis of magnetic nanoparticles and their application in magnetic hyperthermia. Magnetic nanoparticles with narrow size distribution, high saturation magnetization and optimal anisotropy have demonstrated excellent heating ability in an AC magnetic field. In term of synthetic strategy, hydrothermal synthesis and thermal decomposition usually produce nanoparticles with great control over particle size, distribution, shape, morphology, and chemical compositions. Biocompatibility is one of the most important properties that must be achieved if MNPs are going to be used not only for hyperthermia but also for other bio-applications. We discussed the various methods that can help improve biocompatibility of MNPs. We also discussed the different types of materials that could be used to functionalize MNPs and the advantages and disadvantages provided by these materials. Organic and inorganic materials have the capability to help improve biocompatibility as well as other properties such as magnetic property and colloidal stability. Finally, the adaption of magnetic hyperthermia into clinical cancer treatment was reviewed. Based on the data presented in this review, it is the authors' conclusion that magnetic hyperthermia will have an increasingly bright future in treating metastasizing solid tumors as a main treatment modality. We have only just begun to implement hyperthermia-based treatment methods in the clinic.

References

1. (a) S. Nagarajan and Z. Yong, *Recent Pat. Biomed. Eng.*, 2008, **1**, 34–42; (b) http://www.medicinenet.com/hyperthermia/article.html.
2. J. van der Zee, *Ann. Oncol.*, 2002, **13**, 1173–1184.
3. (a) P. Wust, B. Hildebrandt, G. Sreenivasa, B. Rau, J. Gellermann, H. Riess, R. Felix and P. Schlag, *Lancet Oncol.*, 2002, **3**, 487–497; (b) J. L. Roti Roti, *Int. J. Hyperthermia*, 2004, **20**, 109–114.
4. (a) A. Jordan, *EuroNanoForum Proc.*, 2005, 76–80; (b) K. Tanaka, T. Ito, T. Kobayashi, T. Kawamura, S. Shimada, K. Matsumoto, T. Saida and H. Honda, *J. Biosci. Bioeng.*, 2005, **100**, 112–115; (c) G. Vincze, O. Szasz and A. Szasz, *Open J. Biophys.*, 2015, **5**, 97–114.
5. R. K. Gilchrist, R. Medal, W. D. Shorey, R. C. Hanselman, J. C. Parrott and C. B. Taylor, *Ann. Surg.*, 1957, **146**, 596–606.
6. (a) A. Jordan, P. Wust, H. Fählin, W. John, A. Hinz and R. Felix, *Int. J. Hyperthermia*, 1993, **9**, 51–68; (b) A. Jordan, R. Scholz, P. Wust, H. Fahling and R. Felix, *J. Magn. Magn. Mater.*, 1999, **201**, 413–419; (c) E. A. Perigo, G. Hemery, O. Sandre, D. Ortega, E. Garaio, F. Plazaola and F. J. Teran, *Appl. Phys. Rev.*, 2015, **2**, 041302.
7. (a) S. Laurent, D. Forge, M. Port, A. Roch, C. Robic, L. V. Elst and R. N. Muller, *Chem. Rev.*, 2008, **108**, 2064–2110; (b) D. Ho, X. Sun and S. Sun, *Acc. Chem. Res.*, 2011, **44**, 875–882.
8. D. Bahadur and J. Giri, Biomaterials and magnetism, *Sadhana*, 2003, **28**, 639–656.
9. (a) Y. Matsumura, *Adv. Drug Delivery Rev.*, 2011, **63**, 184–192; (b) H. Maeda, J. Fang, T. Inutsuka and Y. Kitamoto, *Int. Immunopharmacol.*, 2003, **3**, 319–328; (c) F. Yuan, M. Dellian, D. Fukumura, M. Leunig, D. A. Berk, V. P. Torchilin and R. K. Jain, *Cancer Res.*, 1995, **55**, 3752–3756; (d) H. Maeda, *Adv. Enzyme Regul.*, 2001, **41**, 189–207; (e) H. Maeda, L. W. Seymour and Y. Miyamoto, *Bioconjugate Chem.*, 1992, **3**, 351–362.

10. (a) H. Koo, M. S. Huh, I. Sun, S. H. Yuk, K. Choi, K. Kim and I. C. Kwon, *Acc. Chem. Res.*, 2011, **44**, 1018–1028; (b) J. Xie, K. Chen, H. Lee, C. Xu, A. R. Hsu, S. Peng, X. Chen and S. Sun, *J. Am. Chem. Soc.*, 2008, **130**, 7542–7543; (c) F. M. Kievit and M. Zhang, *Acc. Chem. Res.*, 2011, **44**, 853–862.
11. (a) R. T. Gordon, J. R. Hines and D. Gordon, *Med. Hypotheses*, 1979, **5**, 83–102; (b) M. Shinkai, M. Yanase, H. Honda, T. Wakabayashi, J. Yoshida and T. Kobayashi, *J. Cancer Res.*, 1996, **87**, 1179–1183; (c) G. P. Amorino and M. H. Fox, *J. Membr. Biol.*, 1996, **152**, 217–222; (d) A. Jordan, R. Scholz, P. Wust, H. Schirra, T. Schiestel, H. Schmidt and R. Felix, *J. Magn. Magn. Mater.*, 1999, **194**, 185–196; (e) A. Ito, H. Honda and T. Kobayashi, *Cancer Immunol. Immunother.*, 2006, **55**, 320–328; (f) Y. Xu, M. Mahmood, Z. Li, E. Dervishi, S. Trigwell, V. P. Zharov, N. Ali, V. Saini, A. R. Biris, D. Lupu, D. Boldor and A. S. Biris, *Nanotechnology*, 2008, **19**, 435102.
12. (a) C. S. S. R. Kumar and F. Mohammad, *Adv. Drug Delivery Rev.*, 2011, **63**, 789–808; (b) F. Lu, A. Popa, S. Zhou, J. Zhu and A. C. S. Samia, *Chem. Commun. (Camb.)*, 2013, **49**, 11436–11438; (c) M. Pernia Leal, A. Torti, A. Riedinger, R. La Fleur, D. Petti, R. Cingolani, R. Bertacco and T. Pellegrino, *ACS Nano*, 2012, **6**, 10535–10545; (d) S. D. Kong, W. Zhang, J. H. Lee, K. Brammer, R. Lal, M. Karin and S. Jin, *Nano Lett.*, 2010, **10**, 5088–5092; (e) A. Riedinger, P. Guardia, A. Curcio, M. A. Garcia, R. Cingolani, L. Manna and T. Pellegrino, *Nano Lett.*, 2013, **13**, 2399–2406.
13. U. Gneveckow, A. Jordan, R. Scholz, V. Brüss, N. Waldöfner, J. Ricke, A. Feussner, B. Hildebrandt, B. Rau and P. Wust, *Med. Phys.*, 2004, **31**, 1444–1451.
14. http://www.magforce.de/en/unternehmen/ueber-uns.html.
15. (a) M. Faraji, Y. Yamini and M. Rezaee, *J. Iran. Chem. Soc.*, 2010, **7**, 1–37; (b) G. M. Whitesides, *Nat. Biotechnol.*, 2003, **21**, 1161–1165; (c) I. Brigger, C. Dubernet and P. Couvreur, *Adv. Drug Delivery Rev.*, 2002, **54**, 631–651.
16. (a) J. Giri, P. Pradhan, T. Sriharsha and D. Bahadur, *J. Appl. Phys.*, 2005, **97**, 10Q916; (b) J. P. Fortin, C. Wilhelm, J. Servais, C. Menager, J. C. Bacri and F. Gazeau, *J. Am. Chem. Soc.*, 2007, **129**, 2628–2635.
17. L. H. Reddy, J. L. Arias, J. Nicolas and P. Couvreur, *Chem. Rev.*, 2012, **112**, 5818–5878.
18. (a) H. Wang, T. B. Shrestha, M. T. Basel, R. K. Dani, G. Seo, S. Balivada, M. M. Pyle, H. Prock, O. B. Koper, P. S. Thapa, D. Moore, P. Li, V. Chikan, D. L. Troyer and S. H. Bossmann, *Beilstein J. Nanotechnol.*, 2012, **3**, 444–455; (b) M. E. Caldorera-Moore, W. B. Liechty and N. A. Peppas, *Acc. Chem. Res.*, 2011, **44**, 1061–1070.
19. R. Hergt, S. Dutz, R. Muller and M. Zeisberger, *J. Phys.: Condens. Matter*, 2006, **18**, S2919–S2934.
20. R. E. Rosensweig, *J. Magn. Magn. Mater.*, 2002, **252**, 370–374.
21. (a) R. Hergt, R. Hiergeist, I. Hilger, W. A. Kaiser, Y. Lapatnikov, S. Margel and U. Richter, *J. Magn. Magn. Mater.*, 2004, **270**, 345–357; (b) R. Hergt and S. Dutz, *J. Magn. Magn. Mater.*, 2007, **311**, 187–192; (c) E. A. Perigo, G. Hemery, O. Sandre, D. Ortega, E. Garaio, F. Plazaola and F. J. Teran, *Appl. Phys. Rev.*, 2015, **2**, 041302.

22. Y. Jun, Y. Huh, J. Choi, J. Lee, H. Song, S. Kim, S. Yoon, K. Kim, J. Shin, J. Suh and J. Cheon, *J. Am. Chem. Soc.*, 2005, **127**, 5732–5733.
23. E. C. Abenojar, S. Wickramasinghe, J. Bas-Concepcion and A. C. S. Samia, *Prog. Nat. Sci.: Mater. Int.*, 2016, **26**, 440–448.
24. L. M. Bauer, S. F. Situ, M. A. Griswold and A. C. S. Samia, *Nanoscale*, 2016, **8**, 12162–12169.
25. (a) Z. Nemati, J. Alonso, L. M. Martinez, H. Khurshid, E. Garaio, J. A. Garcia, M. H. Phan and H. Srikanth, *J. Phys. Chem. C*, 2016, **120**, 8370–8379; (b) R. Das, J. Alonso, Z. Nemati Porshokouh, V. Kalappattil, D. Torres, M. Phan, E. Garaio, J. A. Garcia, J. L. Sánchez Llamazares and H. Srikanth, *J. Phys. Chem. C*, 2016, **120**, 10086–10093; (c) P. Hugounenq, M. Levy, D. Alloyeau, L. Lartigue, E. Dubois, V. Cabuil, C. Ricolleau, S. Roux, C. Wilhelm, F. Gazeau and R. Bazzi, *J. Phys. Chem. C*, 2012, **116**, 15702–15712.
26. J. Lee, J. Jang, J. Choi, S. H. Moon, S. Noh, J. Kim, J. Kim, I. Kim, K. I. Park and J. Cheon, *Nat. Nanotechnol.*, 2011, **6**, 418–422.
27. V. K. LaMer and R. H. Dinegar, *J. Am. Chem. Soc.*, 1950, **72**, 4847–4854.
28. (a) R. Massart, *IEEE Trans. Magn.*, 1981, **17**, 1247–1248; (b) R. Massart and V. Cabuil, *J. Chim. Phys.*, 1987, **84**, 967–973.
29. D. Langevin, *Annu. Rev. Phys. Chem.*, 1992, **43**, 341–369.
30. B. K. Paul and S. P. Moulik, *Curr. Sci.*, 2001, **80**, 990–1001.
31. A. K. Gupta and M. Gupta, *Biomaterials*, 2005, **26**, 3995–4021.
32. C. Liu, B. Zou, A. J. Rondinone and Z. J. Zhang, *J. Phys. Chem. B*, 2000, **104**, 1141–1145.
33. A. K. Gupta and S. Wells, *IEEE Trans. Nanobiosci.*, 2004, **3**, 66–73.
34. J. Vidal-Vidal, J. Rivas and M. A. López-Quintela, *Colloids Surf., A*, 2006, **288**, 44–51.
35. M. A. Willard, L. K. Kurihara, E. E. Carpenter, S. Calvin and V. G. Harris, in *Encyclopedia of Nanoscience and Nanotechnology*, ed. H. S. Nalwa, American Scientific Publishers, Valencia, CA, 2004, vol. 1, p. 815.
36. T. Hyeon, S. S. Lee, J. Park, Y. Chung and H. B. Na, *J. Am. Chem. Soc.*, 2001, **123**, 12798–12801.
37. S. Ge, X. Shi, K. Sun, C. Li, C. Uher, J. R. Baker Jr, M. M. Banaszak Holl and B. G. Orr, *J. Phys. Chem. C*, 2009, **113**, 13593–13599.
38. Y. Kolen'ko, M. Bañobre-López, C. Rodriguez-Abreu, E. Carbo-Argibay, A. Sailsman, Y. Piñeiro-Redondo, M. F. Cerqueira, D. Petrovykh, K. Kovnir, O. Lebedev and J. Rivas, *J. Phys. Chem. C*, 2014, **118**, 8691–8701.
39. T. J. Daou, G. Pourroy, S. Begin-Colin, C. Greneche, J. M. Ulhaq-Bouillet, P. Legare, P. Bernhardt, C. Leuvrey and G. Rogez, *Chem. Mater.*, 2006, **18**, 4399–4404.
40. R. Das, J. Alonso, Z. N. Porshokouh, V. Kalappattil, D. Torres, M. Phan, E. Garaio, J. Á. García, J. L. S. Llamazares and H. Srikanth, *J. Phys. Chem. C*, 2016, **120**, 10086–10093.
41. L. Wu, A. Mendoza-Garcia, Q. Li and S. Sun, *Chem. Rev.*, 2016, **116**, 10473–10512.
42. S. H. Sun and H. Zeng, *J. Am. Chem. Soc.*, 2002, **124**, 8204–8205.

43. D. Kim, N. Lee, M. Park, B. H. Kim, K. An and T. Hyeon, *J. Am. Chem. Soc.*, 2009, **131**, 454–455.
44. P. Guardia, R. Di Corato, L. Lartigue, C. Wilhelm, A. Espinosa, M. Garcia-Hernandez, F. Gazeau, L. Manna and T. Pellegrino, *ACS Nano*, 2012, **6**, 3080–3091.
45. H. Wang, T. B. Shrestha, M. T. Basel, M. Pyle, Y. Toledo, A. Konecny, P. Thapa, M. Ikenberry, K. L. Hohn, V. Chikan, D. L. Troyer and S. H. Bossmann, *J. Mater. Chem. B*, 2015, **3**, 4647–4653.
46. J. T. Jang, H. Nah, J. H. Lee, S. H. Moon, M. G. Kim and J. Cheon, *Angew. Chem., Int. Ed.*, 2009, **48**, 1234–1238.
47. T. Hyeon, S. S. Lee, J. Park, Y. Chung and H. B. Na, *J. Am. Chem. Soc.*, 2001, **123**, 12798–12801.
48. J. Park, E. Lee, N. M. Hwang, M. S. Kang, S. C. Kim, Y. Hwang, J. G. Park, H. J. Noh, J. Y. Kini, J. H. Park and T. Hyeon, *Angew. Chem., Int. Ed.*, 2005, **44**, 2872–2877.
49. S. Peng, C. Wang, J. Xie and S. H. Sun, *J. Am. Chem. Soc.*, 2006, **128**, 10676–10677.
50. L. Lacroix, N. F. Huls, D. Ho, X. Sun, K. Cheng and S. Sun, *Nano Lett.*, 2011, **11**, 1641–1645.
51. F. Dumestre, B. Chaudret, C. Amiens, P. Renaud and P. Fejes, *Science*, 2004, **303**, 821–823.
52. L. Lacroix, S. Lachaize, A. Falqui, M. Respaud and B. Chaudret, *J. Am. Chem. Soc.*, 2009, **131**, 549–557.
53. R. A. Bohara, N. D. Thorat and S. H. Pawar, *RSC Adv.*, 2016, **6**, 43989–44012.
54. A. V. Bychkova, O. N. Sorokina, M. A. Rosenfeld and A. L. Kovarski, *Russ. Chem. Rev.*, 2012, **81**, 1026–1050.
55. N. T. K. Thanh and L. A. W. Green, *Nano Today*, 2010, **5**, 213–230.
56. X. Chu, J. Yu and Y. Hou, *Chin. Phys. B*, 2015, **24**, 014704.
57. Y. Sahoo, A. Goodarzi, M. T. Swihart, T. Y. Ohulchanskyy, N. Kaur, E. P. Furlani and P. N. Prasad, *J. Phys. Chem. B*, 2005, **109**, 3879–3885.
58. S. Mohapatra, N. Pramanik, S. K. Ghosh and P. Pramanik, *J. Nanosci. Nanotechnol.*, 2006, **6**, 823–829.
59. L. Maurizi, H. Bisht, F. Bouyer and N. Millot, *Langmuir*, 2009, **25**, 8857–8859.
60. R. M. Patil, P. B. Shete, N. D. Thorat, S. V. Otari, K. C. Barick, A. Prasad, R. S. Ningthoujam, B. M. Tiwale and S. H. Pawar, *J. Magn. Magn. Mater.*, 2014, **355**, 22–30.
61. M. Lewin, N. Carlesso, C. H. Tung, X. W. Tang, D. Cory, D. T. Scadden and R. Weissleder, *Nat. Biotechnol.*, 2000, **18**, 410–414.
62. H. Pardoe, W. Chua-anusorn, T. G. St Pierre and J. Dobson, *J. Magn. Magn. Mater.*, 2001, **225**, 41–46.
63. C. Fang, N. Bhattarai, C. Sun and M. Zhang, *Small*, 2009, **5**, 1637–1641.
64. M. Chastellian, A. Petri and H. Hofman, *J. Colloid Interface Sci.*, 2004, **278**, 353–360.

65. N. K. Prasad, K. Rathinasamy, D. Panda and D. Bahadur, *J. Biomed. Mater. Res., Part B*, 2008, 409–416.
66. N. D. Thorat, V. M. Khot, A. B. Salunkhe, A. I. Prasad, R. S. Ningthoujam and S. H. Pawar, *J. Phys. D*, 2013, **46**, 105003.
67. A. Hardiansyah, L. Huang, M. Yang, T. Liu, S. Tsai, C. Yang, C. Kuo, T. Chan, H. Zou, W. Lian and C. Lin, *Nanoscale Res. Lett.*, 2014, 497–509.
68. S. A. Meenach, J. Z. Hilt and K. W. Anderson, *Acta Biomater.*, 2010, **6**, 1039–1046.
69. D. H. Kim, D. E. Nikles, D. T. Johnson and C. S. Brazel, *J. Magn. Magn. Mater.*, 2008, **19**, 2390–2396.
70. S. H. Im, T. Herricks, Y. T. Lee and Y. Xia, *Chem. Phys. Lett.*, 2005, **401**, 19–23.
71. H. Yang, S. Zhang, X. Chen, Z. Zhuang, J. Xu and X. Wang, *Anal. Chem.*, 2004, **76**, 1316–1321.
72. Q. Liu, Z. Xu, J. A. Finch and R. Egerton, *Chem. Mater.*, 1998, **10**, 3936–3940.
73. L. Wang, J. Luo, Q. Fan, M. Suzuki, I. S. Suzuki, M. H. Engelhard, Y. Lin, N. Kim, J. Q. Wang and C. Zhong, *J. Phys. Chem. B*, 2005, **109**, 21593–21601.
74. J. L. Lyon, D. A. Fleming, M. B. Stone, P. Schiffer and M. E. Williams, *Nano Lett.*, 2004, **4**, 719–723.
75. P. Gong, H. Li, X. He, K. Wang, J. Hu, W. Tan, S. Zhang and X. Yang, *Nanotechnology*, 2007, **18**, 285604.
76. R. Y. Hong, S. Z. Zhang, G. Q. Di, H. Z. Li, Y. Zheng, J. Ding and D. G. Wei, *Mater. Res. Bull.*, 2008, **43**, 2457–2468.
77. C. R. Thomas, D. P. Ferris, J. Lee, E. Choi, M. H. Cho, E. S. Kim, J. F. Stoddart, J. Shin, J. Cheon and J. I. Zink, *J. Am. Chem. Soc.*, 2010, **132**, 10623–10625.
78. F. M. Martín-Saavedra, E. Ruíz-Hernández, A. Boré, D. Arcos, M. Vallet-Regí and N. Vilaboa, *Acta Biomater.*, 2010, **6**, 4522–4531.
79. H. C. Kolb, M. G. Finn and K. B. Sharpless, *Angew. Chem., Int. Ed. Engl.*, 2001, **40**, 2004–2021.
80. K. K. Alharbi and Y. Al-Sheikh, *Saudi J. Biol. Sci.*, 2014, **21**, 109–117.
81. A. H. Lu, E. L. Salabas and F. A. Schüth, *Chem. Int. Ed.*, 2007, **46**, 1222–1244.
82. H. Arami, A. Khandhar, D. Liggitt and K. M. Krishnan, *Chem. Soc. Rev.*, 2015, **44**, 8576–8607.
83. V. I. Shubayev, T. R. Pisanic and S. Jin, *Adv. Drug Delivery Rev.*, 2009, **61**, 467–477.
84. Y. Liu, S. Jia, Q. Wu, J. Ran, W. Zhang and S. Wu, *Catal. Commun.*, 2011, **12**, 717–720.
85. A. M. G. C. Dias, A. Hussain, A. S. Marcos and A. C. A. Roque, *Biotechnol. Adv.*, 2011, **29**, 142–155.
86. J. W. M. Bulte and D. L. Kraitchman, *NMR Biomed.*, 2004, **17**, 484–499.
87. W. J. M. Mulder, G. J. Strijkers, G. A. F. van Tilborg, A. W. Griffioen and K. Nicolay, *NMR Biomed.*, 2006, **19**, 142.

88. J. Gao, G. Liang, J. S. Cheung, Y. Pan, Y. Kuang, F. Zhao, B. Zhang, X. Zhang, E. X. Wu and B. Xu, *J. Am. Chem. Soc.*, 2008, **130**, 11828.
89. D. Shao, K. Xu, X. Song, J. Hu, W. Yang and C. Wang, *J. Colloid Interface Sci.*, 2009, **336**, 526–532.
90. C. L. Du, W. Du, B. Wang, W. Y. Feng, Z. Wang and Y. L. Zhao, *Chin. J. Anal. Chem.*, 2010, **38**, 902.
91. Y. R. Chemla, H. L. Grossman, Y. Poon, R. McDermott, R. Stevens, M. D. Alper and J. Clarke, *Proc. Natl. Acad. Sci.*, 2000, **97**, 14268–14272.
92. S. J. Denardo, G. L. Denardo, L. A. Miers, A. Natarajan, A. R. Foreman, C. Gruettner, G. N. Adamson and R. Ivkov, *Clin. Cancer Res.*, 2005, **11**, 7087s–7092s.
93. M. Johannsen, U. Gneveckow, L. Eckelt, A. Feussner, N. Waldöfner, R. Scholz, S. Deger, P. Wust, S. A. Loening and A. Jordan, *Int. J. Hyperthermia*, 2005, **21**, 637–647.
94. K. Maier-Hauff, R. Rothe, R. Scholz, U. Gneveckow, P. Wust, B. Thiesen, A. Feussner, A. Deimling, N. Waldoefner, R. Felix and A. Jordan, *J. Neurooncol.*, 2007, **81**, 53–60.
95. K. Maier-Hauff, F. Ulrich, D. Nestler, H. Niehoff, P. Wust, B. Thiesen, H. Orawa, V. Budach and A. Jordan, *J. Neurooncol.*, 2011, **103**, 317–324.
96. F. Yang, P. Chen, W. He, N. Gu, X. Zhang, K. Fang, Y. Zhang, J. Sun and J. Tong, *Small*, 2010, **6**, 1300–1305.
97. F. Assa, H. Jafarizadeh-Malmiri, H. Ajamein, N. Anarjan, H. Vaghari, Z. Sayyar and A. Berenjian, *Nano Res.*, 2016, **9**, 2203–2225.
98. American Cancer Society, *Cancer Facts and Figures 2016*, American Cancer Society, Atlanta, Ga, 2016.
99. T. Hehr, P. Wust, M. Bamberg and W. Budach, *Oncol. Res. Treat.*, 2003, **26**, 295–302.
100. J. Beik, Z. Abed, F. S. Ghoreishi, S. Hosseini-Nami, S. Mehrzadi, A. Shakeri-Zadeh and S. K. Kamrava, *J. Controlled Release*, 2016, **235**, 205–221.
101. L. S. Goldstein, M. W. Dewhirst, M. Repacholi and L. Kheifets, *Int. J. Hyperthermia*, 2003, **19**, 373–384.
102. M. Bañobre-López, A. Teijeiro and J. Rivas, *Rep. Pract. Oncol. Radiother.*, 2013, **18**, 397–400.
103. G. P. Raaphorst, M. L. Freeman and W. C. Dewey, *Radiat. Res.*, 1979, **79**, 390–402.
104. https://www.cancer.gov/about-cancer/treatment/types/surgery/hyperthermia-fact-sheet#r1.
105. S. Kossatz, J. Grandke, P. Couleaud, A. Latorre, A. Aires, K. Crosbie-Staunton, R. Ludwig, H. Dähring, V. Ettelt, A. Lazaro-Carrillo, M. Calero, M. Sader, J. Courty, Y. Volkov, A. Prina-Mello, A. Villanueva, Á. Somoza, A. L. Cortajarena, R. Miranda and I. Hilger, *Breast Cancer Res.*, 2015, **17**, 66.
106. G. Sreenivasa, J. Gellermann, B. Rau, J. Nadobny, P. Schlag, P. Deuflhard, R. Felix and P. Wust, *Int. J. Radiat. Oncol., Biol., Phys.*, 2003, **55**, 407–419.

107. K. L. Mcnerny, Y. Kim, D. E. Laughlin and M. E. Mchenry, *J. Appl. Phys.*, 2010, **107**, 09A312–14.
108. A. Jordan, *Int. J. Hyperthermia*, 2009, **25**, 512–516.
109. A. Jordan, P. Wust, H. Fähling, W. John, A. Hinz and R. Felix, *Int. J. Hyperthermia*, 1993, **9**, 51–68.
110. C. Iacovita, A. Florea, R. Dudric, E. Pall, A. I. Moldovan, R. Tetean, R. Stiufiuc and C. M. Lucaciu, *Molecules*, 2016, **21**, 1357.

CHAPTER 3

Magnetic Nanoparticles in Catalysis

JIANMING LIU*[a] AND WEI SUN*[b]

[a]Henan Normal University, School of Chemistry and Chemical Engineering, No. 46, East of Construction Road, Xinxiang, 453007, China; [b]Lanzhou Institute of Chemical Physics, Chinese Academy of Sciences, State Key Laboratory for Oxo Synthesis and Selective Oxidation, No. 18, Tianshui Middle Road, 730000, Lanzhou, China
*E-mail: jmliu@htu.cn, wsun@licp.cas.cn

3.1 Introduction

In traditional catalytic chemistry, homogeneous catalysts have been attracting much attention owing to high selectivities, high yields, and easy process optimization.[1] Despite these advantages in applications, there are obvious disadvantages of homogeneous catalysts, such as difficult separation of the final product(s) from the reaction mixture, and difficulties in commercialization of homogeneous processes because of the limited availability and high costs of the required noble metals. In comparison with homogeneous catalysts, a great deal of attention has been devoted to emerging novel heterogeneous catalysts, because they are easily recovered and reused due to their good stability. In fact, heterogeneous catalysts could overcome all of the above-discussed limitations of homogeneous systems. As already discussed, magnetic nanosize catalysts facilitate the easy separation of the final products from the catalyst.[2] To date, heterogeneous catalysts are not widely used

Smart Materials No. 26
Magnetic Nanomaterials: Applications in Catalysis and Life Sciences
Edited by Stefan H. Bossmann and Hongwang Wang

Published by the Royal Society of Chemistry, www.rsc.org

in chemical transformations owing to their low efficiency and, consequently, inferior reaction yields. The solution of this dilemma is to use magnetic nanoparticles, because they can combine the advantages of homogeneous catalysts and heterogeneous catalysts.

Magnetic nanoparticles can be utilized in many catalytic reactions as nanocatalysts for heterogeneous catalysts.[3–10] These magnetic nanoparticles could easily be prepared by straightforward synthetic methods with high activities, stabilities, and chemical- or regio-selectivity, taking advantage of their special shapes and small sizes.[11] Magnetic nanocatalysts have not only the characteristics of magnetic materials, catalytic materials, and nanometer materials, but also show the following features:

(a) Size Effect. The magnetic nanocatalysts show high activity owing to their size and feature an enormous surface on the catalyst.
(b) Surface Effect and Volume Effect. As the size of particles decreases to the nanoscale, the particles can show many special characteristics, such as the surface effect and the volume effect.
(c) Catalytic Activity. The magnetic nanometer catalysts can smoothly promote the reaction due to the nano-effect.

Nanocatalysts have much greater catalytic activity and selectivity than those of traditional heterogeneous catalysts, because of their large surface area and high surface activity. However, because of their small size, separation by simple filtration is difficult to achieve, thus increasing the complexity of successful separation and recycling, especially when scaled up to larger processes. The recycling problem must be overcome before nanocatalytic processes can successfully be scaled-up. Magnetic nanocatalysts can be recycled with the aid of an external magnetic field (Figure 3.1). Therefore, it is definitely possible to develop an inexpensive and easy route for the preparation of magnetic nanocatalysts for numerous applications in the catalytic process.

3.2 Application of Magnetic Nanoparticles in Catalysis

3.2.1 Transition Metal Loading onto the Surface of Nano-Magnetite-Supported Catalysts

Ruthenium hydroxide catalyst on magnetite (Ru(OH)x/Fe_3O_4) prepared by Mizuno and co-workers promoted the aerobic oxidation of alcohols to generate aldehydes and ketones, as well as the aerobic oxidation of amines to afford the corresponding nitriles, and reduction of carbonyl compounds to produce alcohols (Scheme 3.1).[12] The catalysts were well characterized by XPS, XRD and BET. The results showed that the Ru species was highly dispersed on Fe_3O_4. A series of substrates bearing aromatic, aliphatic, and heterocyclic groups were well tolerated in generating the desired products with good to excellent yields without the necessity of any additives.

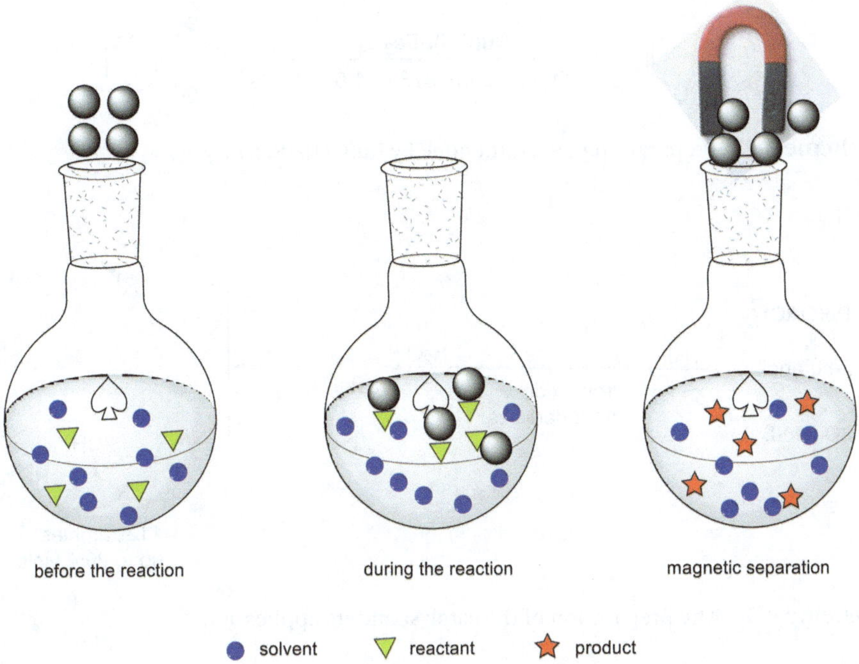

Figure 3.1 Catalytic separation by magnetic attraction.

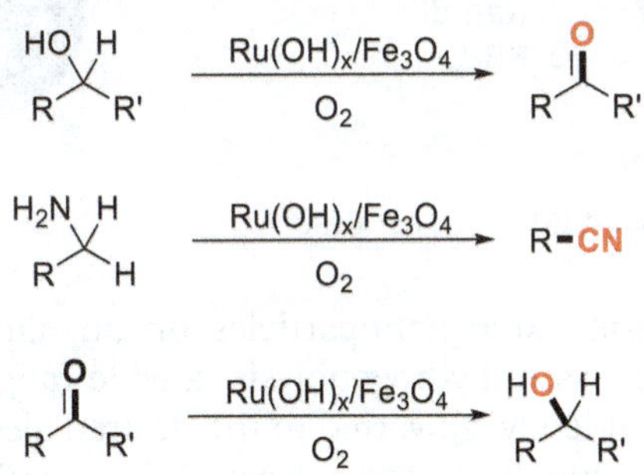

Scheme 3.1 Ru(OH)x/Fe_3O_4 catalyzed: the oxidation of alcohols and amines, the reduction of carbonyl compounds.

It is noteworthy that xanthene is quantitatively converted into the corresponding ketone of 9-xanthenone under oxygen atmosphere catalyzed by Ru(OH)x/Fe_3O_4 (Scheme 3.2). The recovered Ru(OH)x/Fe_3O_4 catalysts were found to retain their catalytic performance after several cycles.

In 2007, Park *et al.* reported a highly efficient and selective epoxide hydrogenolysis catalyzed by a magnetically separable palladium catalyst under 1.0 atm H_2 at room temperature (Scheme 3.3).[13] The palladium catalyst was prepared by a sol–gel process incorporating palladium nanoparticles and

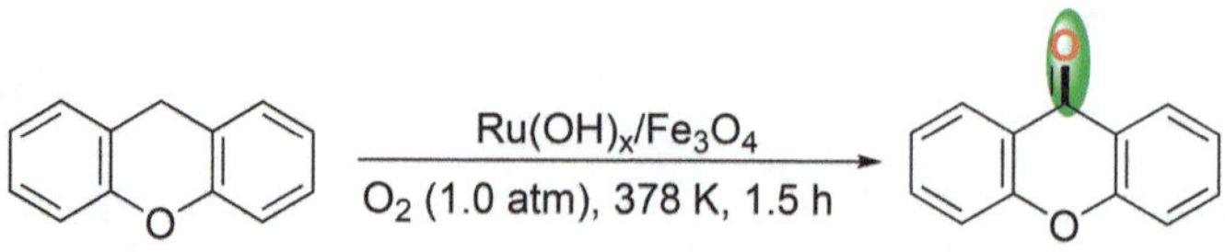

Scheme 3.2 Preparation of 9-xanthenone by Ru(OH)x/Fe_3O_4.

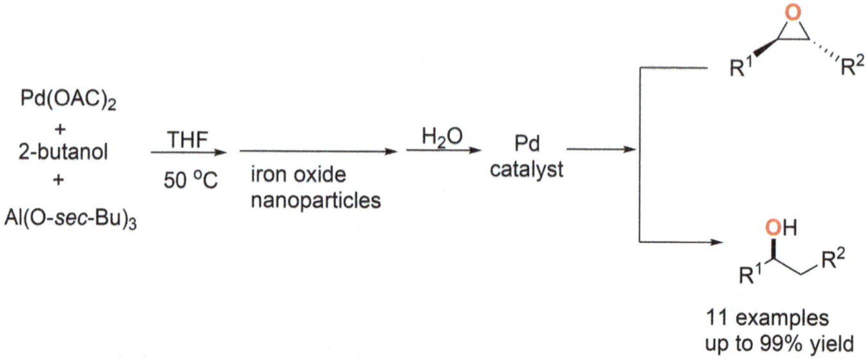

Scheme 3.3 The preparation of the catalyst and its application.

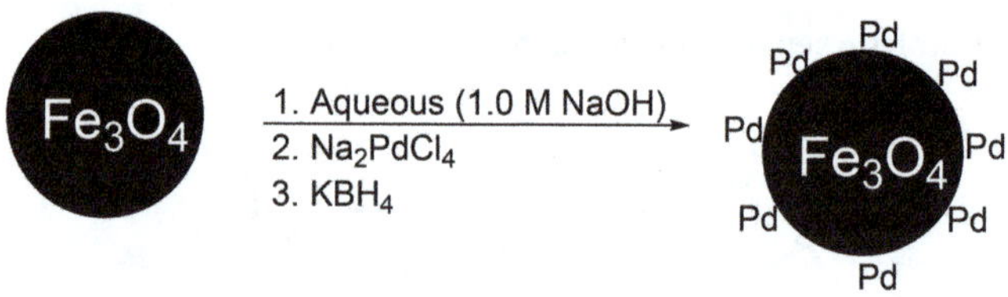

Scheme 3.4 Synthesis of Pd/Fe_3O_4 catalyst.

superparamagnetic iron oxide nanoparticles on an aluminum oxyhydroxide matrix. In the process of hydrogenolysis, a wide range of epoxides were well tolerated and completely converted to the desired alcoholic compounds. Moreover, the catalyst could be reused at least 25 times without loss of activity.

In 2008, Sun and Xia designed a Pd/Fe_3O_4 catalyst prepared by a wet impregnation process incorporating palladium nanoparticles and superparamagnetic Fe_3O_4 nanoparticles, which were reduced by KBH_4 (Scheme 3.4). This nanocatalyst was successfully applied in the carbonylative Sonogashira coupling reaction of aryl iodides with terminal alkynes under phosphine-free conditions.[14a] A series of aryl iodides with terminal alkynes were well tolerated, as shown in Scheme 3.5. The results of XPS and TEM indicated that palladium nanoparticles of 5 nm entrapped in iron oxide were formed during catalyst synthesis. Pd(II) could then be reduced to metallic Pd(0). Finally, the Pd/Fe_3O_4 catalyst could be reused seven times without appreciable loss of the catalytic performance in the carbonylative Sonogashira coupling of

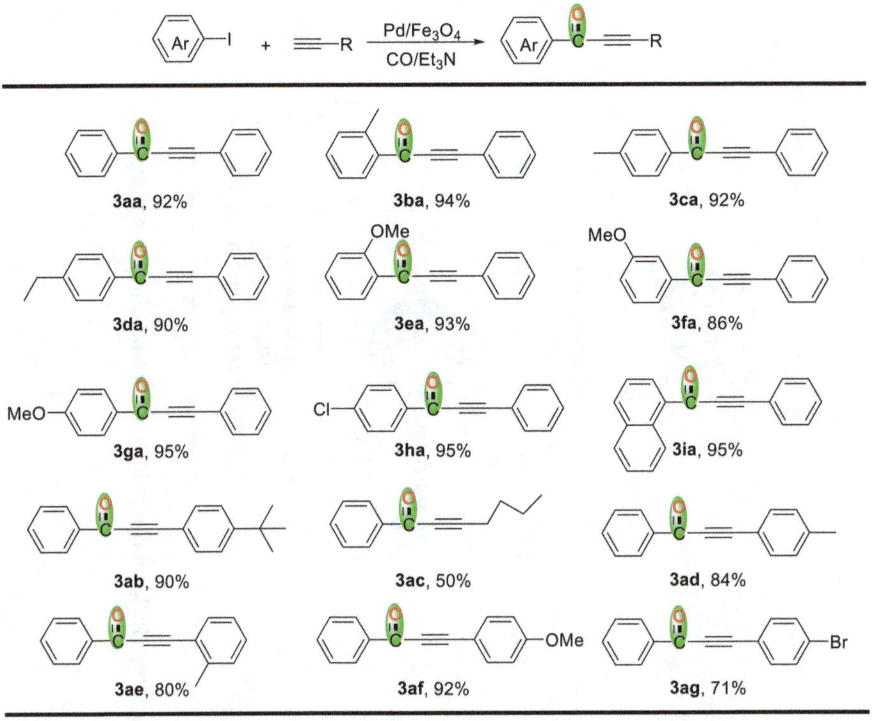

Scheme 3.5 Reactions were carried out in 5.0 mL of toluene under 2.0 MPa pressure of CO at 130 °C for 4.0 h with 2.5 mmol aryl iodides, 3.0 mmol phenylacetylene alkynes, 7.2 mmol Et_3N, and Pd/Fe_3O_4 catalyst (Pd, 1.04 wt%), 50 mg (Pd/substrate).

iodobenzene and phenylacetylene. In the above process, the catalytic active species, which is Pd(0), partially dissolved from the magnetite materials, but is partially or completely reprecipitated on the surface of Fe_3O_4. In 2014, Song and Cao prepared noble-metal-nanoparticle-loaded carbon nanotube (CNT) composites to be applied in the carbonylative Sonogashira coupling reaction and in the catalytic hydrodegradation of methylene blue.[14b] A variety of substrates reacted smoothly to give the desired products. After the fifth run, the MWCNTs–Fe_3O_4–Pd catalysts were characterized by XRD and ICP-AES. The XRD pattern of MWCNTs–Fe_3O_4–Pd catalysts after the fifth run demonstrated that the crystal structures of Fe_3O_4 and Pd were still maintained. ICP-AES analysis revealed the reason for the observed decrease in catalytic yields of the recycled catalysts: the Pd content of MWCNTs–Fe_3O_4–Pd was lost in the carbonylative Sonogashira-coupling reaction.

Plucinski and co-workers have reported that a series of easily-separable magnetic nanoparticle-supported Pd catalysts with diameters of 7–17 nm were applied in Suzuki coupling, Heck coupling, hydrogenation of benzaldehyde and amination reactions of bromobenzene with morpholine (Scheme 3.6). These nanocatalysts were characterized by transmission electron microscopy

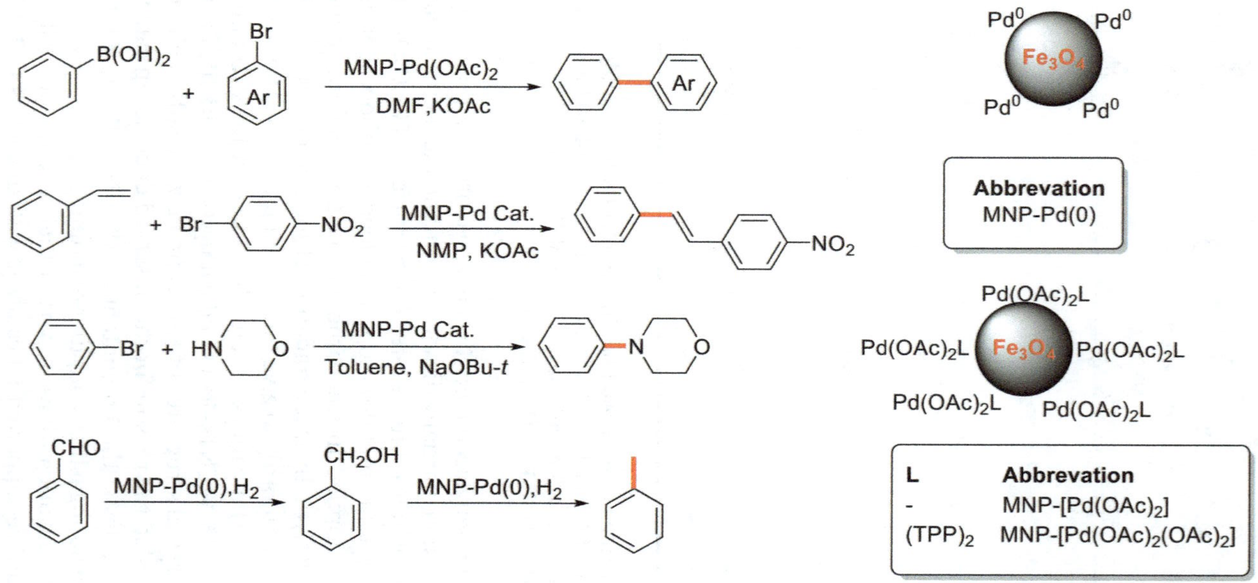

Scheme 3.6 Nanoparticle-supported Pd catalysts catalyze Suzuki coupling, Heck coupling, hydrogenation and amination.

(TEM), atomic absorption spectrophotometry (AAS) analysis and X-ray diffraction (XRD).[15] Pseudo-first-order kinetics were observed in all cases, which is of mechanistic importance, because it indicates that catalytic atomic Pd centers or small Pd clusters exist on the surface. The results of palladium-leaching investigations indicate that the catalyst instability was critically dependent on the presence of the aryl halide during Heck coupling. The catalyst could be easily separated in a magnetic field and its catalytic efficiency remains unaltered even after several cycles although the rate of reaction obviously decreased on reuse.

A magnetically separable palladium catalyst was synthesized simply through a three-step process: (a) the preparation of Fe_3O_4 carrier particles, (b) surface modification of the magnetite particles through the carbonization of glucose, and (c) generation of Pd(0) nanoparticle decorated MFC composites (Scheme 3.7).[16] The morphology, inner structure, and magnetic properties of the Pd/MFC catalyst were carefully investigated by TEM, XRD, Fourier infrared spectroscopy (FT-IR), XPS, and a vibrating sample magnetometer. Pd/MFC catalysts were successfully applied to demonstrate their catalytic efficiency in Suzuki and Heck coupling reactions (Scheme 3.8). The Pd/MFC catalyst recovered after Suzuki and Heck reactions showed no obvious loss of catalytic performance even after five repeated cycles. In the above process, the stability of the catalyst was investigated and the reaction mixture was also detected by ICP-MS. The result showed that the Pd loss amounted only to traces: observed Pd-loss was 0.2% and 0.5% in Suzuki and Heck reactions, respectively.

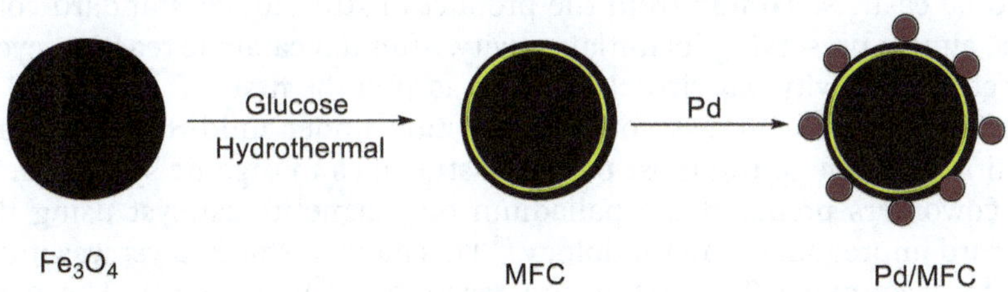

Scheme 3.7 Synthesis route to Pd/MFC nanocomposites.

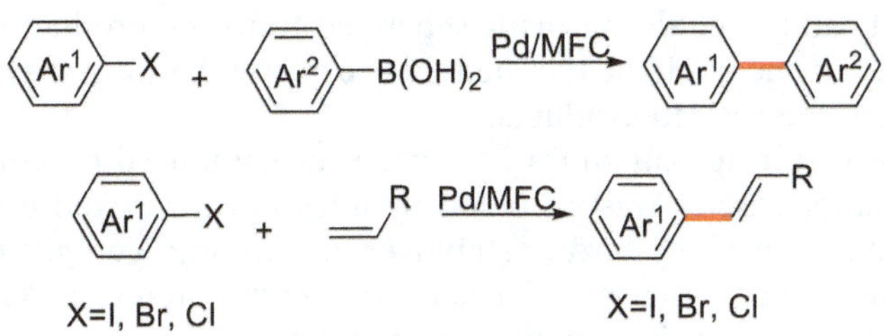

Scheme 3.8 Pd/MFC catalyzes the Suzuki and Heck reactions.

Hyeon and co-workers have reported a new method to synthesize Pd–Fe_3O_4 heterodimer nanocrystals by means of a precisely controlled one-pot thermolysis of a mixture solution composed of iron acetylacetonate, palladium acetylacetonate, oleylamine, and oleic acid.[17,18] These heterodimer nanocrystals were also successfully applied to carry out Suzuki, Heck, and Sonogashira coupling reactions, direct C–H arylation, and Wacker oxidation (Scheme 3.9). Following the demonstration of catalytic efficacy, the authors investigated the mechanisms of the Suzuki coupling and Wacker oxidation reactions employing the Pd–Fe_3O_4 nanoparticle catalysts. General kinetic studies, hot filtration tests, and three-phase tests were carefully carried out. The results of these mechanistic studies confirmed that the reactions most probably follow a solution-phase mechanism, as indicated in Scheme 3.9.

Recently, Yus *et al.* reported an inexpensive and easily obtainable $Ru(OH)_3$–Fe_3O_4 catalyst prepared by a simple wet impregnation process. This catalyst was fully characterized by means of X-ray and TEM.[19] The impregnated ruthenium catalyst was found to be highly active and efficient in the selective *N*-monoalkylation of amino derivatives with poor nucleophilic character, such as aromatic and heteroaromatic amines, sulfonamides, sulfinamides, and nitroarenes (Scheme 3.10). In the above processes, alcohols were applied as the initial source of the electrophile. The reaction was accomplished through a hydrogen autotransfer process (Scheme 3.11). Interestingly, the use of chiral sulfinamides and secondary alcohols successfully proceeded to afford the alkylated compound with a diastereomeric ratio of 92:8. In addition, the *N*-monoalkylated amines were obtained in the alkylation of amines using potassium hydroxide. Related imines were also prepared by the alkylation of amines using sodium hydroxide. The ruthenium catalyst could be easily separated from the product mixture under standard conditions, almost preserving its initial activity. After ten catalytic reaction cycles, the catalytic activity was virtually as high as after the first.

Performing a C–C cross coupling reaction under mild Suzuki–Miyaura conditions is one of the most powerful strategies in organic synthesis. Yus and coworkers prepared the palladium on magnetite catalyst using their standard impregnation methodology.[20] This nanocrystal catalyst was indeed able to accomplish Suzuki–Miyaura reactions (Scheme 3.12). The use of this particular nanocatalyst does not only avoid expensive ligands and the additional drawback of having palladium metal or cations contaminating the final products, it also shows remarkably high activity. Furthermore, the nanocrystal catalyst could be easily separated from the product mixture by using a magnet and could be recycled three times with only a slight decrease in the yield of the desired products.

A catalyst featuring palladium nanoparticles supported on cobalt ferrite magnetic nanoparticles was prepared by ultrasound assisted co-precipitation in the absence of any surface stabilizers or capping agents.[21] TEM of the catalyst reveals the formation of palladium-incorporated $CoFe_2O_4$ NPs (40–50 nm). The palladium-incorporated cobalt catalyst was found suitable for Suzuki coupling reactions in ethanol under ligand free condition (Scheme 3.13). In the above process, only low catalyst loading (1.6 mol%) could be

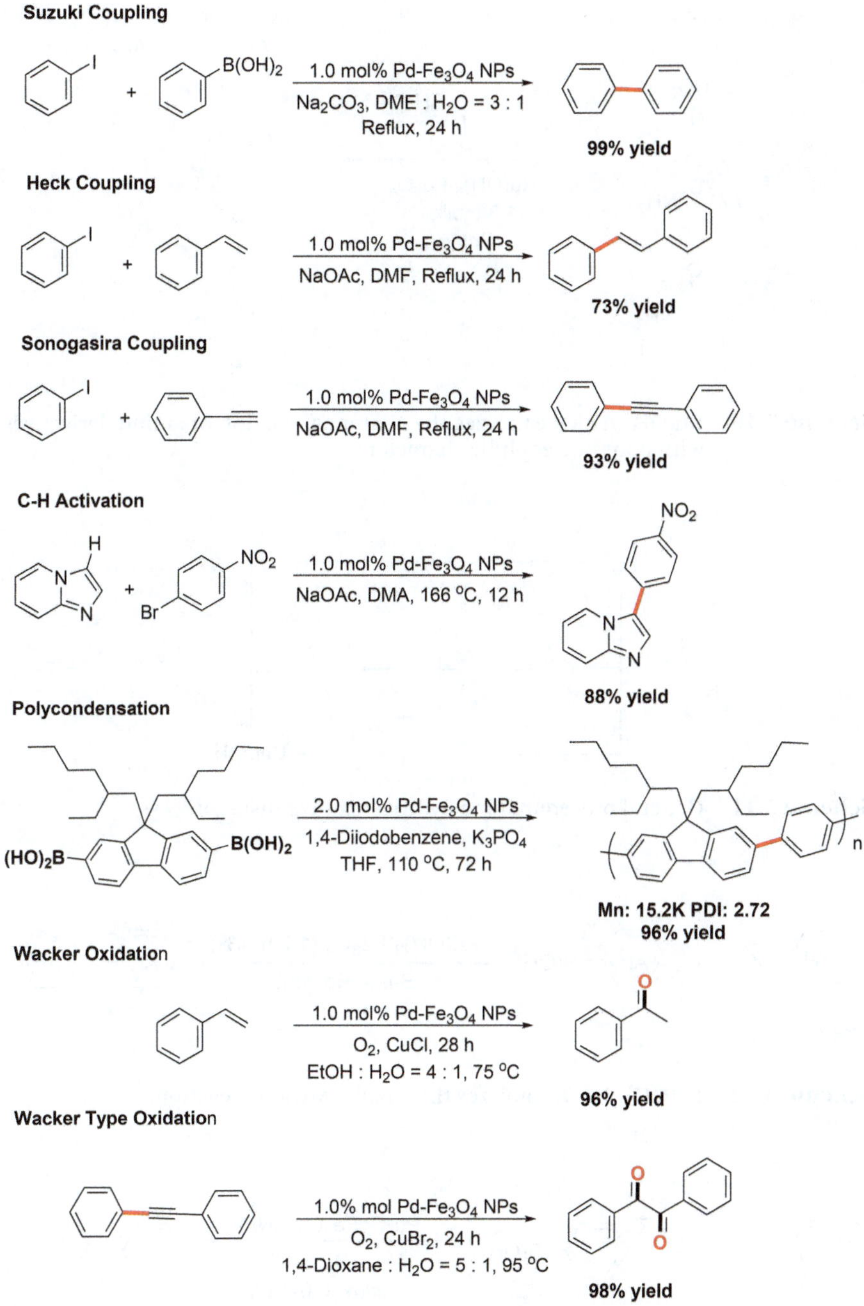

Scheme 3.9 Different applications of Pd/Fe_3O_4 nanocrystals.

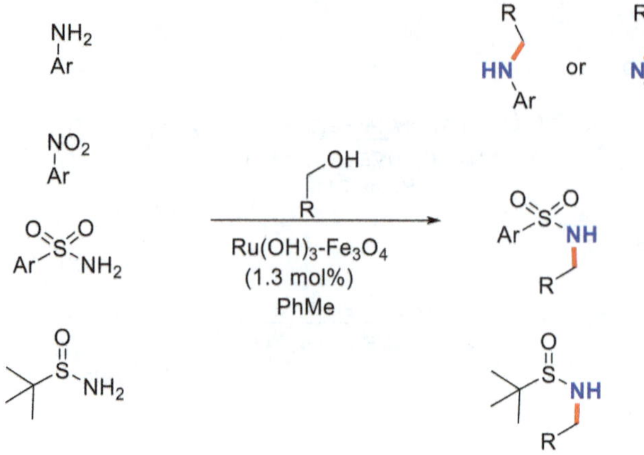

Scheme 3.10 $Ru(OH)_3/Fe_3O_4$ catalyzes the *N*-monoalkylation of amino derivatives with poor nucleophilic character.

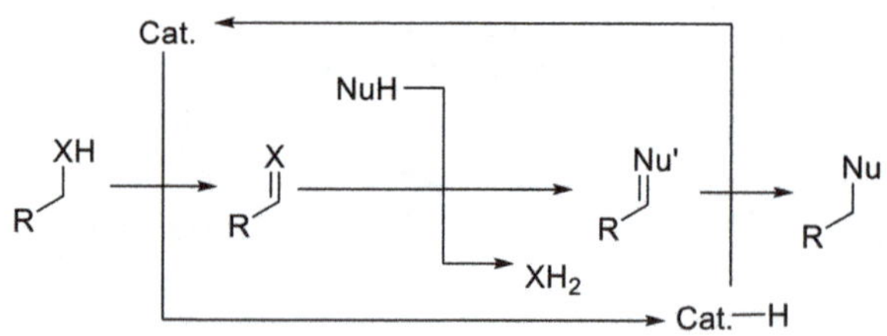

Scheme 3.11 General procedure for a hydrogen autotransfer process.

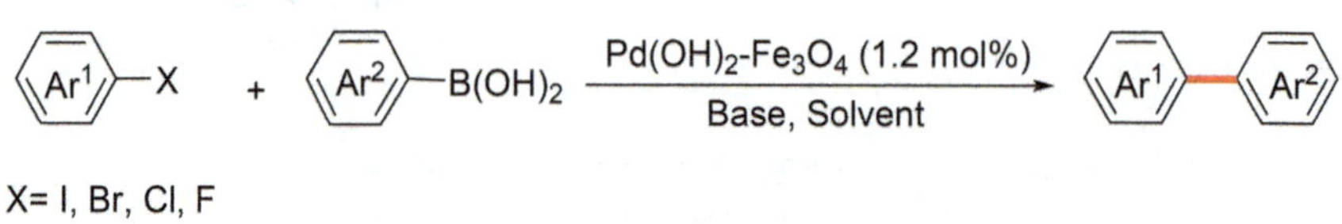

Scheme 3.12 $Pd(OH)_2/Fe_3O_4$ catalyzes the Suzuki–Miyaura reaction.

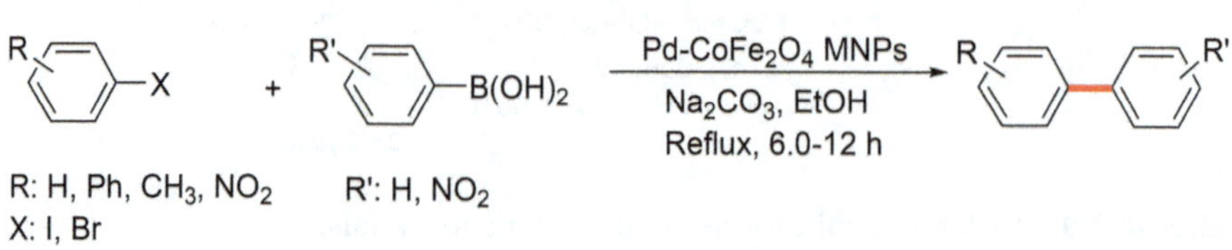

Scheme 3.13 Suzuki reaction catalyzed by $Pd\text{-}CoFe_2O_4$ MNPs.

efficiently employed for Suzuki cross coupling reactions. After the recycling of the magnetic catalyst reaction, its activity was not significantly decreased in up to four cycles.

Gawande and co-workers have developed new, efficient, recyclable and inexpensive ferrite–nickel magnetic-nanoparticles (Fe_3O_4–Ni MNPs), which were applied in the reduction of nitro arenes and carbonyl compounds by using glycerol as a hydrogen source (Scheme 3.14).[22] Fe_3O_4–Ni MNPs synthesized by the wet impregnation method and subsequent chemical reduction were fully characterized by XRD, ICP-AES, TEM and SIMS. Then, the catalytic performances of Fe_3O_4–Ni MNPs were investigated by the reduction of nitro arenes and carbonyl compounds to afford the desired products in excellent yields. This process not only solved the problem of selecting a green solvent, but also avoided the use of a precious metal and expensive ligands to permit the reuse of the catalyst.

Branco *et al.* designed their Nanocat-Fe-Pd catalyst synthesized by means of wet impregnation using $PdCl_2$ as a palladium source, followed by chemical reduction (Scheme 3.15).[23] ICP-AES analysis indicated that no traces of a

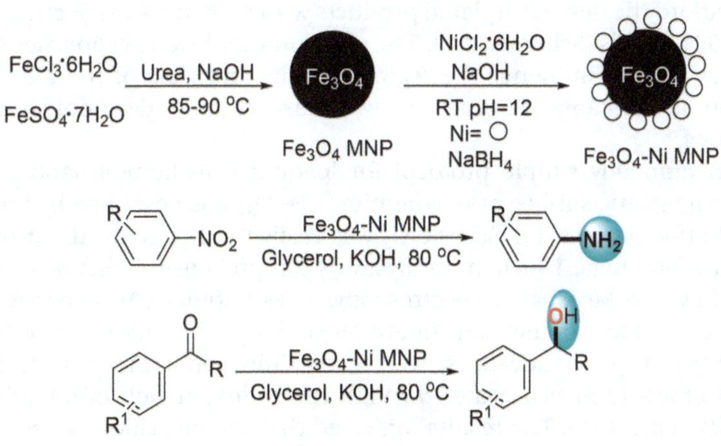

Scheme 3.14 Synthesis Fe_3O_4–Ni MNP and Fe_3O_4–Ni MNP catalyzes the reduction of nitro arenes and carbonyl compounds.

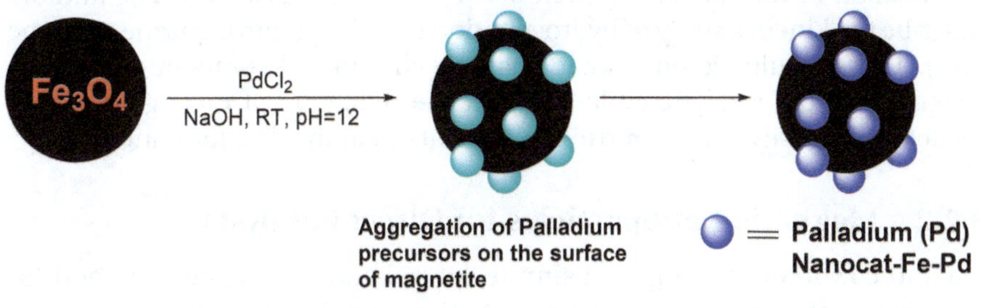

Scheme 3.15 Synthesis of Nanocat-Fe–Pd MNPs.

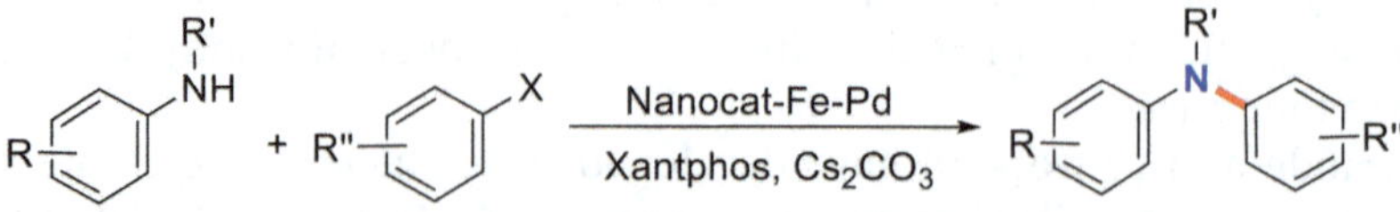

Scheme 3.16 Buchwald–Hartwig C–N cross coupling reaction using Nanocat-Fe–Pd.

crystalline Pd phase were found in the XRD spectrum of Nanocat-Fe–Pd. The catalyst showed activity in the Buchwald–Hartwig reaction for arylation of amines and amides, as shown in Scheme 3.16. The functional amines and amides were obtained in moderate to excellent yields. Notably, only a negligible amount of Pd (5.4×10^{-7} M) was detected *via* ICP analysis in the hot filtered reaction mixture.

The direct C–H arylation under mild conditions is one of the most powerful strategies to prepare the aryl-heteroaryl compounds. Kim and Lee have achieved C–H arylation of a heteroarene with aryl bromides in the presence of Pd–Fe_3O_4 heterodimer nanocrystals (1.0 mol% of palladium).[24] Various aryl bromides were suitable substrates to react smoothly with imidazo[1,2-*a*]pyridine to afford the desired arylated products with exclusive C3-selectivity from 42% to 95% yields (Scheme 3.17). The electronic and steric properties of aryl bromides did not influence the regioselectivity and yield of the direct C–H arylation. Remarkably, the reactions were carried out under phosphine-free conditions.

An operationally simple protocol for loading palladium(0) nanoparticles onto the magnetic surface of maghemite (γ-Fe_2O_3) was described by Roucoux *et al.*[25] In this process, Pd(0)@γ-Fe_2O_3 was easily synthesized without the use of organic modifiers. This nanocatalyst was comprehensively characterized by TEM, XRD, ^{57}Fe Mössbauer spectroscopy, X-ray photoelectron spectroscopy, and superconducting quantum interference device measurements (SQUID). The Pd(0)@γ-Fe_2O_3 nanocatalyst was successfully applied to provide hydrogenated products in neat water through the hydrogenation of olefinic substrates (Scheme 3.18). The results indicated that styrene, chlorobenzene, and nitrobenzene were suitable substrates in this hydrogenation reaction. In this process, the presence of hydrochloric acid decreases the catalytic activity due to progressive destabilization of the Pd(0)@γ-Fe_2O_3 suspension. A complete deactivation of the catalytic system occurred in the fourth run. The authors describe the kinetics study of hydrogenation of 4-chloronitrobenzene (Scheme 3.19). These results demonstrated that the chlorinated compound could be generated in the catalytic cycle and that the reduction of nitro groups proceeded more easily in the control experiments than the dechlorination step.

3.2.2 Magnetic Nanoparticles for Direct Catalysis

Xia and coworkers developed a simple sol–gel autocombustion method for the preparation of magnetic nanocrystalline spinel-type cobalt ferrite complex oxides.[26] These catalysts were synthesized in a variety of molar ratios of

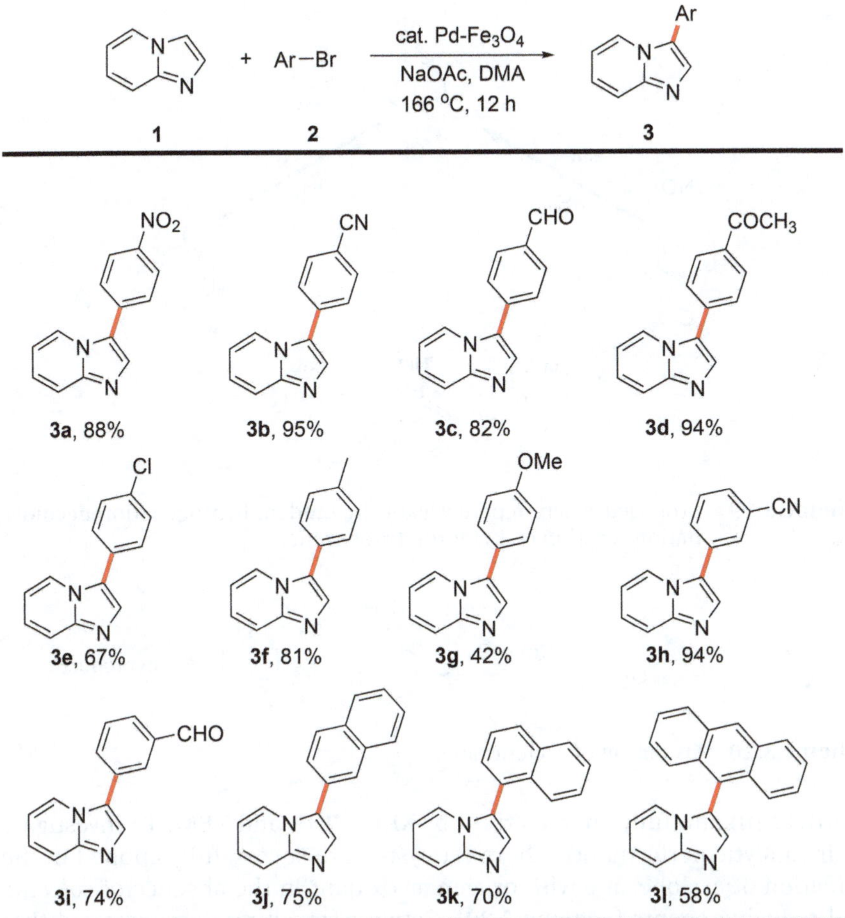

Scheme 3.17 Scope of the direct arylation of imidazo[1,2-*a*]pyridine with aryl bromides. All reactions were carried out with imidazo[1,2-*a*]pyridine (0.75 mmol), aryl bromide (0.50 mmol), NaOAc (1.0 mmol), and Pd/Fe_3O_4 (0.005 mmol, 1.0 mol% of Pd) in DMA (0.25 M) at 166 °C for 12 h.

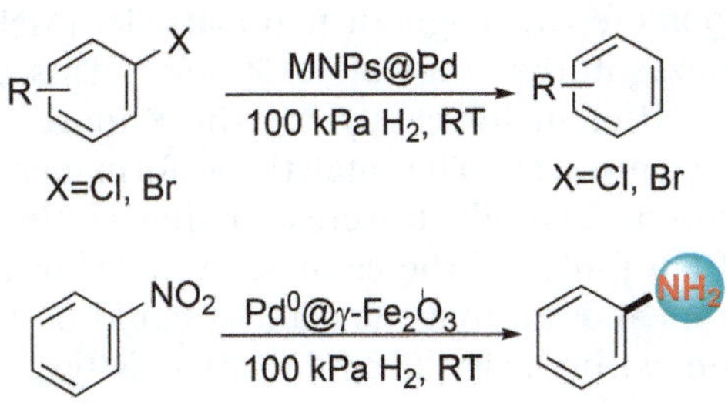

Scheme 3.18 The reduction of olefinic substrates catalyzed by Pd(0)@γ-Fe_2O_3.

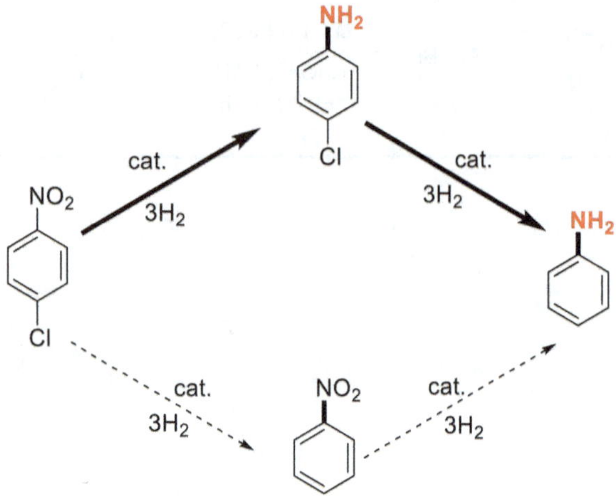

Scheme 3.19 Proposed reaction pathway for the tandem hydrogenation-dechlorination reaction of 4-chloronitrobenzene.

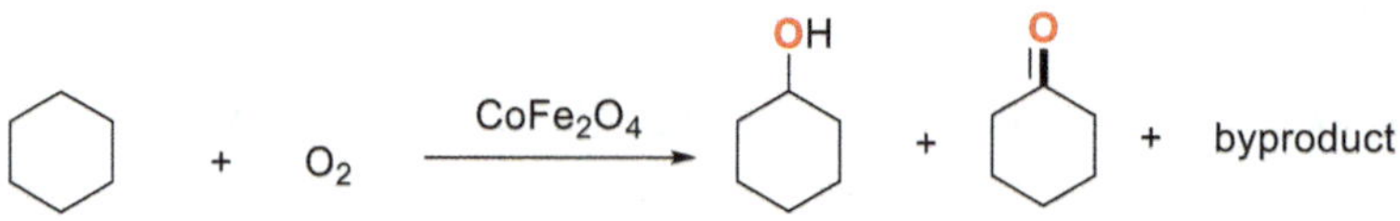

Scheme 3.20 Oxidation of cyclohexane.

Co(II):Fe(III) and fully characterized by XRD, FT-IR and TEM. To investigate their catalytic performance, these catalysts were successfully applied in the oxidation of cyclohexane with oxygen as oxidant in the absence of solvents and reducing agents (Scheme 3.20). Catalyst screening demonstrated that excellent results with a turnover number of 288 and 92.4% selectivity for the products cyclohexanone and cyclohexanol were observed under 1.6 MPa of dioxygen pressure at 145 °C after 6.0 h of reaction time and use of $CoFe_2O_4$ as catalyst. The recycling of this catalyst showed that no obvious loss of activity was observed after five consecutive runs. In addition, oxidation of linear alkanes was realized as well.

In 2010, spinel cobalt ferrite magnetic nanoparticles (MNPs) were synthesized with average sizes in the range of 40–50 nm.[27] This nanocatalyst was found to be highly efficient for catalyzing the Knoevenagel reaction in aqueous ethanol (Scheme 3.21). The catalytic performance showed that the corresponding Knoevenagel products were prepared within a very short time in the presence of 5.0 mol% of the catalyst at 50 °C in high yields. The recycling experiments showed that the catalyst could be readily recovered and reused four times without significant loss of activity.

The biaryl compound represents an important structural motif found in numerous natural products, pharmaceuticals and agrochemicals. Traditionally, biaryl structures could be prepared by transition metal catalyzed cross coupling. Compared to traditional routes, the direct arylation of aromatic C–H bonds with aryl halides in the presence of catalytic transition metals is an attractive alternative strategy, and promising in terms of sustainability and practical applications. Sun *et al.* prepared magnetic Fe–Ni nanoparticle alloys to explore their activity for the arylation of aromatic C–H bonds with aryl halides.[28] The direct arylation of benzene with various aryl halides was conducted after optimizing reaction conditions. The yields of the isolated target products are summarized in Scheme 3.22. The aryl iodides bearing

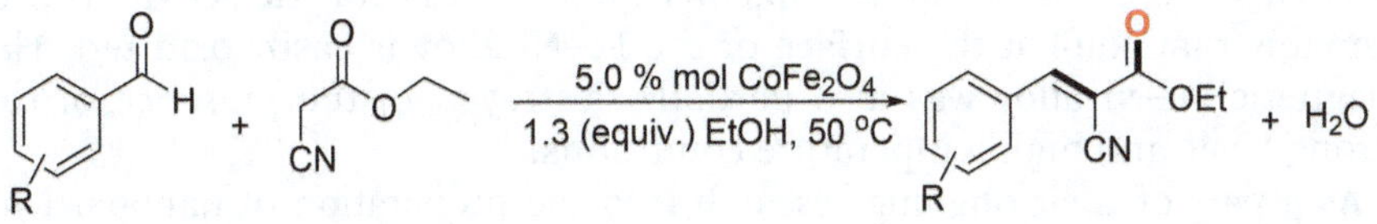

Scheme 3.21 Application of spinel cobalt ferrite magnetic nanoparticles in Knoevenagel reaction.

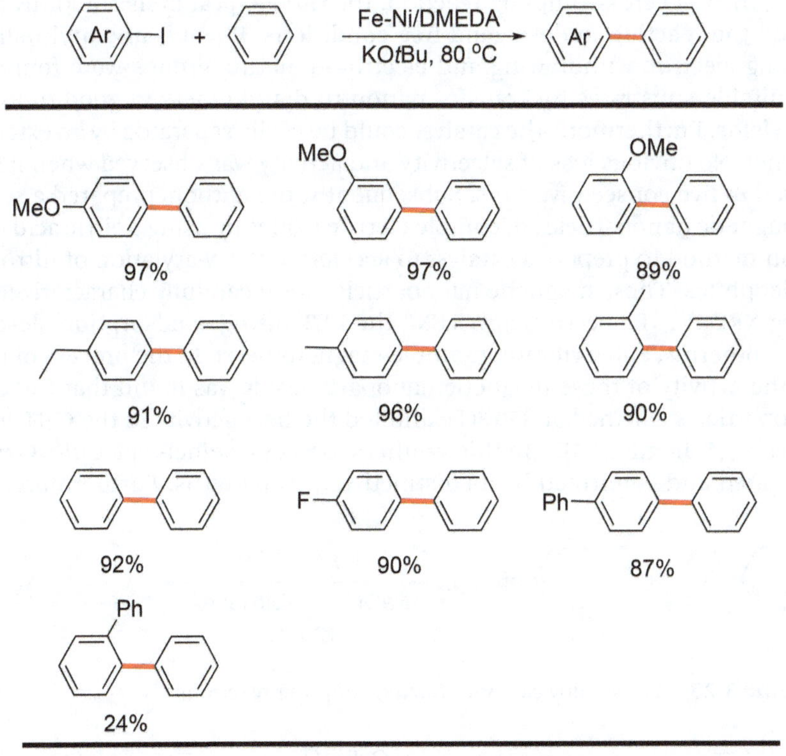

Scheme 3.22 Arylation of benzene with various aryl halides catalyzed by the Fe–Ni alloy. Reaction conditions: aryl halide (0.50 mmol), Fe–Ni (20 mol%), KOtBu (3.0 equiv), benzene (8.0 mL), DMEDA (30 mol%), 48 h. (Isolated yield.)

electron-withdrawing and electron-donating groups in the *ortho*-, *meta*-, and *para*- positions afforded the corresponding products moderate to excellent isolated yields. The efficiency of this catalyst remained unchanged even after eight runs. Subsequently, this research group employed this magnetic Fe–Ni alloy as a nanocatalyst for the Suzuki coupling reaction in the absence of a reducing agent (Scheme 3.23).[29] Most notably, this catalyst successfully utilized the coupling reaction of aryl halides with phenylboronic acid to afford the target products in good to excellent yields. However, the recycling of the catalyst demonstrated that the efficiency of the catalyst markedly decreased after each step. The reason for this observed loss in catalytic activity is presumably that Ni(0) at the surface of the Fe–Ni alloy is easily oxidized. The magnetic Fe–Ni alloy was also partially destroyed in the presence of the strong basic and high temperature conditions.

As a part of their ongoing research into the preparation of nanoparticle-based catalysts, Sun *et al.* synthesized $CuFe_2O_4$ nanoparticles as a magnetically separable catalyst for the C–O cross coupling reaction of substituted phenols and aryl halides (Scheme 3.24).[30] The $CuFe_2O_4$ nanoparticles were carefully characterized by XRD, TEM and FT-IR. In the application of catalyzing the C–O cross coupling reaction, the $CuFe_2O_4$ catalysis smoothly promoted the reaction under ligand-free conditions. Phenols and aryl halides bearing electron-withdrawing and electron-donating groups were found to be suitable substrates for transformation to diaryl ethers in good to excellent yields. Furthermore, the catalyst could be easily separated by an external magnet. No obvious loss of selectivity and activity was observed when it was reused in five consecutive runs. Subsequently, the authors prepared a series of magnetic nanoparticles of complex ferrite oxides by using a citric acid chelation method to prepare a catalyst to accelerate the *N*-arylation of nitrogen nucleophiles. These magnetic nanoparticles were carefully characterized by using XRD, FT-IR spectroscopy, TEM, the BET nitrogen adsorption–desorption isotherms, and a vibrating sample magnetometer. In the process of testing the activity of these magnetic nanoparticles, it was found that $CuFe_2O_4$ nanoparticles calcined at 850 °C exhibited the best activity in the C–N cross coupling (Scheme 3.25).[31] In this synthetic process, spinel-type $CuFe_2O_4$ was generated and confirmed by its distinctive XRD patterns. Furthermore, the

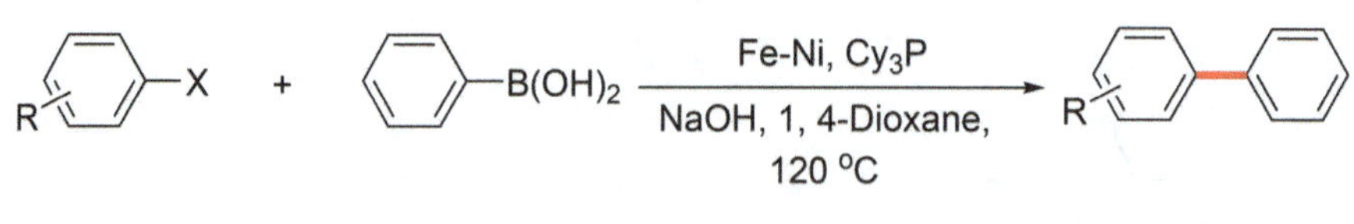

Scheme 3.23 Fe–Ni alloy catalyzes Suzuki coupling reactions.

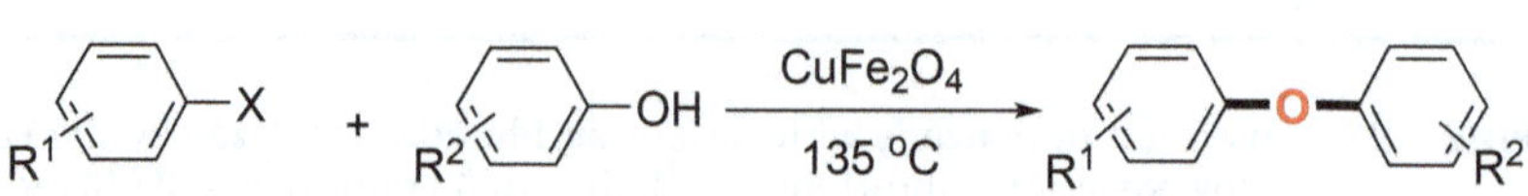

Scheme 3.24 $CuFe_2O_4$-catalyzed diaryl ether formation reaction.

data of the crystal faces of the $CuFe_2O_4$ spinel were used to confirm that the catalyst was indeed able to transform the C–N coupling under ligand-free conditions. These experimental results clearly demonstrated that the C–N coupling was catalyzed by $CuFe_2O_4$.

Readily available and magnetically separable Fe_3O_4 nanoparticles were applied to catalyze the reduction of nitroarenes to afford the respective amines.[32] A set of reactions including both nitroarenes and nitroalkanes was carried out to investigate the activity of Fe_3O_4 nanoparticles (Scheme 3.26). Nitroarenes bearing electron-donating and electron-withdrawing groups were suitable substrates to transform the corresponding amines. Based on the paradigm that hydrazine hydrate acts as a two-electron

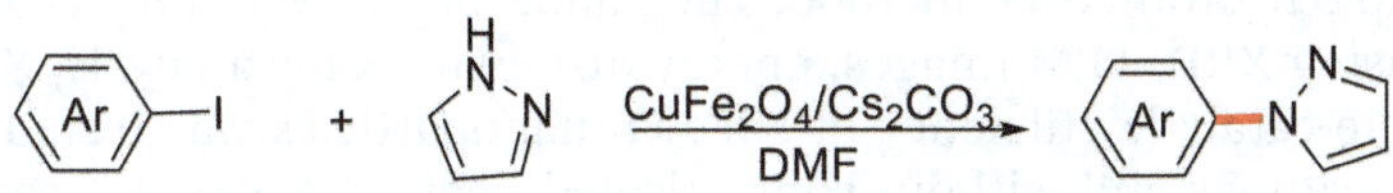

Scheme 3.25 $CuFe_2O_4$-catalyzed diaryl ether formation reaction.

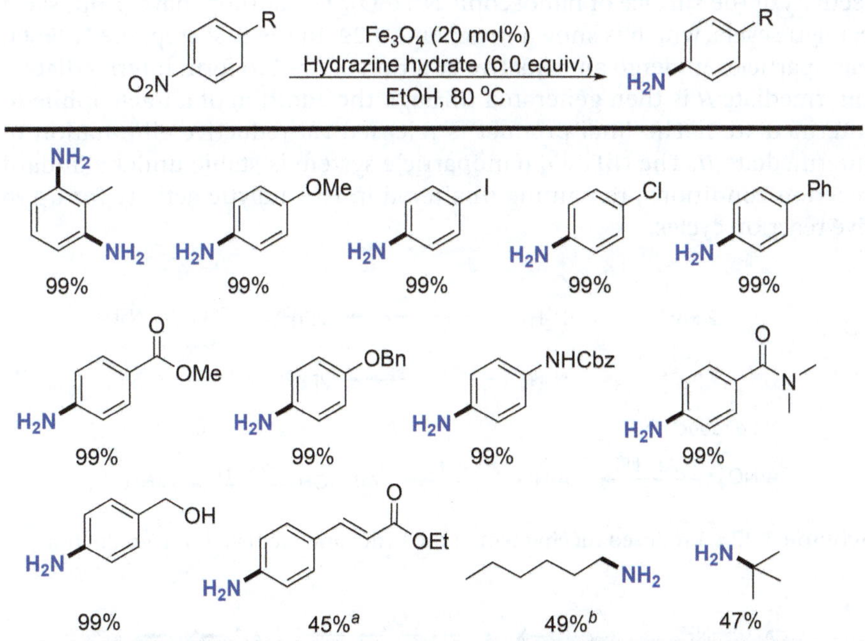

Scheme 3.26 Substrate scope of Fe_3O_4 catalyzed nitro group reductions. Reaction conditions: 1.0 mmol of nitro compound, 6.0 equiv. of hydrazine monohydrate, 6.0 mL of ethanol at 80 °C. All of the yields were determined by GC analysis using anisole as an internal standard. [a]55% of product was ethyl 3-(4-aminophenyl) propanoate. [b]10 equiv. of hydrazine monohydrate was used.

reducing agent, a plausible mechanism for the reduction of the nitro group at the Fe_3O_4 catalyst is shown in Scheme 3.27. Notably, nitroso- and hydroxylamine were not observed in the mixture by GC in the Fe_3O_4-catalyzed nitro reduction. Furthermore, neither double bonds nor imides were reductive enough to afford the respective products. However, the Fe_3O_4-catalyzed system was capable of reducing double bonds together with a nitro group. In addition, TEM images of the Fe_3O_4 nanoparticles before use and after ten cycles showed that they appeared almost unchanged in their size and shape.

In 2014, Das *et al.* developed nickel ferrite magnetic nanoparticles to catalyze the formation of C–N, C–O and C–C bonds under ligand-free conditions (Scheme 3.28).[33] The synthesis of $NiFe_2O_4$ nanoparticles was achieved by a simple hydrothermal method. The nanocatalysts were fully characterized by using XRD, TEM images, energy dispersive X-ray analysis, XPS and FT-IR. The catalytic efficiency of $NiFe_2O_4$ nanoparticles was tested at low catalyst loadings and without sophisticated ligands. Notably, a variety of substrates can be utilized, such as nitro, and methoxy-, acid- and base-sensitive heteroaromatic groups. These results showed that these reactions can be carried out with excellent functional group tolerance and chemoselectivity at the surface of nanoscopic $NiFe_2O_4$. The authors have proposed a catalytic cycle, which is shown in Scheme 3.29. In the first step, the $NiFe_2O_4$ nanoparticles undergo an oxidative addition of Ar–X to form intermediate *I*. Intermediate *II* is then generated through the addition of a nucleophile to intermediate *I*. The final product is released *via* reductive elimination of intermediate *II*. The $NiFe_2O_4$ nanoparticle system is stable under standard reaction conditions, remaining unaltered in its catalytic activity for up to five reaction cycles.

$$2\,Fe^{3+} + N_2H_4 \longrightarrow 2Fe^{2+} + 2H^+ + N_2H_2$$

$$2\,Fe^{3+} + N_2H_2 \longrightarrow 2Fe^{2+} + 2H^+ + N_2$$

Nitro reduction

$$ArNO_2 \xrightarrow{2e^-,\,H^+} ArNO \xrightarrow{2e^-,\,H^+} ArNHOH \xrightarrow{2e^-,\,H^+} ArNH_2$$

Scheme 3.27 Proposed mechanism of the hydrazine-mediated nitro-reduction.

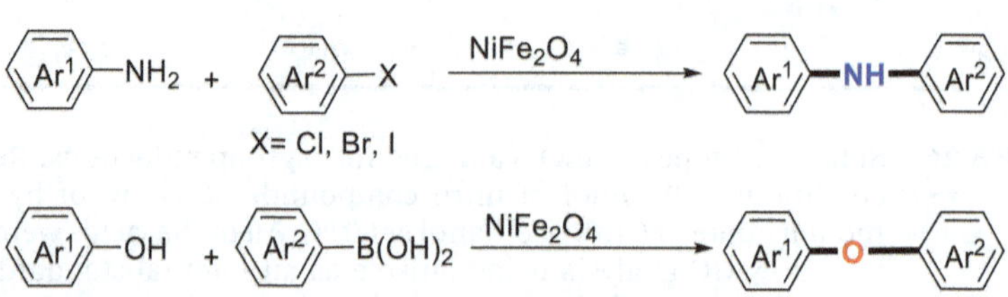

Scheme 3.28 $NiFe_2O_4$-catalyzed C–N and C–O coupling.

3.2.3 Nano-Magnetite Supported Metal- and Organocatalysts

In 2005, Gao *et al.* designed core–shell superparamagnetic nanoparticles consisting of a highly crystalline γ-Fe_2O_3 core and a very thin polymeric shell wall.[34] Then 1-methylimidazole was loaded onto the surface of the core–shell nanoparticles by reflux in toluene solution. Finally, a palladium catalyst was combined with the N-heterocyclic carbenes (NHCs) at the surface (Scheme 3.30). These magnetic nanoparticle-supported Pd/NHC complexes were utilized to promote the Suzuki cross coupling of aryl halides with arylboronic acids. Most notably, it was shown that 1.52×10^{-2} mol% Pd at the iron oxide nanoparticles was sufficient to catalyze the Suzuki cross coupling efficiently. After the completion of the reactions, the nanoparticle supported catalysts were magnetically concentrated at a side wall of the flask by an external permanent magnet. Moreover, the recovered palladium nanocatalyst was reconditioned by washing it several times with organic solvents, followed by air oxidation. More than 97% of active palladium–iron oxide nanoparticles were recycled based on the weight of particles.

The same research group also prepared another maghemite nanoparticle-supported NHC–Pd complex.[35] The catalyst was synthesized by the addition of the complex of Pd and N-heterocyclic carbenes (NHCs) to load onto magnetic maghemite (γ-Fe_2O_3) cores to form iron oxide–Pd. Iron oxide–Pd catalysts were suitable for Suzuki, Heck and Sonogashira cross coupling

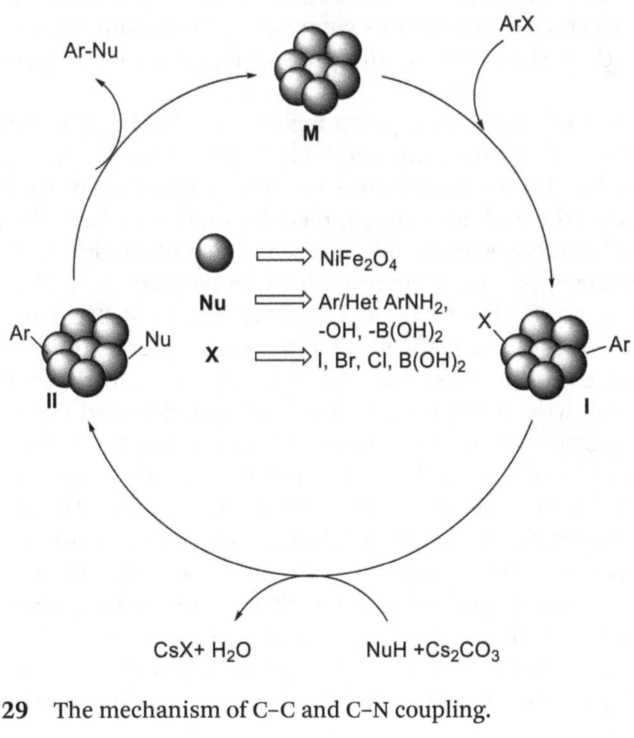

Scheme 3.29 The mechanism of C–C and C–N coupling.

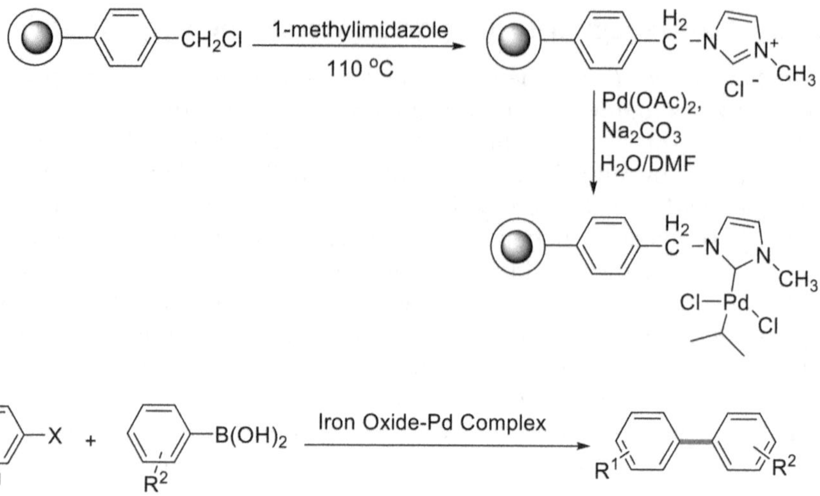

Scheme 3.30 Iron oxide–Pd complex catalyzed Suzuki cross coupling.

reactions (Scheme 3.31). Since the iron oxide–Pd catalyst was immobilized on the nanocluster surfaces, the substrates were interacted with the catalyst's surface to form the final products. This catalyst was able to overcome the typical limitations of heterogeneous support matrixes, where the active sites are often hardly accessible by the substrate. More importantly, comparative kinetic studies show that reactions catalyzed by iron oxide-Pd are promoted much faster than the corresponding reactions that are catalyzed by solid phase-Pd.

Manorama *et al.* have developed a facile route for Pd(0) immobilization on the surface of amine-terminated Fe_3O_4 and $NiFe_2O_4$ nanoparticles.[36] $NiFe_2O_4$-DA-Pd and Fe_3O_4-DA-Pd catalysts were synthesized by the sonication and refluxing ferrite and dopamine molecules on the surface of $NiFe_2O_4$ and Fe_3O_4. The Pd catalyst was loaded on the surface of modified $NiFe_2O_4$ and Fe_3O_4 and reduced by hydrazine monohydrate (Scheme 3.32). The catalysts were fully characterized by FT-IR, ICP-AES, XPS and TEM. The $NiFe_2O_4$-DA-Pd and Fe_3O_4-DA-Pd catalysts were applied to achieve a series of hydrogenation reactions and could be reused in ten repeated cycles, without any significant loss in catalytic efficiency. No loss of Pd was detected due to leaching during the hydrogenation reaction, which is most likely the reason for the persistent activity of the catalyst. TEM results indicated that the morphology of the catalysts remained unaltered after five reactions. The authors then employed these particles in various Pd-catalyzed cross coupling reactions of chloro-derivatives (Scheme 3.33). Suzuki and Heck couplings were successful with these catalysts and led to the formation of the target compound in good to excellent yields. Notably, the enhanced catalytic activity that was observed for the chloro-derivatives remained very high when utilizing the corresponding bromo- and iodo-derivatives, due to the very good dispersion

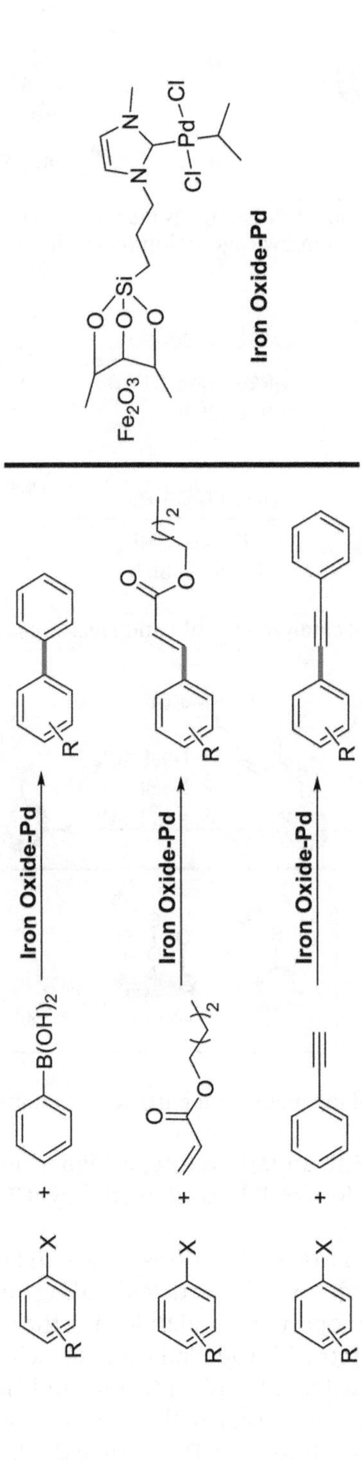

Scheme 3.31 Pd catalysts using superparamagnetic nanoparticles as novel soluble supports for Suzuki, Heck, and Sonogashira cross coupling reactions.

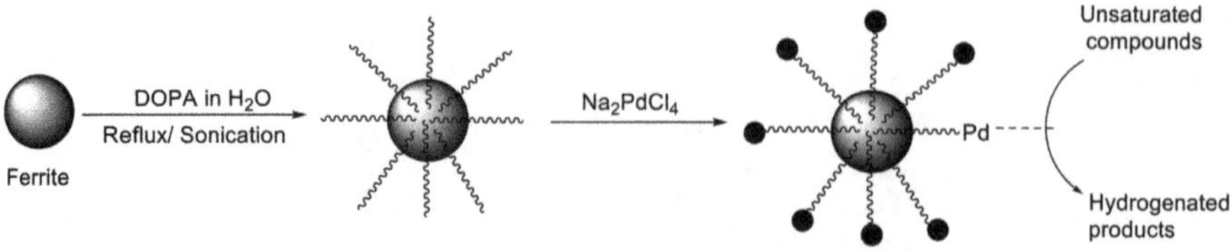

Scheme 3.32 The preparation of $NiFe_2O_4$-DA-Pd and Fe_3O_4-DA-Pd catalysts and their application in hydrogenation reaction.

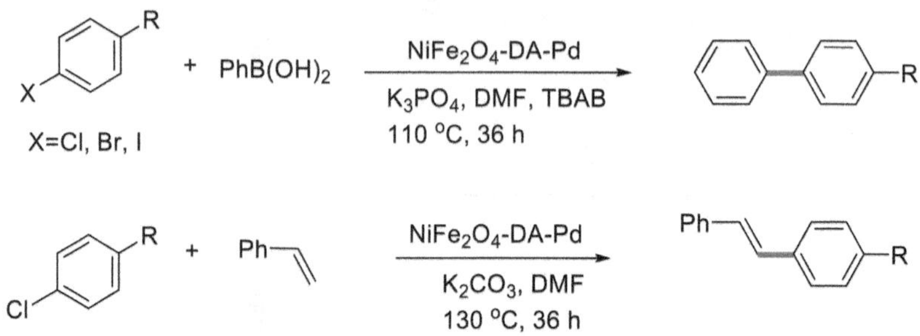

Scheme 3.33 $NiFe_2O_4$-DA-Pd catalyzed Suzuki and Heck cross coupling.

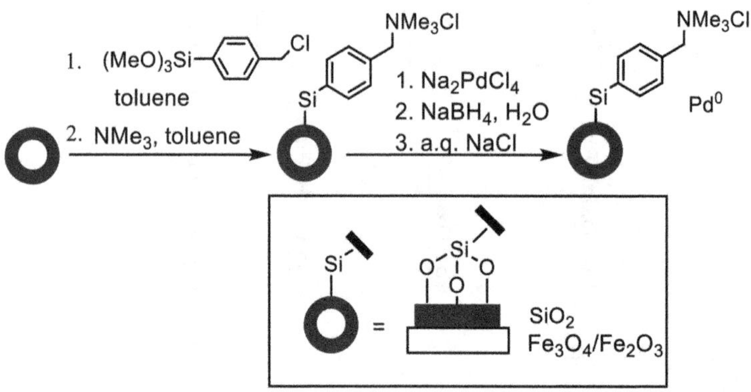

Scheme 3.34 Preparation of magnetic nanoparticles functionalized with Pd^0.

of the catalyst system in DMF and, therefore, a higher number of available active sites. Here, minimal loss of Pd was detected by ICP-AES studies after three repeated reactions.

Similarly, magnetic nanoparticles functionalized with Pd(0) were prepared by Kirschning *et al. via* palladium nanoparticles loading on the surface of the functionalized magnetic nanoparticles and following the reductive precipitation of ammonium-bound tetrachloropalladate salts (Scheme 3.34).[37] These particles were employed to achieve Suzuki–Miyaura and Heck coupling reactions. (Scheme 3.35). It was noted that palladium particles show good catalytic activity under flow conditions. ICP-MS analytic data confirmed that

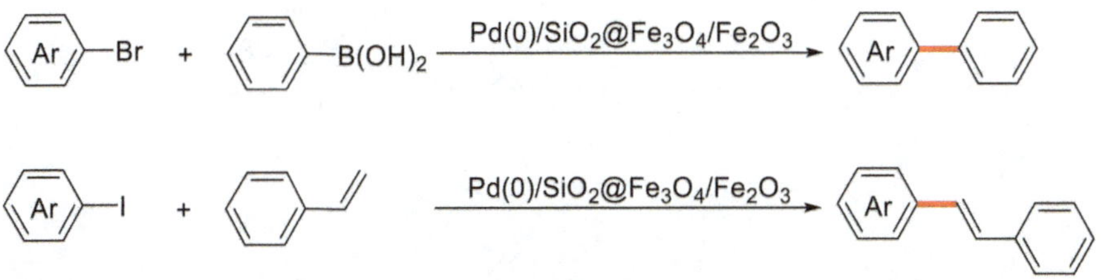

Scheme 3.35 Suzuki–Miyaura and Heck coupling reactions under flow conditions.

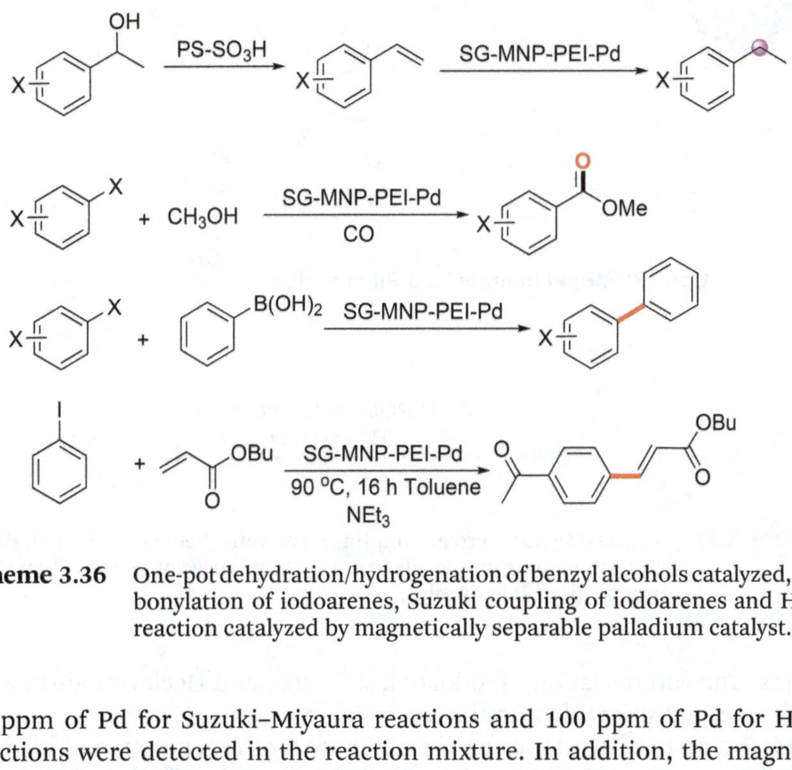

Scheme 3.36 One-pot dehydration/hydrogenation of benzyl alcohols catalyzed, carbonylation of iodoarenes, Suzuki coupling of iodoarenes and Heck reaction catalyzed by magnetically separable palladium catalyst.

34 ppm of Pd for Suzuki–Miyaura reactions and 100 ppm of Pd for Heck reactions were detected in the reaction mixture. In addition, the magnetic nanoparticles functionalized with Pd(0) were reusable three times without a decrease in activity.

In 2008, a magnetically separable palladium-based sol–gel catalyst had been prepared by the following routes: the palladium catalyst could be encapsulated in silica sol–gel-modified polyethylenimine composite with the aid of the surface of magnetic nanoparticles modified with ionic liquid groups.[38] These systems were utilized to promote the one-pot reactions of dehydration–hydrogenation of benzyl alcohols to afford the desired products in good yields (Scheme 3.36). In the above process, benzyl alcohols first underwent dehydration to afford the substituted olefins. Alkylated arenes were obtained *via* hydrogenation of the double bonds formed. 4-Methylbenzenesulfonic acid was able to accelerate the reactions of benzyl alcohols to substituted

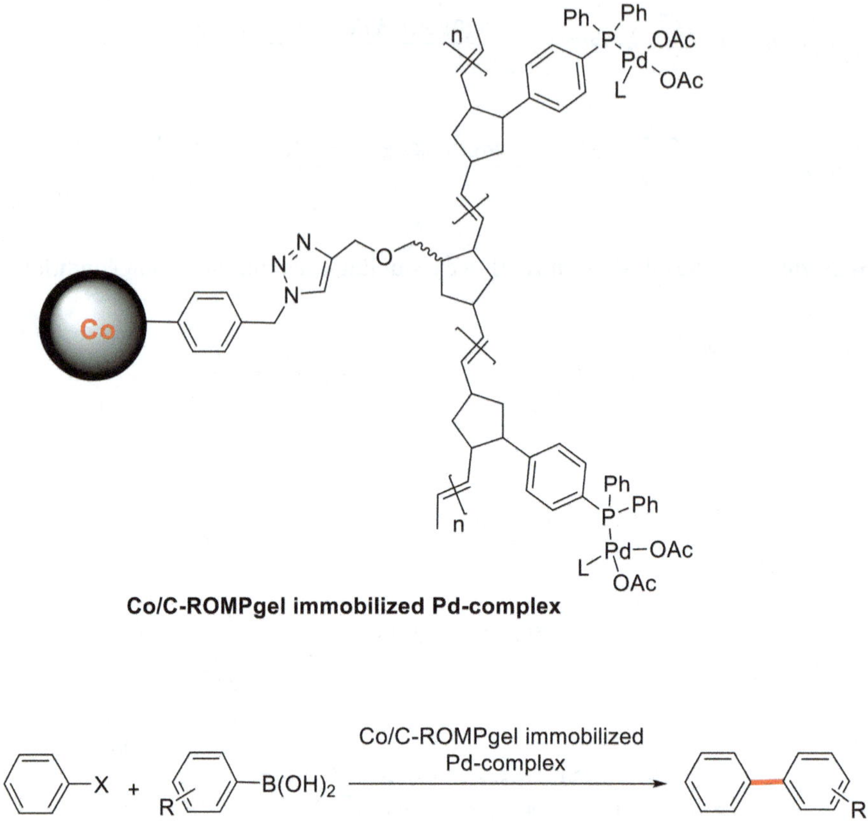

Scheme 3.37 Suzuki–Miyaura cross coupling reactions between phenylhalides and phenylboronic acids catalyzed by recyclable Co/C-ROMPgel immobilized Pd-complex.

olefins. The carbonylation of iodoarenes, Suzuki, and Heck couplings were also proceeded smoothly with these systems (Scheme 3.36).

Reiser *et al.* have developed an elegant route for the generation of a hybrid material as support for a recyclable palladium catalyst.[39] The catalyst followed a new strategy of a click reaction to produce a norbornene tag (Nb-tag) at the surface of carbon coated cobalt nanoparticles. The subsequent polymerization of a mixture of Nb-tagged ligand and palladium on the surface of carbon coated cobalt afforded hybrid magnetic nanoparticles. The performance of Co/C-ROMPgel immobilized Pd-complex has been investigated for Suzuki–Miyaura cross coupling reactions between phenylhalides and phenylboronic acids (Scheme 3.37). After recycling the catalyst, it was confirmed by AAS-analysis that a rather insignificant Pd-content (1.9 ppm) was observed in the aqueous layer.

A Pd-complex on the surface of silica-coated Fe_3O_4 (Pd-SiO_2@Fe_3O_4) was reported by Jin *et al.* (Scheme 3.38).[40] Inductive plasma atomic emission

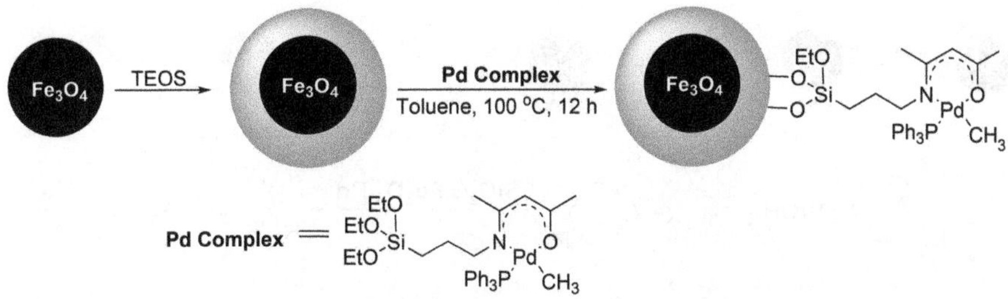

Scheme 3.38 Synthesis of magnetite-supported Pd catalyst.

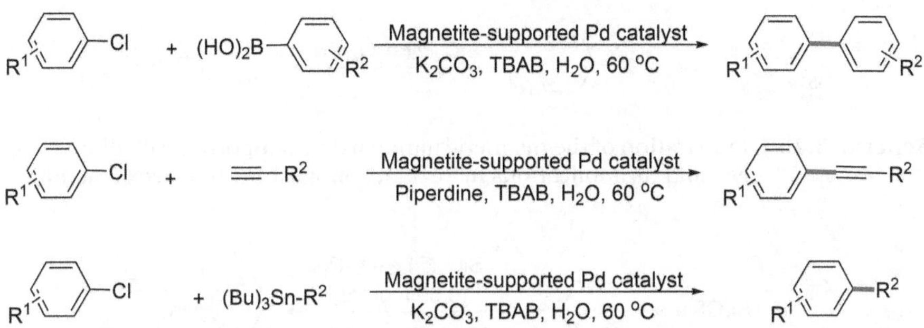

Scheme 3.39 Magnetite-supported Pd catalyst catalyzed C–C coupling reaction.

spectrometry (ICP-AES) indicated that 0.21 mmol of palladium per gram was loaded on the surface of silica-coated Fe_3O_4. TEM revealed a uniform thickness on the core–shell structure of the particles of 7.0 nm. The synthesized Pd-SiO_2@Fe_3O_4 catalyst was found to be highly active, easily recoverable, and practically heterogeneous for the Suzuki, Sonogashira and Stille coupling reactions (Scheme 3.39). A wide range of otherwise unreactive aryl chlorides were transformed successfully into their respective target compounds under mild conditions. Owing to the monodentate electron-rich phosphine and bulky imine ligands on the palladium atom, oxidative activation of less reactive aryl chlorides to the active Pd(0) species was observed. The superior activity fueled the speculation that these two independent ligands cooperatively not only facilitated the oxidative addition, but also accelerated the reductive elimination.

Wang *et al.* have reported a highly efficient, air- and moisture-stable and easily recoverable magnetic nanoparticle-supported palladium catalyst (SiO_2@Fe_3O_4-Pd)[41a] following the procedure summarized in Scheme 3.40. The SiO_2@Fe_3O_4-Pd catalysts were characterized by ICP-AES, XRD and TEM. This catalyst also showed high activity in Suzuki, Sonogashira and Heck reactions. In the process of recycling experiments, the reaction mixture was analyzed for palladium metal using ICP-AES. The results indicated that less than 0.036% of the starting palladium catalyst had leached out from the catalyst surface. Meanwhile, the reaction mixture was also analyzed for iron metal

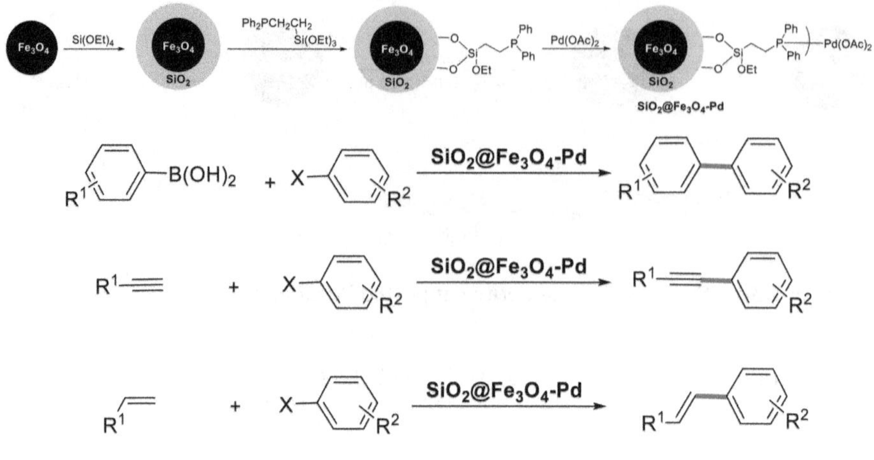

Scheme 3.40 Preparation of the magnetic nanoparticle-supported palladium catalyst and its applications in Suzuki, Sonogashira and Heck reactions.

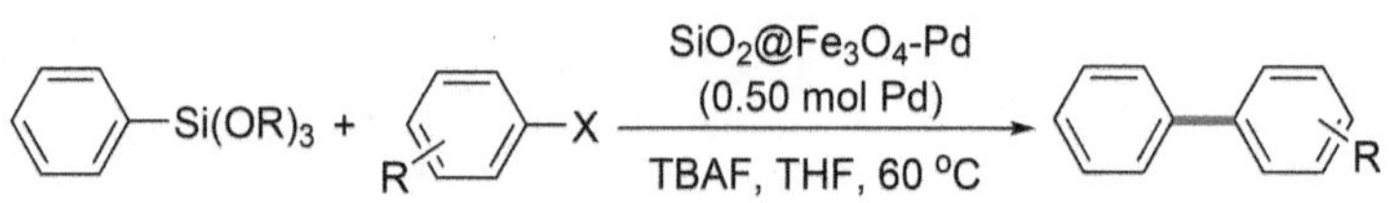

Scheme 3.41 Hiyama cross-coupling reaction of aryl halides with aryltrimethoxysilanes.

using ICP-AES. The experiment showed that the Fe content was less than 0.10 ppm in the reaction mixture. The result indicated that the corrosion on the silica layer did not occur under anhydrous weak basic reaction conditions. The same research group successfully utilized this catalyst to accelerate the Hiyama reaction of aryltrialkoxysilanes (Scheme 3.41).[41b] Various aryltrialkoxysilanes coupled well with organic halides in the presence of 0.5 mol% supported palladium catalyst, thus affording the corresponding products in good yields.

Recently, Heuzé *et al.* synthesized dendritic structures for the grafting of core–shell γ-Fe_2O_3/polymer 300 nm superparamagnetic nanoparticles functionalized with diphosphino–palladium complexes.[42] The chelating diphosphine groups on the dendron played key roles in stability and catalytic efficiency. The catalytic performance of these nanocatalysts was investigated in Suzuki cross couplings (Scheme 3.42). A variety of aryl boronic acids reacted with deactivated aryl bromide substrates to give the desired products in good yields. Even after 25 cycles, no significant loss of reactivity was observed.

In 2005, Ying and co-workers synthesized Pd-nanoclusters on silica-coated Fe_2O_3 support (Scheme 3.43).[43] The Pd/SiO_2/Fe_2O_3 catalysts were characterized by XRD, TEM and ICP-MS. The catalyst systems were successfully utilized to promote the hydrogenation of nitrobenzene (Scheme 3.44). Notably, the turnover number (TON) reached 2000 for 100% conversion. In the process of recycling, grain growth of the Pd-nanoclusters led to reduced conversion

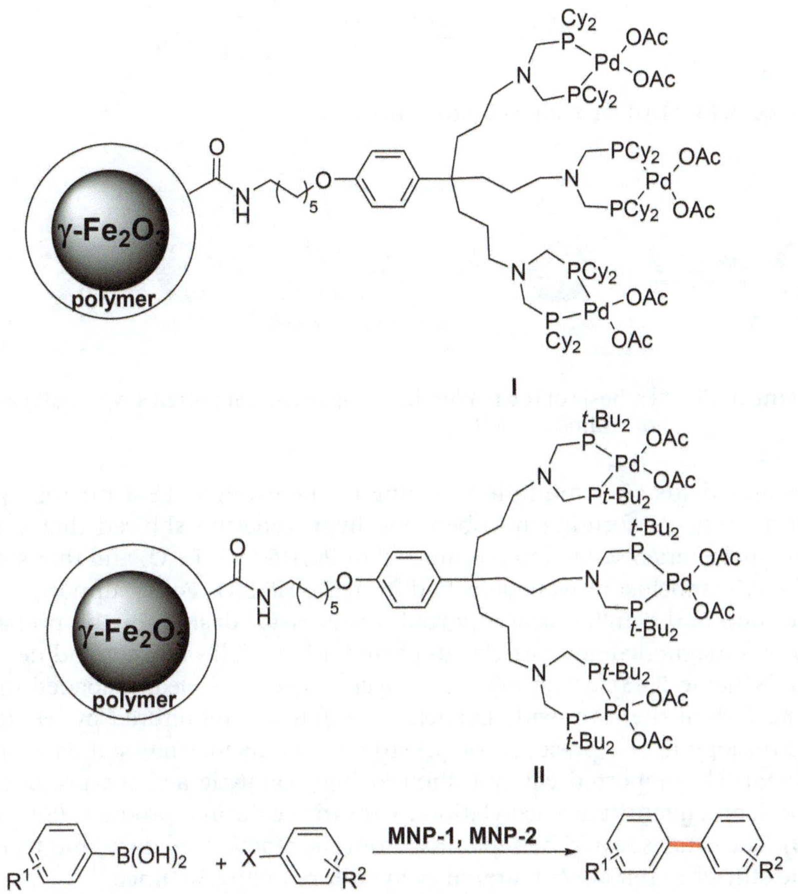

Scheme 3.42 Suzuki cross couplings using grafted MNPs.

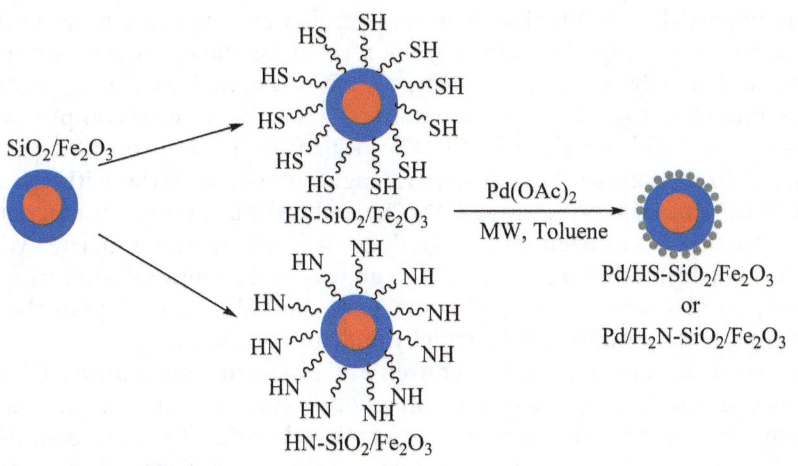

Scheme 3.43 Preparation of the magnetic nanoparticle-supported palladium catalyst.

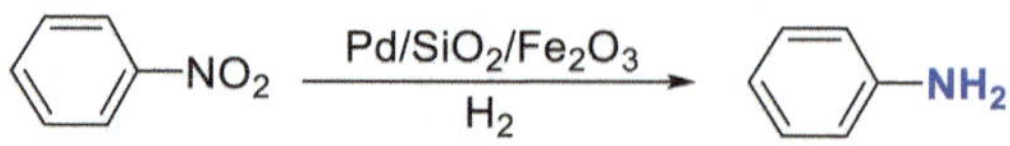

Scheme 3.44 Hydrogenation of nitrobenzene.

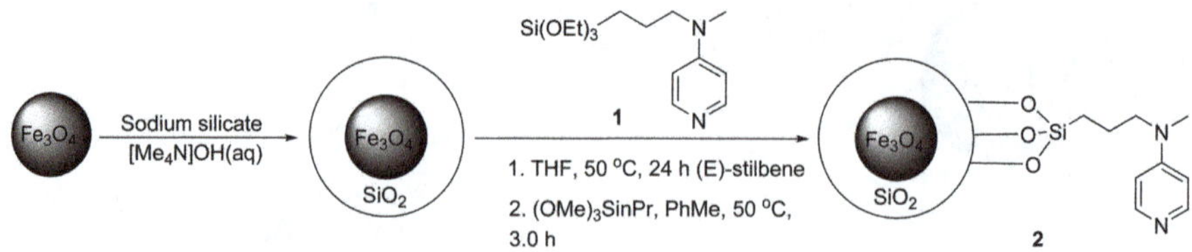

Scheme 3.45 Synthesis of the magnetic-nanoparticle-supported 4-*N*,*N*-dialkylaminopyridine catalyst.

rates and yields after multiple recycling of the catalyst. TEM micrographs after 14 runs performing nitrobenzene hydrogenation showed that some Pd-nanoclusters became interconnected in Pd/HS-SiO_2/Fe_2O_3 and that some of the Pd-nanoclusters were detached from the HS-SiO_2/Fe_2O_3 support.

Connon and Gun'ko have reported a successful design for the preparation of a magnetic-nanoparticle-supported 4-*N*,*N*-dialkylaminopyridine catalyst (Scheme 3.45).[44] The organocatalyst moiety was cleanly loaded onto the surface of the iron oxide particle core and was monitored by ^{1}H NMR spectroscopy in the presence of (E)-stilbene as an internal standard. The nanoparticle-supported catalyst showed high catalytic activity, promoting smooth and quantitative acetylation to afford the desired products (Scheme 3.46). The catalyst could be separated from the reaction mixture, and its catalytic efficiency remains unaltered even after recycling 30 times.

Asymmetric catalysis is one of the most powerful strategies for the preparation of an important product and intermediates. Many chiral motifs are found in natural products, pharmaceuticals, and agrochemicals. Magnetite nanoparticles modified with chiral ligands were employed in asymmetric catalysis, solving the difficulty in recovering nanoparticle-supported asymmetric catalysts. Lin and coworkers have designed novel magnetically recoverable heterogeneous chiral catalysts.[45] A ruthenium(II) complex with phosphonic acid-substituted BINAP [Ru(BINAP-PO_3H_2)(DPEN)Cl_2] was tethered to the surface of a superparamagnetic nanomaterial with the aid of a phosphate group (Scheme 3.47). This chiral Ru-catalyst on magnetite nanoparticles was utilized for the highly enantioselective asymmetric hydrogenation of aromatic ketones with high activity and enantioselectivity (up to 98% ee and 99% conversion) (Scheme 3.48). Notably, the bulk phosphonic acid groups mainly affected the activity and enantioselectivity.

Reiser *et al.* have reported a convenient procedure for grafting Cu(II)–azabis(oxazoline)-complexes onto magnetite nanoparticles coated with amorphous silica by means of copper-catalyzed azide/alkyne cycloaddition (CuAAC) as a tagging strategy.[46] The immobilized complexes showed good

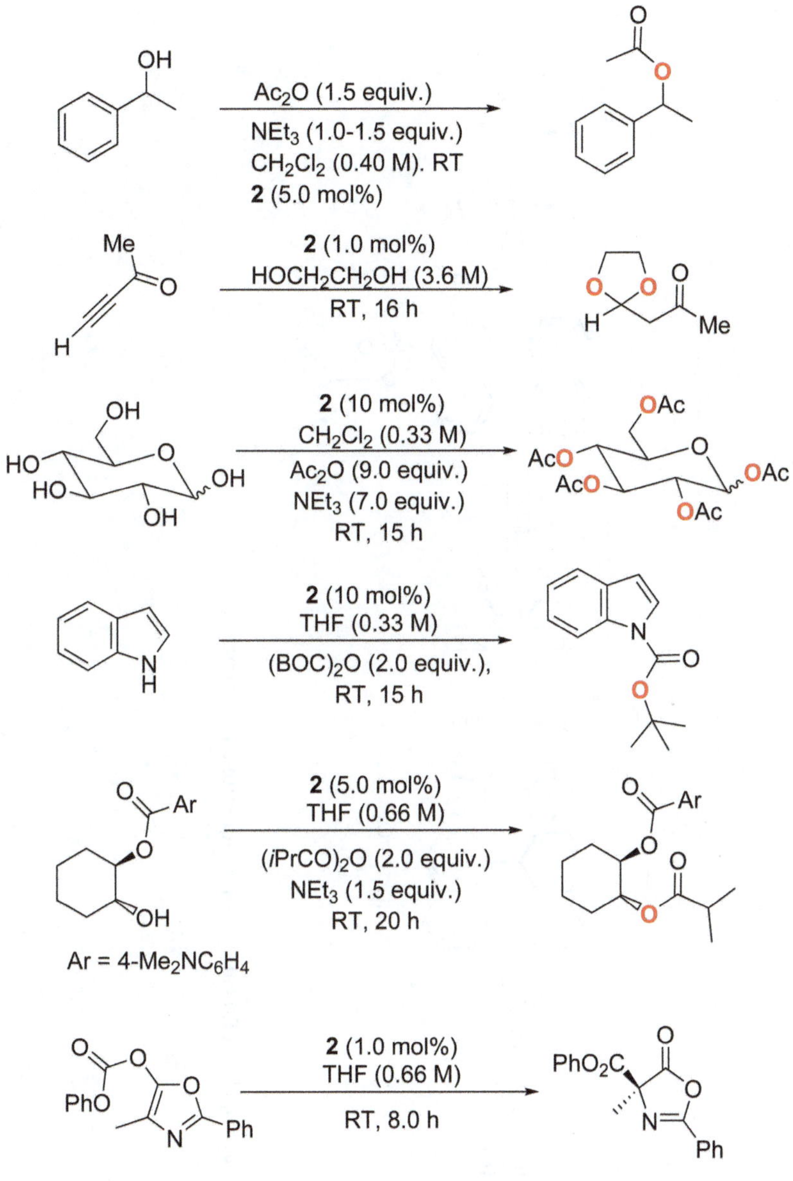

Scheme 3.46 Application of nanoparticle-supported catalyst in acetylation.

activity and selectivity in the desymmetrization of racemic 1,2-diols through asymmetric benzoylation (Scheme 3.49). Fe_3O_4@SiO_2 demonstrated higher activity and selectivity in comparison to other common polymer resin supports, such as MeOPEG or the Merrifield resin.

In 2010, Glorius *et al.* have performed the synthesis of N-heterocyclic carbenes (NHCs) as chiral modifiers of Fe_3O_4/Pd nanoparticles (Scheme 3.50).[47]

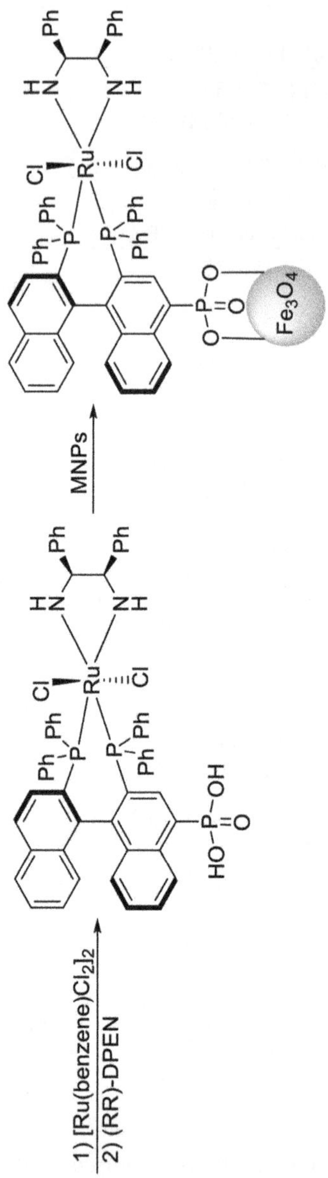

Scheme 3.47 Immobilization of chiral Ru catalyst on magnetite nanoparticles.

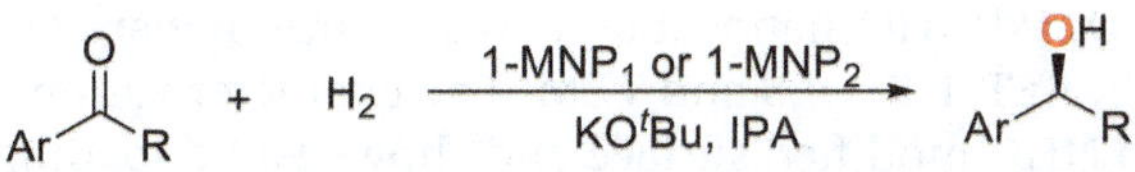

Scheme 3.48 Enantio excess values (% ee) for the heterogeneous hydrogenation of aromatic ketones.

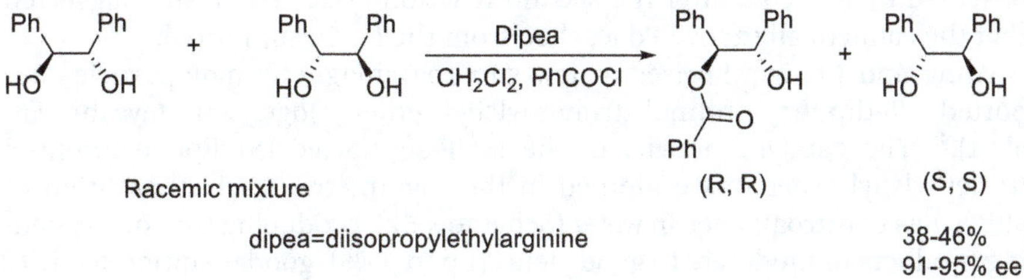

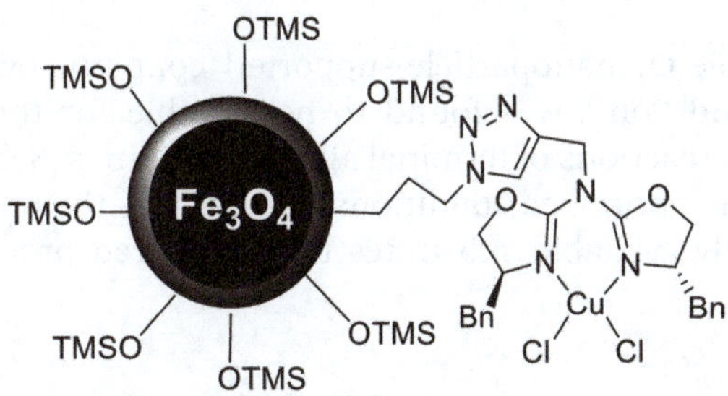

Scheme 3.49 Asymmetric benzoylation using magnetically separable azabis(oxazoline)copper (II) catalyst.

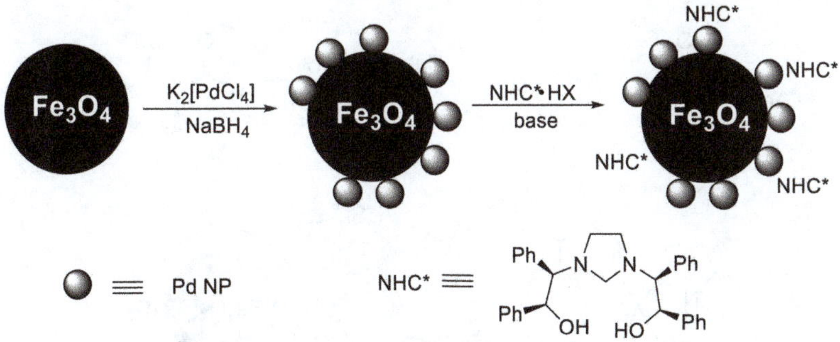

Scheme 3.50 Preparation of Fe_3O_4/Pd NPs modified by chiral NHC; for clarity, the sizes are not represented proportionally.

Then, the Fe_3O_4/Pd/NHC nanocatalyst was comprehensively characterized by XPS, ATR-IR, BET, ICP-OES and TEM. The characterization suggested the presence of an NHC-modified surface and showed a Pd-loading of 0.92 wt%. The activity of the Fe_3O_4/Pd/NHC nanocatalyst was investigated in the asymmetric α-arylation between 2-methyl-1-tetralone using chloro- and bromobenzene, affording the desired products in 72% yield (48% ee) and in 56% yield (60% ee), respectively (Scheme 3.51). A low Pd content of 0.232 ppm was observed by ICP-OES after the second reaction cycle. The result suggested that the catalytically active Pd leached from the Fe_3O_4 supported.

Wang and Ma synthesized a new superparamagnetic nanoparticle supported (*S*)-diphenylprolinol trimethylsilyl ether (Jøgensen–Hayashi catalyst).[48] The catalytic activity of the MNP-supported (*S*)-diphenylprolinol trimethylsilyl ether was examined in the asymmetric Michael addition of aldehydes to nitroalkenes in water (Scheme 3.52), producing the corresponding products in moderate to good yields (up to 96%), good enantioselectivity (up to 90% ee) and diastereoselectivities (up to 99 : 1). The activity, enantioselectivity and diastereoselectivity remained unaltered after four recycling cycles.

Recently, a Fe_3O_4 nanoparticle-supported copper(I) pybox catalyst, prepared by Li and Song, was found to be suitable for the enantioselective direct-addition reactions of terminal alkynes to imines (Scheme 3.53).[49] After establishing the optimized conditions, the authors then converted a variety of commercially available substrates to the desired propargylamines with

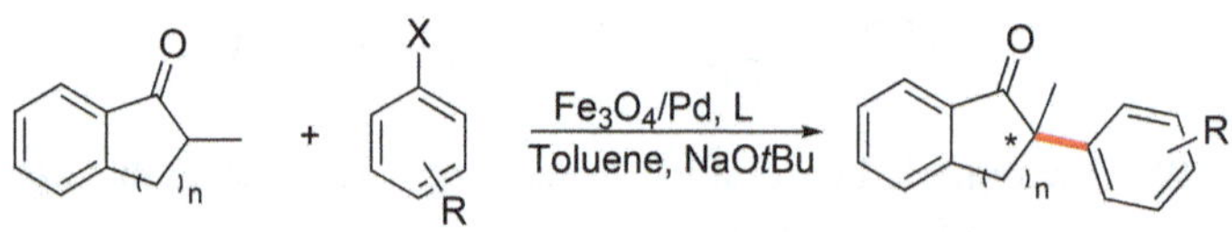

Scheme 3.51 Asymmetric α-arylation of ketones with aryl halides catalyzed by Fe_3O_4/Pd/L.

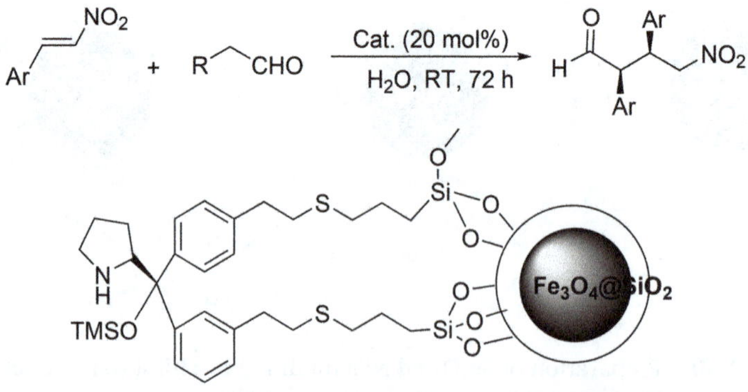

Scheme 3.52 Asymmetric Michael addition of aldehydes to nitroalkenes using heterogeneous catalyst.

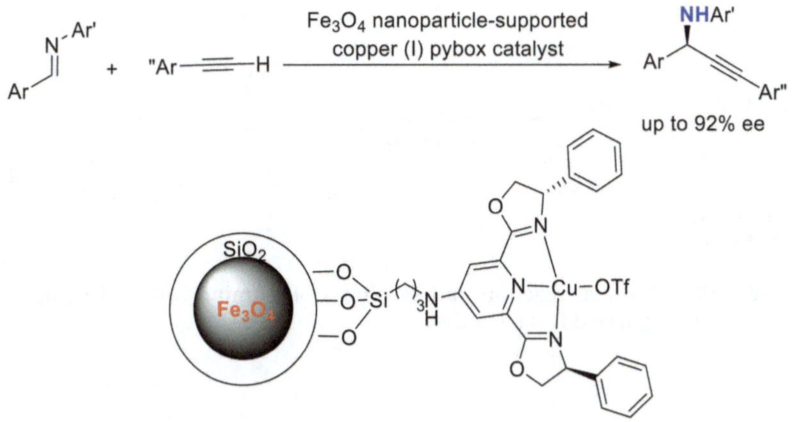

Scheme 3.53 Magnetite supported copper(I) pybox catalyzed enantioselective direct-addition reactions of terminal alkynes to amines.

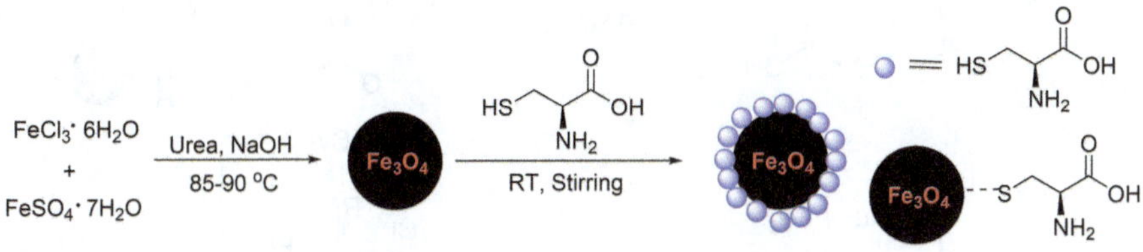

Scheme 3.54 Synthesis of Fe_3O_4–Cys MNPs.

high yields and good enantioselectivity. The optical rotation of the product showed the same sign as the one obtained from the homogeneous catalysts. The catalyst could be reused at least six times and the optical activity and enantioselectivity remained unchanged.

Branco and Gawande developed a facile synthesis of cysteine–ferrite magnetic nanoparticles without any additive or additional source of linkers (Scheme 3.54).[50] The authors performed the Mannich reaction of aniline with benzaldehyde and cyclohexanone with high chemoselectivity (Scheme 3.55). The stability of Fe_3O_4–Cys was evaluated. No significant loss in activity and chemoselectivity was observed after nine recycling cycles.

Yang and coworkers immobilized the Hoveyda–Grubbs catalysts on the surface of magnetic nanoparticles with amino groups by means of organosilane bifunctional groups (Scheme 3.56).[51] For evaluation of the efficiency and reactivity of magnetic nanoparticle-supported Hoveyda–Grubbs catalysts in the RCM reaction, a series of cyclic olefins reacted smoothly to the corresponding products in high yields (Scheme 3.57). Interestingly, the catalyst was reused 22 times in RCM reactions under standard conditions with no loss of the catalytic efficiency.

Robinson and co-workers reported an olefin metathesis precatalyst on iron oxide magnetic particles, obtained by the immobilization of a quaternary

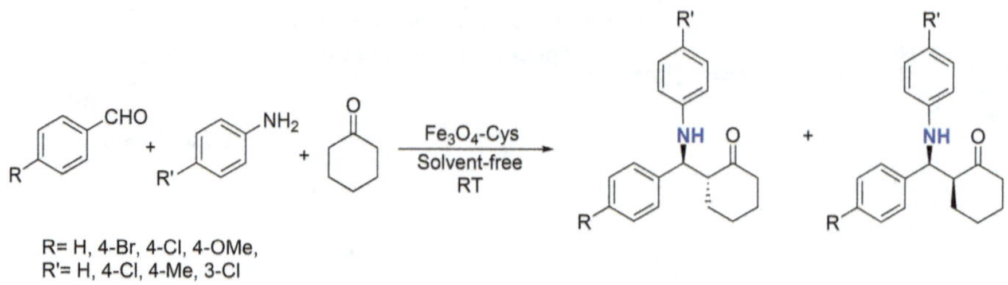

Scheme 3.55 Mannich reaction and synthesis of β-amino carbonyl compounds catalyzed by Fe_3O_4–Cys.

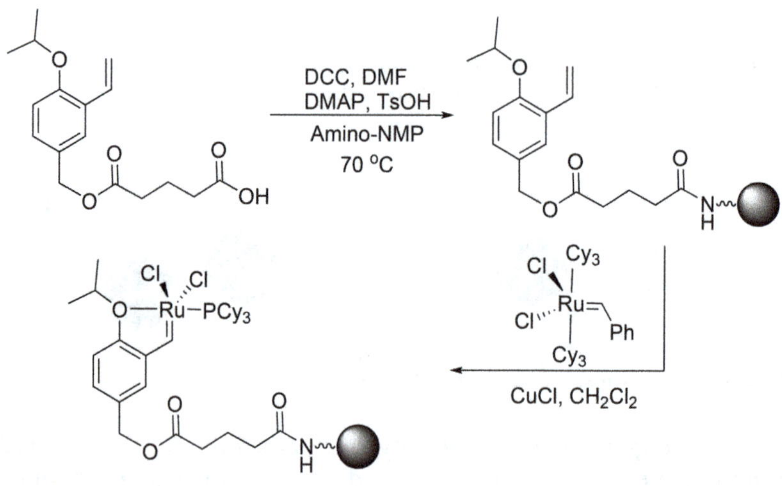

Scheme 3.56 Synthesis of magnetic nanoparticle-supported catalyst.

ammonium Hoveyda–Grubbs catalyst on sulfonic acid-functionalized silica-coated iron oxide magnetic particles with (Scheme 3.58).[52] Ring closing metathesis of diethyl diallylmalonate was selected as the model reaction to evaluate the activity of this olefin metathesis catalyst (Scheme 3.59). It was noted that >95% conversion to cyclopentene was obtained and a clear solution after magnetic separation of the catalyst particles was also observed.

Varma *et al.* have developed a convenient method for the preparation of a nanoferrite-supported glutathione-copper (nano-FGT-Cu) catalyst.[53] The TEM revealed the formation of single-phase Fe_3O_4 nanoparticles, with a size range of 10–25 nm. XRD confirmed that the Cu species was highly dispersed on ferrites. The nano-FGT-Cu catalyst was suitable for catalyzing a one-pot multi-component reaction using MW irradiation to synthesize the 1, 2, 3-triazoles (Scheme 3.60). The Cu content in the completion reaction was not detected by ICP-AES analysis. The result demonstrated that glutathione not only provided enough binding sites on the surface of Fe_3O_4 nanoparticles but also acted as a ligand for the coordination with Cu.

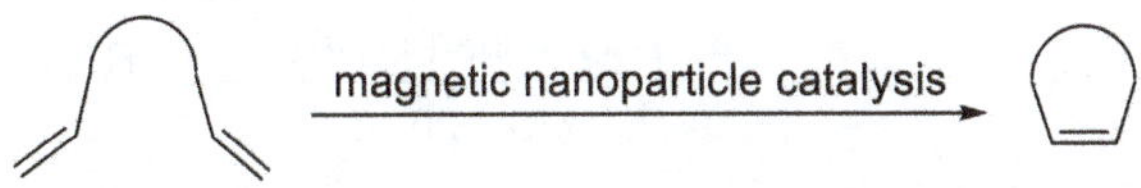

Scheme 3.57 Activity of magnetic catalyst in RCM reactions.

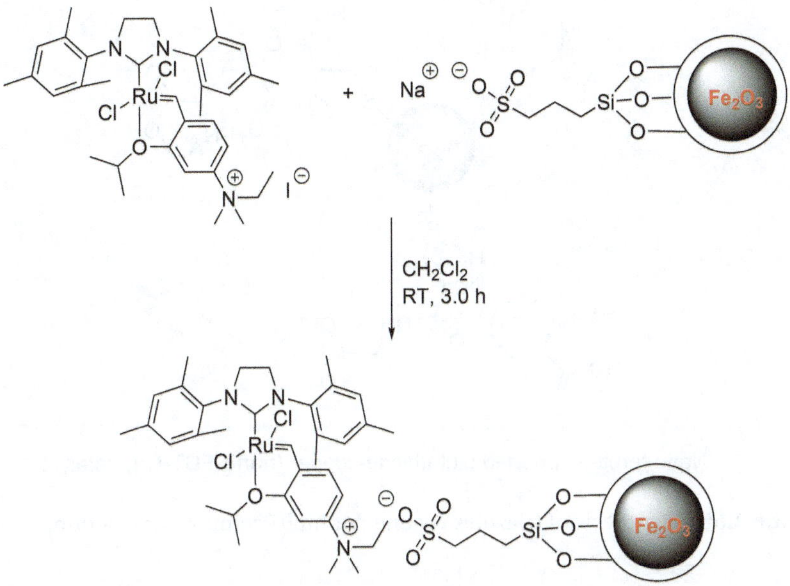

Scheme 3.58 Synthesis of Fe_3O_4–Cys MNPs.

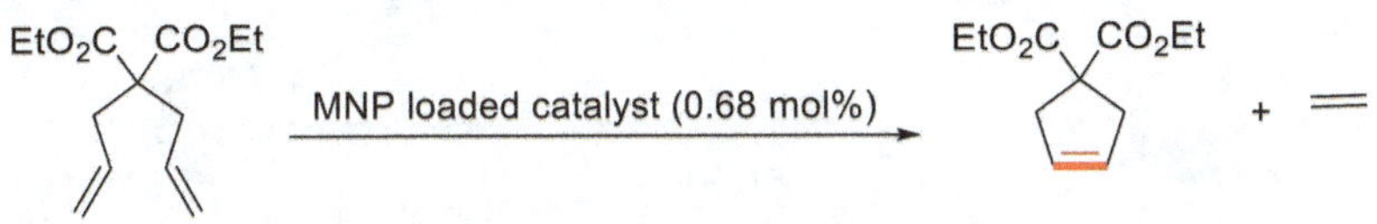

Scheme 3.59 The RCM reaction by MNP loaded catalyst.

The same group of researchers has prepared a magnetically recoverable heterogeneous Cu catalyst by sonicating nanoferrites with dopamine hydrochloride and adding $CuCl_2$ at basic pH (Scheme 3.61).[54] Then the authors investigated the activity of the nano-Fe_3O_4-DOPA-Cu catalyst in the C–S coupling reaction using MW (microwave) irradiation (Scheme 3.62). The reactions of a wide variety of commercially available aryl iodides and aryl bromides with thiophenols proceeded smoothly to generate the corresponding diaryl sulfides in good to excellent yields. TEM revealed that the nano-FeDOPACu remained unaltered in morphology and size during the catalyzed reaction.

Magnetic silica supported Ru-hydroxide nanoparticles were synthesized by ruthenium hydroxide immobilization on magnetic silica (Scheme 3.63).[55] The XRD revealed that the Ru and Si species were well dispersed on the

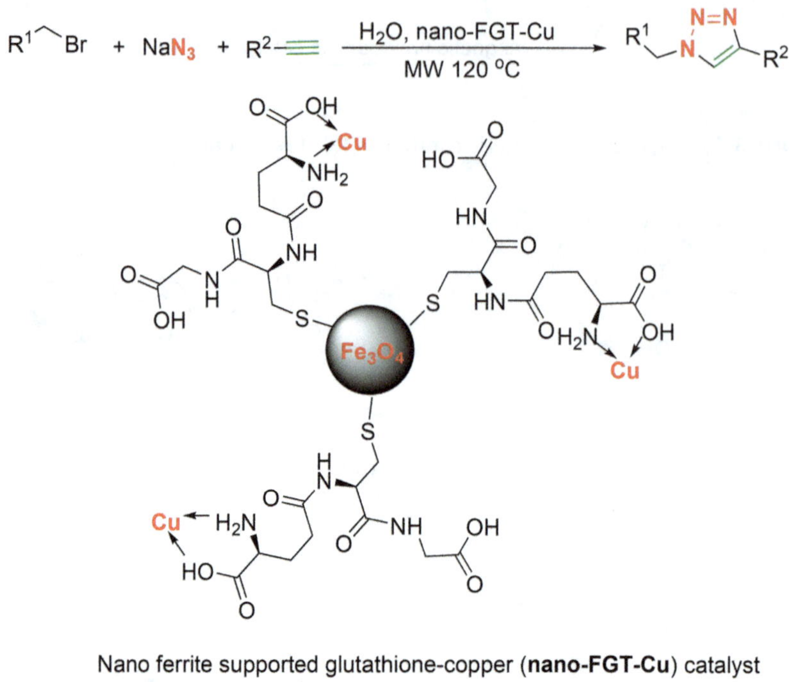

Scheme 3.60 Synthesis of triazoles *via* one-pot multi-component reaction.

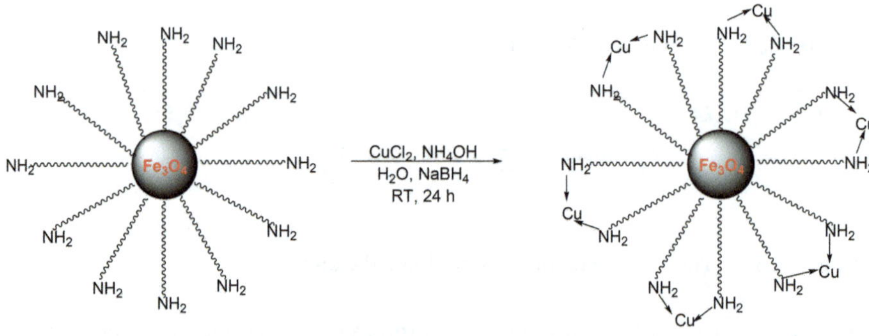

Scheme 3.61 Synthesis of nano-Fe_3O_4-DOPA-Cu catalyst (nano-FeDOPACu).

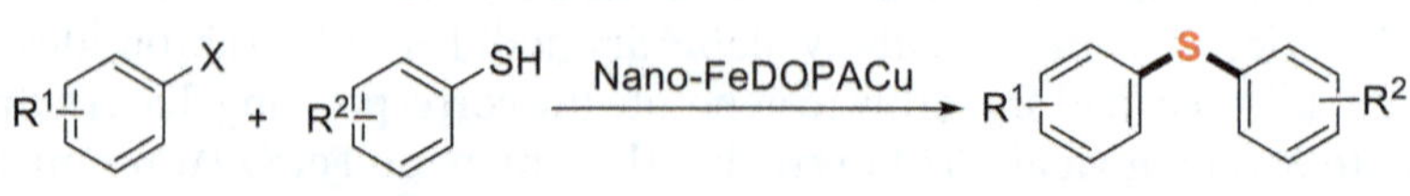

Scheme 3.62 Cross coupling of aryl halides and thiols.

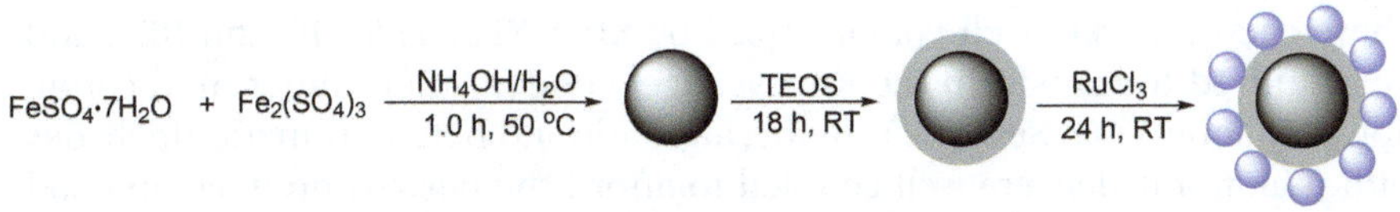

Scheme 3.63 One-pot synthesis of the nano-Fe@SiO_2Ru catalyst.

Scheme 3.64 Nano-Fe@SiO_2Ru-catalyzed hydration of nitriles.

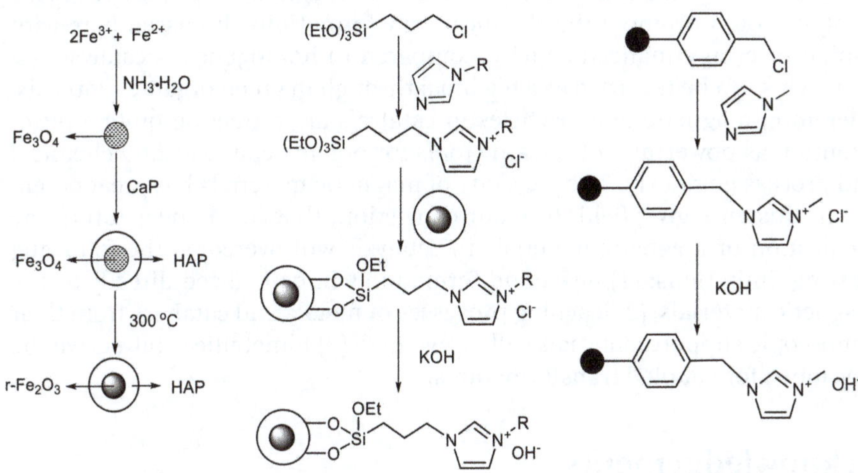

Scheme 3.65 Preparation of HAP-γ-Fe_2O_3 and SBILs.

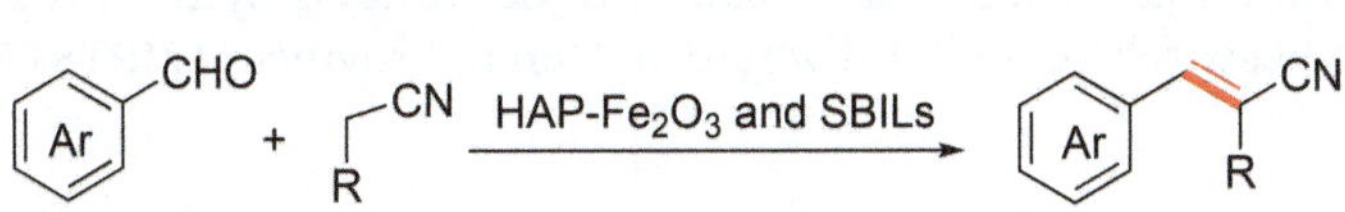

Scheme 3.66 Knoevenagel condensation reaction.

ferrites. TEM confirmed that nano-Fe@SiO_2Ru with spherical morphology and a size range of 15–30 nm was formed. Nano-Fe@SiO_2Ru was utilized for catalyzing the hydration of nitriles to form the corresponding amides with high yields (Scheme 3.64). The reaction was conducted in an aqueous medium under neutral conditions. The content of Ru was detected by ICP-AES analysis with a loss of 0.09%.

Xia and colleagues reported a series of basic ionic-liquid-functionalized hydroxyapatite-encapsulated-Fe_2O_3 magnetic nanoparticles (Scheme 3.65).[56]

These catalysts were well characterized by XRD, XPS, TEM, IR and BET, and were utilized to transform the Knoevenagel condensation under mild conditions (Scheme 3.66). Several aromatic, aliphatic and heteroaromatic aldehydes with malononitrile were well coupled to afford the desired products in good yields. It was noted that cooperativity between the basic magnetic nanoparticle carrier and SBILs had an important influence on the catalytic activity.

3.3 Conclusion

This chapter summarized a "snapshot" of recent impressive advances in utilizing magnetic nanoparticles in catalysis. A series of bond formation reactions, such as C–C, C–O and C–S cross coupling, Mannich reactions, carbonylation, alkylation, reduction and asymmetric synthesis, were realized by catalysts on magnetic nanoparticles. The magnetic nanoparticle-based catalysts could promote the chemical transformations discussed here with similar or even enhanced activity, compared to homogeneous catalyst systems. This can be regarded as a big breakthrough in green organic synthesis. Therefore, magnetic nanoparticles in catalysis are attracting quite a bit of attention as powerful and versatile tools for organic chemists and chemical and process engineers. The recycling of magnetic materials has great potential in this emerging field. It is our conviction, that in the near future the application of magnetic materials in catalysis will overcome the following existing limitations: (1) new bond formation will be realized directly by the magnetic materials; (2) leaching processes of metals and catalysts from their nanoscopic support materials will be avoided; (3) bimetallic catalysts will be developed for coupled transformations.

Acknowledgements

We are grateful for financial support from National Natural Science Foundation of China (21573057, 21473226), project funded by the Program for Innovative Research Team in University of Henan Province (17IRTSTHN001).

References

1. (a) D. J. Cole-Hamilton, *Science*, 2003, **299**, 1702; (b) R. T. Baker and W. Tumas, *Science*, 1999, **284**, 1477.
2. (a) C. Copéret, M. Chabanas, R. P. Saint-Arroman and J.-M. Basset, *Angew. Chem., Int. Ed.*, 2003, **42**, 156; (b) N. Mizuno and M. Misono, *Chem. Rev.*, 1998, **98**, 199; (c) F. Lefebvre and J.-M. Basset, *J. Mol. Catal. A: Chem.*, 1999, **146**, 3.
3. S. Roy and M. A. Pericàs, *Org. Biomol. Chem.*, 2009, **7**, 2669.
4. R. B. N. Baig and R. S. Varma, *Chem. Commun.*, 2013, **49**, 752.
5. T. Y. Cheng, D. C. Zhang, H. X. Li and G. H. Liu, *Green Chem.*, 2014, **16**, 3401.
6. V. Polshettiwar, R. Luque, A. Fihri, H. B. Zhu, M. Bouhrara and J.-M. Basset, *Chem. Rev.*, 2011, **111**, 3036.

7. M. B. Gawande, P. S. Branco and R. S. Varma, *Chem. Soc. Rev.*, 2013, **42**, 3371.
8. Á. Molnár, *Chem. Rev.*, 2011, **111**, 2251.
9. K. V. S. Ranganath and F. Glorius, *Catal. Sci. Technol.*, 2011, **1**, 13.
10. S. Shylesh, V. Schünemann and W. R. Thiel, *Angew. Chem., Int. Ed.*, 2010, **49**, 3428.
11. S. E. Glacía-Garrido, J. Francos, V. Cadierno, J.-M. Basset and V. Polshettiwar, *ChemSusChem*, 2011, **4**, 104.
12. M. Kotani, T. Koike, K. Yamaguchi and N. Mizuno, *Green Chem.*, 2006, **8**, 735.
13. M. S. Kwon, I. S. Park, J. S. Jang, J. S. Lee and J. Park, *Org. Lett.*, 2007, **9**, 3417.
14. (a) J. M. Liu, X. G. Peng, W. Sun, Y. W. Zhao and C. G. Xia, *Org. Lett.*, 2008, **10**, 3393; (b) S.-L. Yang, C.-Y. Cao, F.-F. Wei, P.-P. Huang, Y.-B. Sun and W.-G. Song, *ChemCatChem*, 2014, **6**, 1868.
15. U. Laska, C. G. Frost, G. J. Price and P. K. Plucinski, *J. Catal.*, 2009, **268**, 318.
16. M. Y. Zhu and G. W. Diao, *J. Phys. Chem. C*, 2011, **115**, 24743.
17. Y. Jang, J. Chung, S. Kim, S. W. Jun, B. H. Kim, D. W. Lee, B. M. Kim and T. Hyeon, *Phys. Chem. Chem. Phys.*, 2011, **13**, 2512.
18. S. Byun, J. Chung, J. Kwon and B. M. Kim, *Chem.–Asian J.*, 2015, **10**, 982.
19. R. Cano, D. J. Ramón and M. Yus, *J. Org. Chem.*, 2011, **76**, 5547.
20. R. Cano, D. J. Ramón and M. Yus, *Tetrahedron*, 2011, **67**, 5432.
21. K. K. Senapati, S. Roy, C. Borgohain and P. Phukan, *J. Mol. Catal. A: Chem.*, 2012, **352**, 128.
22. M. B. Gawande, A. K. Rathi, P. S. Branco, I. Nogueira, A. Velhinho, J. J. Shrikhande, U. U. Indulkar, R. V. Jayaram, C. A. A. Ghumman, N. Bundaleski and O. M. N. D. Teodoro, *Chem.–Eur. J.*, 2012, **18**, 12628.
23. S. Sá, M. B. Gawande, A. Velhinho, J. P. Veiga, N. Bundaleski, J. Trigueiro, A. Tolstogouzov, O. M. N. D. Teodoro, R. Zboril, R. S. Varma and P. S. Branco, *Green Chem.*, 2014, **16**, 3494.
24. J. Lee, J. Chung, S. M. Byun, B. M. Kim and C. Lee, *Tetrahedron*, 2013, **69**, 5660.
25. C.-H. Pélisson, A. Denicourt-Nowicki, C. Meriadec, J.-M. Greneche and A. Roucoux, *ChemCatChem*, 2015, **7**, 309.
26. J. H. Tong, L. L. Bo, Z. Li, Z. Q. Lei and C. G. Xia, *J. Mol. Catal. A: Chem.*, 2009, **307**, 58.
27. K. K. Senapati, C. Borgohain and P. Phukan, *J. Mol. Catal. A: Chem.*, 2011, **339**, 24.
28. R. Z. Zhang, C.-X. Miao, S. F. Wang, C. G. Xia and W. Sun, *ChemCatChem*, 2012, **4**, 192.
29. R. Z. Zhang, C.-X. Miao, S. F. Wang, C. G. Xia and W. Sun, *ChemCatChem*, 2013, **5**, 142.
30. R. Z. Zhang, J. M. Liu, S. F. Wang, J. Z. Niu, C. G. Xia and W. Sun, *ChemCatChem*, 2011, **3**, 146.
31. R. Z. Zhang, C. X. Miao, Z. Q. Shen, S. F. Wang, C. G. Xia and W. Sun, *ChemCatChem*, 2012, **4**, 824.

32. S. Kim, E. Kim and B. M. Kim, *Chem.–Asian J.*, 2011, **6**, 1921.
33. S. Paul, K. Pradhan, S. Ghosh, S. K. De and A. R. Das, *Adv. Synth. Catal.*, 2014, **356**, 1301.
34. P. D. Stevens, J. Fan, H. M. R. Gardimalla, M. Yen and Y. Gao, *Org. Lett.*, 2005, 7, 2085.
35. P. D. Stevens, G. F. Li, J. D. Fan, M. Yen and Y. Gao, *Chem. Commun.*, 2005, 4435.
36. (a) D. Guin, B. Baruwati and S. V. Manorama, *Org. Lett.*, 2007, **9**, 1419; (b) B. Baruwati, D. Guin and S. V. Manorama, *Org. Lett.*, 2007, **9**, 5377.
37. S. Ceylan, C. Friese, C. Lammel, K. Mazac and A. Kirschning, *Angew. Chem., Int. Ed.*, 2008, **47**, 8950.
38. R. Abu-Reziq, D. S. Wang, M. Post and H. Alper, *Chem. Mater.*, 2008, **20**, 2544.
39. A. Schätz, T. R. Long, R. N. Grass, W. J. Stark, P. R. Hanson and O. Reiser, *Adv. Funct. Mater.*, 2010, **20**, 4323.
40. M.-J. Jin and D.-H. Lee, *Angew. Chem., Int. Ed.*, 2010, **49**, 1119.
41. (a) P. H. Li, L. Wang, L. Zhang and G.-W. Wang, *Adv. Synth. Catal.*, 2012, **354**, 1307; (b) L. Zhang, P. H. Li, H. J. Li and L. Wang, *Catal. Sci. Technol.*, 2012, **2**, 1859.
42. D. Rosario-Amorin, M. Gaboyard, R. Clérac, L. Vellutini, S. Nlate and K. Heuzé, *Chem.–Eur. J.*, 2012, **18**, 3305.
43. D. K. Yi, S. S. Lee and J. Y. Ying, *Chem. Mater.*, 2006, **18**, 2459.
44. C. Ó. Dálaigh, S. A. Corr, Y. Gun'ko and S. J. Connon, *Angew. Chem., Int. Ed.*, 2007, **46**, 4329.
45. A. G. Hu, G. T. Yee and W. B. Lin, *J. Am. Chem. Soc.*, 2005, **127**, 12486.
46. A. Schätz, M. Hager and O. Reiser, *Adv. Funct. Mater.*, 2009, **19**, 2109.
47. K. V. S. Ranganath, J. Kloesges, A. H. Schäfer and F. Glorius, *Angew. Chem., Int. Ed.*, 2010, **49**, 7786.
48. B. G. Wang, B. C. Ma, Q. Wang and W. Wang, *Adv. Synth. Catal.*, 2010, **352**, 2923.
49. T. Q. Zeng, L. Yang, R. Hudson, G. H. Song, A. R. Moores and C.-J. Li, *Org. Lett.*, 2011, **13**, 442.
50. M. B. Gawande, A. Velhinho, I. D. Nogueira, C. A. A. Ghumman, O. M. N. D. Teodoro and P. S. Branco, *RSC Adv.*, 2012, **2**, 6144.
51. C. Che, W. Z. Li, S. Y. Lin, J. W. Chen, J. Zheng, J.-C. Wu, Q. X. Zheng, G. Q. Zhang, Z. Yang and B. W. Jiang, *Chem. Commun.*, 2009, 5990.
52. M. J. Byrnes, A. M. Hilton, C. P. Woodward, W. R. Jackson and A. J. Robinson, *Green Chem.*, 2012, **14**, 81.
53. R. B. N. Baig and R. S. Varma, *Green Chem.*, 2012, **14**, 625.
54. R. B. N. Baig and R. S. Varma, *Chem. Commun.*, 2012, **48**, 2582.
55. R. B. N. Baig and R. S. Varma, *Chem. Commun.*, 2012, **48**, 6220.
56. (a) Y. Zhang, Y. W. Zhao and C. G. Xia, *J. Mol. Catal. A: Chem.*, 2009, **306**, 107; (b) Y. Zhang and C. G. Xia, *Appl. Catal., A*, 2009, **366**, 141.

CHAPTER 4

Sustainable Magnetic Nanocatalysts in Heterogeneous Catalysis

AYOMI S. PERERA*[a]

[a]Department of Chemical Engineering and Centre for Nature Inspired Engineering, University College London, Torrington Place, London, WC1E 7JE, UK
*E-mail: ayomi.perera@ucl.ac.uk

4.1 Introduction

4.1.1 What Are Sustainable Catalysts?

Sustainability is a concept that is essentially tied to catalysis, as the very purpose of a catalyst is to make a chemical process more efficient, thereby saving resources, time and cost. However, a catalyst itself needs to fulfill certain criteria if it is to be classified as a sustainable material. By broad definition, an environmentally-friendly or sustainable material is one that has no harmful environmental impact from the extraction of its raw material constituents, to production, consumption and finally, disposal.[1–3] Such materials have the additional characteristics of high recyclability, social and ecological significance and customer satisfaction as well as offering significant improvements to socio-ecological problems and showing continuous

Smart Materials No. 26
Magnetic Nanomaterials: Applications in Catalysis and Life Sciences
Edited by Stefan H. Bossmann and Hongwang Wang

Published by the Royal Society of Chemistry, www.rsc.org

improvement to keep up with such expectations.[2] Heterogeneous catalysts which are categorized as "sustainable" should encompass most, if not all, of the characteristics above.

For instance, a catalyst which is highly active and cost-effective is not a sustainable material, if it is non-recyclable, energy-inefficient or is harmful to living organisms. A perfect example is sulfuric acid which is used to catalyze cellulose dehydrogenation, where either concentrated acid is used[4] or dilute acid solutions are used at high temperatures[5,6] which are non-recyclable and corrosive. Other examples include hydrogen peroxide, which is considered a "green" catalyst for alkene oxidation at an industrial level, due to its remarkable efficiency and the formation of water as the only benign byproduct.[7] However, its production, handling and transportation are highly expensive, complex and dangerous, making it a non-sustainable material.[8–11] A material that is considered environmentally-benign at one point may no longer be so considered when new scientific evidence comes to light about its long-term harmful effects. The most recognizable example is chlorofluorocarbons (CFCs), which at their discovery were thought to be highly sustainable, before their detrimental impact on ozone depletion was brought to attention especially during the 1980s.[12,13] Similarly, hydrogen fuel cells, introduced as highly sustainable alternatives to fossil fuel cells, emit molecular hydrogen, which can damage multiple layers of the atmosphere.[14] Hence, it is crucial that sustainable material or catalyst development technologies evolve continuously in order to keep up with current environmental and health protocols.

4.1.2 The Role of Magnetic Nanomaterials in Sustainable Heterogeneous Catalysis

Heterogeneous catalysis has become a crucial component in modern industrial processes and has enormous economic and environmental implications.[15–18] For decades it has shaped chemical manufacture and energy generation and counteracted environmental pollution.[19] Recent developments in rational design, advances in synthesis and characterization, together with mechanistic insights have resulted in greatly increasing the efficiency of heterogeneous catalysts, and have consequently solidified their role as pivotal ingredients in modern human life.[20–22] With such significance comes greater demand in both quantity and quality. It is therefore important that sustainable technologies simultaneously emerge, to produce, utilize and then dispose of heterogeneous catalysts. The advent of efficient heterogeneous catalysts with magnetic properties has contributed significantly to achieving exactly that.

The aim of this chapter is to convey and highlight the sustainable properties of magnetic nanocatalysts using one of their most prominent applications, which is heterogeneous catalysis, by carefully examining a few chosen examples from literature (Figure 4.1).

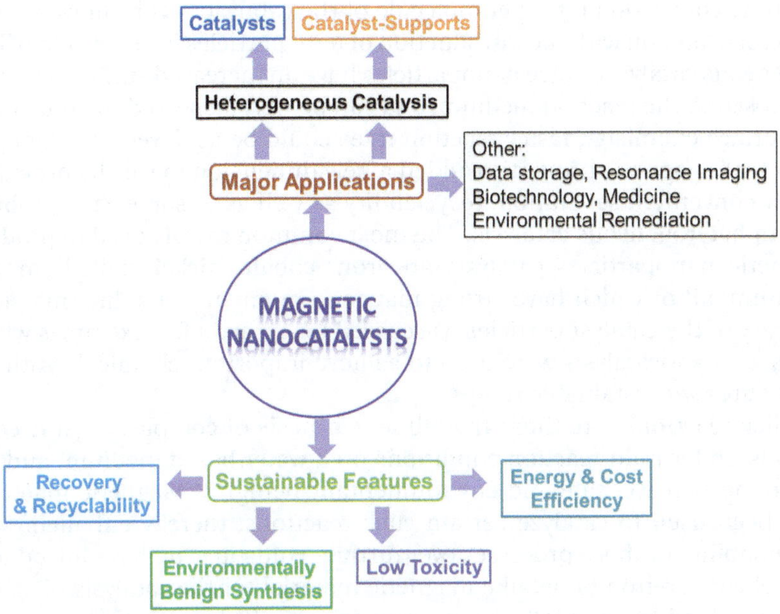

Figure 4.1 Overview of chapter discussion.

4.2 Major Applications of Magnetic Nanomaterials

Magnetic nanomaterials are developed with transition metal elements that are inherently ferromagnetic and highly versatile, and employed in a variety of fields including biotechnology,[23] medicine,[24–26] data storage,[27] environmental remediation[28,29] and resonance imaging.[30–32] Each of the above areas has been greatly improved by the introduction of magnetic materials. This chapter focuses on two of the most chemically and industrially significant applications which are: heterogeneous catalysis and catalyst-supports.

4.2.1 Heterogeneous Catalysis

The development of heterogeneous catalysts was driven by the need to counterattack the drawbacks of homogeneous catalysts, such as difficulty in separation and recyclability, leading to higher costs, which are especially significant for processes that require high-value catalysts.[33] Heterogeneous catalysts are also advantageous from an industrial standpoint as they can be used for both gas-phase and liquid-phase reactions and can be utilized in continuous reactor processes, which are particularly useful in petroleum refining and bulk-chemical production. However, conventional heterogeneous catalysts have lacked certain features that make homogeneous catalysts so desirable to begin with. Since homogeneous catalysts are miscible in the reaction medium, their reaction rates are extremely fast and provide high substrate conversion rates per molecule of the catalyst.[34] This, however, was addressed later on with the introduction of nanoparticles into the field of heterogeneous catalysis. Since nanoparticles have an increased surface area that is exposed to the reaction medium, and can be well-dispersed by controlling the surface chemistry, faster reaction rates could be achieved.[35,36] Nanoparticles with magnetic properties added a new dimension to this by providing highly convenient separation, recyclability as well as in some cases, robustness in heterogeneous catalysis. The most common metals used to produce magnetic nanoparticles catalysts are iron, cobalt, nickel, palladium and platinum, all of which have strong magnetic moments, enabling magnetic recovery of the catalyst particles. Discussed below are a few examples where magnetic nanocatalysts were used to achieve important chemical synthesis procedures *via* sustainable routes.

"Click reactions" are those that allow synthesis of complex organic compounds under mild reaction conditions on a water-based medium, and are hence inexpensive, safe and environmentally-benign.[37] Magnetic materials have been used to catalyze certain click reactions, thereby enhancing the sustainability of those processes ever further. Hudson *et al.* have introduced a novel copper–iron bimetallic magnetic material for the catalysis of azide–alkyne cycloaddition (AAC) click reaction in water (Scheme 4.1).[38]

The AAC reaction is conventionally achieved *via* Cu(I) ions as a homogeneous catalyst. Cu(I) is either added directly into the reaction mixture as a salt solution or Cu(II) is reduced *in situ* to form Cu(I) by sodium ascorbate. Introduction of a solid, magnetic catalyst provides recyclability, thereby considerably reducing the cost of such processes. The novel catalyst synthesis consisted of reduction of $CuSO_4$ to Cu(I), by Fe(0) nanoparticles *via* galvanic reduction, and its subsequent deposition onto the Fe(0) surface, with no additional ligands or reducing agents (Cu@FeNPs). Hence, catalyst synthesis is facile, cost-effective and environmentally-benign. In this case, the magnetic Fe(0) core has three functions; (1) to provide means for recoverability and recyclability (up to 5 times), (2) to act as reducing agent to achieve Cu(I) from Cu(II), (3) to prevent release of Cu(I) ions into the medium by acting as anchor, thereby facilitating a heterogeneous mechanism. Baig and Varma also reported another copper-based magnetic catalyst for the AAC reaction, during this time.[39] Their catalyst was developed by anchoring Glutathione (GT) molecules onto magnetic nanoferrites *via* thiol groups and subsequently depositing copper onto the surface (nano-FGT-Cu). This catalyst was used to achieve the AAC reaction in one pot, by generating the azide *in situ* first, and then followed by the cycloaddition

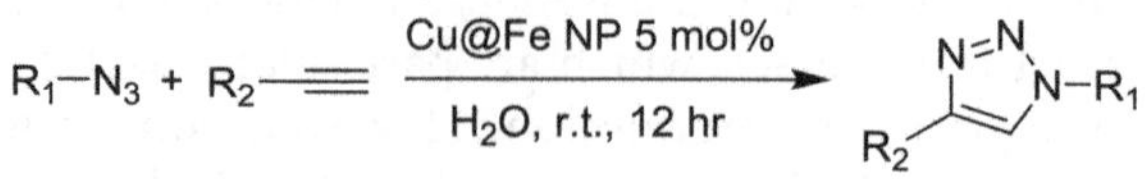

Scheme 4.1 Azide–alkyne click reaction in water, catalyzed by Magnetic copper–iron nanoparticles. Adapted from ref. 38 with permission from the Royal Society of Chemistry.

step in aqueous media. The nano-FGT-Cu catalyst was recyclable up to three times without significant loss of activity. Pourjavadi *et al.* reported yet another copper based novel magnetic heterogeneous catalyst, for click reactions.[37] This catalyst consisted of a Fe_3O_4 core, coated with several layers of copper and poly(ionic liquid) (Figure 4.2). This was prepared by the polymerization of 3-carboxymethyl-1-vinylimidazolium with Fe_3O_4 magnetic nanoparticles, functionalized with 3-(trimethoxysilyl) propylmethacrylate, followed by the coordination of the carboxylate units in the polymer chains with $CuSO_4$. Catalytic characterization of the material was carried out *via* one-pot synthesis of the 1,4-disubstituted 1,2,3-triazoles, by click reactions between primary halides or tosylates and sodium azide and terminal acetylenes. The reactions were carried out in hallmark click reaction conditions at room temperature in aqueous media.

The copper-loaded polymeric catalyst showed good to excellent yields for a diverse range of products for the synthesis of 1,4-disubstituted 1,2,3-triazoles. Apart from being a highly sustainable process due to the lack of toxic reagents and usability in environmentally-benign conditions, the catalyst was also stable and showed no aggregation or leaching of copper, and was highly recyclable. These features offer useful, greener possibilities to carry out click reactions, on an industrial scale.

In addition to click reactions, a myriad of useful organic processes can be catalyzed by magnetic heterogeneous nanocatalysts. Menini *et al.* reported the liquid-phase synthesis of disulfides *via* the aerobic oxidation of thiols, using cobalt-iron composites as a catalyst.[40] Removal of thiols is an important process in the petroleum refining industry, as aromatic thiols in petroleum cause degradation of the metal catalysts used in refining. Moreover, residual thiols in petroleum products cause corrosiveness, bad odor and environmental pollution. Oxidation into disulfides is the most common method used for removal of such thiols. Disulfides themselves are useful chemicals and are industrially used for vulcanization of rubber and elastomers. They are also key intermediates in a broad range of

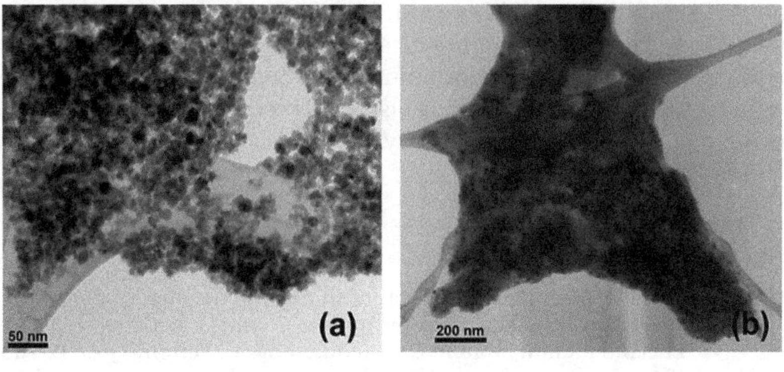

Figure 4.2 TEM images of (a) MNP and (b) MNP@ImAc/Cu. Reproduced from ref. 37 with permission from The Royal Society of Chemistry.

organic processes, hence their production is commercially and industrially relevant. An inexpensive catalyst was developed from an iron oxide-rich red soil. First, the soil was impregnated with cobalt to produce a magnetic material *via* mixing with a sucrose solution containing $CoSO_4{\cdot}7H_2O$, and subsequent solvent removal and thermal treatment in air. This new catalyst was able to efficiently convert thiols into disulfides with only atmospheric oxygen as the oxidant, with no additional alkaline co-catalysts, under mild conditions (Scheme 4.2).

Apart from being cost-effective, this catalyst is also superior to its predecessors; by using weakly basic solvents such as dimethylformamide and dimethylacetamide, the substrate can be activated, and the need for corrosive alkaline co-catalysts, which produce toxic waste, can be eliminated. The economic viability of the catalyst was further solidified by the generation of high yields with very low Co loadings and with no significant leaching. The magnetic nature allows for facile recovery and reusability. This type of sustainable research opens up doors for many other possibilities for the use of abundant natural iron oxides, as heterogeneous catalyst precursors for processes such as the mass production of biomass-based olefins.

Another example of such a sustainable catalyst was introduced by Shahamat *et al.*, by the development of a magnetic carbon nano composite as a novel catalyst for the degradation of phenol from industrial wastewater *via* catalytic ozonation.[41] The catalyst was prepared by impregnation of activated carbon (AC) into Fe_3O_4 by first heating the AC with nitric acid, followed by washing with water, filtering and drying overnight. Then the modified AC was dispersed in aqueous $Fe(NO_3)_3{\cdot}9H_2O$, followed by thermal treatment under nitrogen to form Fe_3O_4 magnetic nanoparticles, incorporated with AC. The catalyst was then used to degrade a solution of pH adjusted phenol under ozone. It was found that the efficiency of phenol removal in the catalytic ozonation process was significantly higher than that of the single ozonation process (98.5% *vs.* 78.7%), and that the highest catalytic potential was achieved at neutral pH. Also the process could be further enhanced by using it in combination with an SBR bioreactor, for effective degradation of phenol. This catalyst offers a practical, efficient and economic means for the treatment of wastewaters containing pollutants such as phenol. Due to its magnetic nature, the catalyst was also easily recycled and reused up to 5 times with no significant loss of activity.

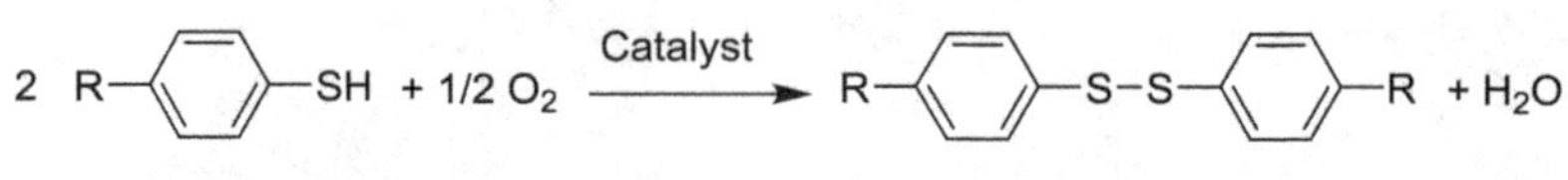

Scheme 4.2 General scheme of oxidation of thiols into disulfides *via* the cobalt-iron catalyst. Adapted from Applied Catalysis A: General, 392(1–2), Luciano Menini, Márcio C. Pereira, André C. Ferreira, José D. Fabris, Elena V. Gusevskaya, Cobalt-iron magnetic composites as heterogeneous catalysts for the aerobic oxidation of thiols under alkali free conditions, 151–157, Copyright (2011) with permission from Elsevier.

4.2.2 Heterogeneous-Catalyst Supports

Perhaps the most prudent use of magnetic nanomaterials in catalysis is found in their use as support materials. In particular, core–shell type materials are a relatively recent addition that have revolutionized the scope of magnetic nano-catalysis. They are generally made of a magnetic core and an outer coating which is either made up of organic or inorganic materials and can be functionalized with various chemical groups, in order to achieve a specific function. Thus, the shell itself is the catalyst and the magnetic core acts either as a reversing/recycling agent or an anchor. Such catalysts have found significant applications in chemical synthesis. Table 4.1 summarizes examples of Fe_3O_4 supported core–shell catalysts used in the synthesis of various organic compounds.

Another prominent type of core–shell magnetic nanocatalyst includes those that contain a silica coating. Lee *et al.* reported a metal-oxide@SiO_2 core–shell magnetic catalyst, where a Pd and $CuFe_2O_4$ dual core was encapsulated within a silica-shell for selective hydrogenation of arylacetylenes.[48] The dual catalyst showed >98% conversion of arylacetylenes into related styrenes with an excellent selectivity of 98%. The high activity of the catalyst was due to the $CuFe_2O_4$ being in close contact with Pd, which promotes the catalytic reaction. Abu-Reziq *et al.* reported a pioneering technique to homogenize a heterogeneous catalyst by growing polyaminoamido (PAMAM) dendrons on silica-coated magnetic nanoparticles.[49] This technique leads to higher solubility of the silica-magnetic catalysts in organic solvents. The dendronized catalyst is subsequently phosphonated, complexed with $[Rh(COD)Cl]_2$, and applied in catalytic hydroformylation reactions and is proven to be highly selective and reactive. Kassaee *et al.* introduced a technique to develop a silica-coated catalyst with an Fe/Fe_3O_4 core, functionalized with sulfamic acid groups, as an efficient, reusable catalyst for one-pot synthesis of α-amino nitriles in water.[50] Wang *et al.* extended this approach to develop a sulfamic acid and sulfonic acid-functionalized, silica-coated, Fe/Fe_3O_4 core–shell magnetic nanocatalyst for efficient synthesis of biodiesel compounds.[51] Esterification of oleic acid and the transesterification of glyceryl trioleate were studied as model reactions that may lead to biodiesel production. The silica core stabilizes the magnetic core while providing a base to carry functional groups.

Ordered mesoporous silica such as SBA-15 provide an excellent option as a shell or shell-template, as they have a large surface area that can facilitate reagent–product diffusion while having the possibility to serve as a template to generate other coating materials with those features. Lu *et al.* reported a magnetic cobalt catalyst with a highly porous carbon coating, generated *via* an SBA-15 template, as a hydrogeneation catalyst.[52] In order to produce the carbon coating, first, mesoporous SBA-15 was infiltrated with furfuryl alcohol. Afterwards, polyfurfuryl alcohol was used to block the pore system of the silica template. This allows the cobalt nanoparticles to be spatially selectively deposited onto the silica–polymer composite surface. Next, the

Table 4.1 Magnetic catalysts with a Fe_3O_4 core used for synthesis of various organic compounds. Adapted from ref. 42–47.

Reaction type	Catalyst	Reaction scheme	Reference
Mannich-type multicomponent reactions of alkynes, aldehydes, and amines	Fe_3O_4–Cu(OH)x	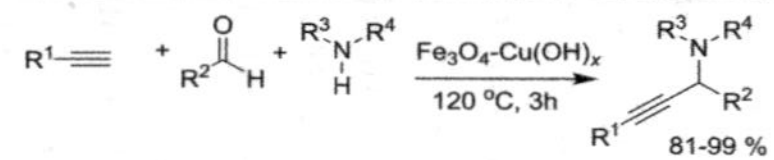	Aliaga *et al.*[42]
Michael-type borylation of olefins	Fe_3O_4–Cu(OH)x	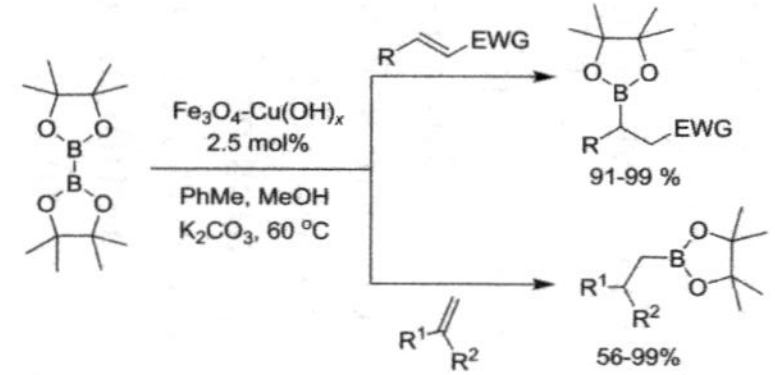	Reprinted from Cano *et al.*,[43] Copyright (2010) American Chemical Society
Aerobic oxidation/ reduction	Ru(OH)$_x$–Fe_3O_4	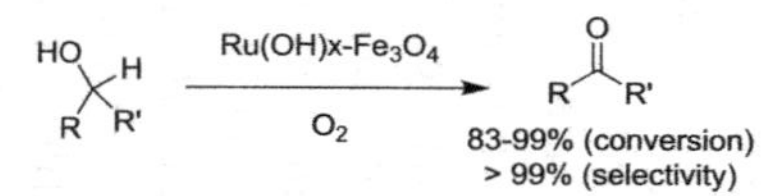 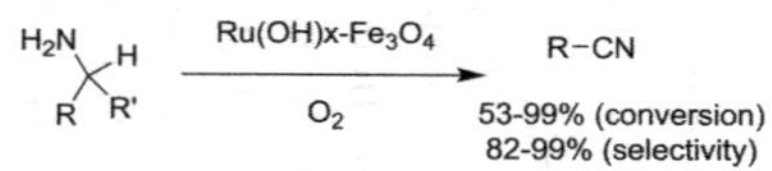 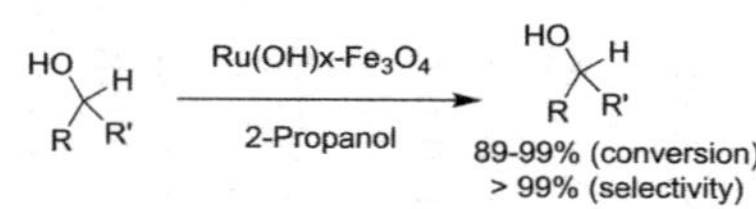	Kotani *et al.*[44]
Synthesis of sulfonamides	Ru/Fe_3O_4	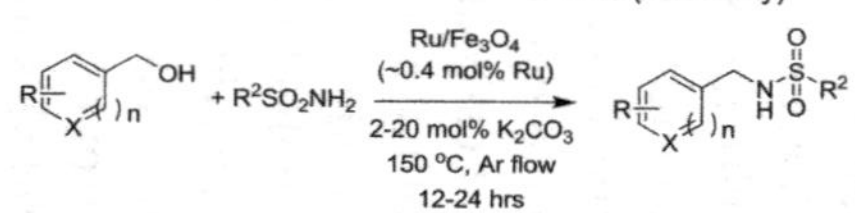	Reprinted with permission from Shi *et al.*,[45] Copyright (2009) American Chemical Society

Sonogashira-cyclization reaction	$CuO–Fe_3O_4$, Pd(0, II)/ $CuO–Fe_3O_4$	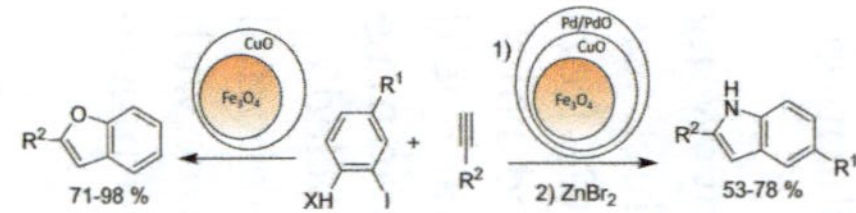	Reprinted from Tetrahedron, **68**(5), R. Cano, M. Yus, D. J. Ramón, impregnated copper or palladium-copper on magnetite as catalysts for the domino and stepwise Sonogashira-cyclization processes: a straightforward synthesis of benzo[*b*] furans and indoles, 1393–1400, Copyright (2012) with permission from Elsevier[46]
N-alkylation of aromatic amines, sulfonamides, sulfinamides, and nitrocompounds	$Ru(OH)_3–Fe_3O_4$	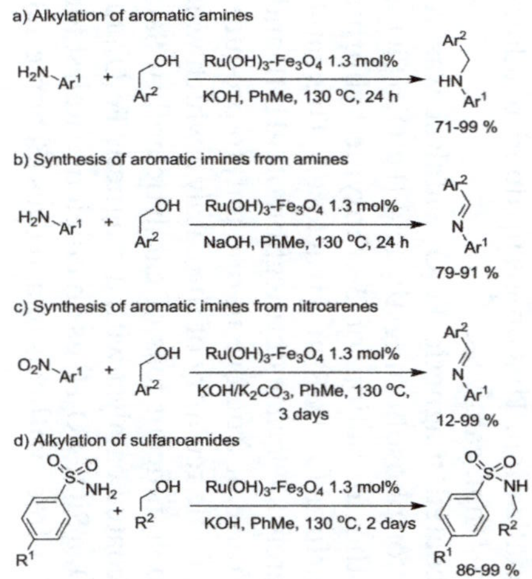	Reprinted with permission from Cano *et al.*,[47] Copyright (2011) American Chemical Society

cobalt containing composite was coated to protect the nanoparticles from corrosion and high-temperature. A small amount of carbon precursor was used to cover the cobalt nanoparticles and was subsequently subject to pyrolysis. This resulted in the encapsulation of the cobalt nanoparticles by a graphitic carbon layer, while simultaneously converting the polyfurfuryl alcohol in the silica shell into carbon. The silica template was leached out later. The resulting carbon is superparamagnetic and is mesoporous, and allows further functionalization of catalytically active species such as palladium. This catalyst was found to be highly stable and was used for hydrogenation of octane, and can be easily separated by applying a magnetic field. Lu *et al.* further extended this work by developing carbon-coated cobalt nanoparticles deposited onto mesoporous silica (Figure 4.3).[53] This was synthesized *via* a reversible pore blocking and opening strategy. The cobalt nanoparticles were deposited on the external surface of the silica particles and were further protected by a thin carbon coating against acid erosion. Thus, the magnetic particles do not block the pore of silica, and the particles are still accessible for further functionalization. This opens up doors to additional applications of the material other than catalysis.

Another silica-coated magnetic Fe_3O_4 catalyst was reported by Wang *et al.* as a support for phosphotungstic acid (Fe_3O_4@SiO_2-HPW), which was used to catalyze the formation of 5-ethoxymethylfurfural (5-EMF) from 5-hydroxymethylfurfural (5-HMF) and fructose.[54] This catalyst showed a high yield of approximately 84% *via* esterification of 5-HMF and a moderate yield of 55% from direct use of fructose in one pot, to produce 5-EMF. The magnetic core allowed facile recovery of the catalyst which was then recycled up to 5 times with no significant loss of catalytic ability. Safari *et al.* reported yet another silica-coated catalyst with a magnetic Fe_3O_4, functionalized with sulfonic acid (Fe_3O_4@SiO_2–SO_3H) as an efficient catalyst for the environmentally-benign, solvent-free synthesis of azo dyes (Scheme 4.3).[55]

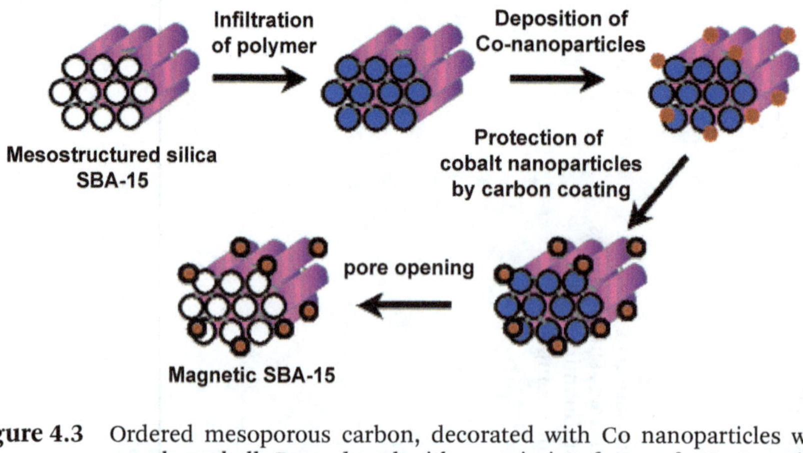

Figure 4.3 Ordered mesoporous carbon, decorated with Co nanoparticles with a carbon shell. Reproduced with permission from ref. 53. Copyright (2004) American Chemical Society.

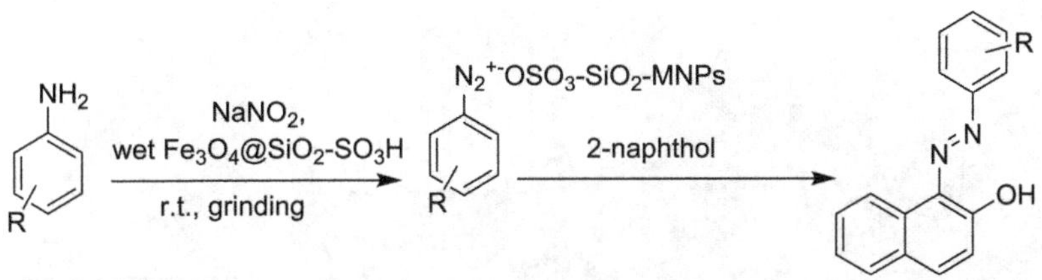

Scheme 4.3 One-pot method for diazotization and diazo coupling reactions using Fe_3O_4@SiO_2–SO_3H under solvent-free conditions. Adapted from ref. 55 with permission from The Royal Society of Chemistry.

The azo dyes were obtained *via* the diazo coupling of aromatic amines with β-naphthol in the presence of sulfonic acid groups on the surface of the magnetic catalyst, by an environmentally-benign grinding method at room temperature. Conventional methods for the synthesis of azo dyes have many drawbacks and limitations such as use of corrosive acids/alkalis, toxic solvents, low temperatures, decomposition of diazonium salts at room temperature, low yields, and long reaction times. Due to the use of the new catalyst and procedure, advantageous and sustainable features such as mild reaction conditions, elimination of the need for toxic solvents, high reagent conversions and product yields, facile product isolation and cost-effectiveness were achieved. Moreover, the use of the magnetic catalyst renders the element of facile recovery and recyclability thereby enhancing the overall efficiency of the process Scheme 4.3.

4.3 Sustainable Features of Magnetic Nanomaterials

Magnetic nanomaterials have multiple features that impart sustainable qualities. Among these are recovery and recyclability, energy and cost-efficiency as well as environmentally-benign synthesis and low toxicity towards living organisms. Discussed below are a few examples of catalysts from each of the above categories. These specific cases were chosen as they offered significant improvements upon existing technologies and showed potential to expand into other processes and applications.

4.3.1 Recovery and Recyclability

One of the most advantageous features of magnetic catalysts is efficient recovery and consequent recyclability, particularly in liquid-phase reactions. Conventionally, the process of heterogeneous catalyst recovery is done *via* filtration or centrifugation which is time consuming and challenging, especially when the catalyst particles are smaller than the micro-meter size range. However, the introduction of magnetic property to catalysts has dramatically improved these issues.[56] Although almost all magnetic catalysts have the features of recovery and recyclability due to their magnetic nature, the below examples showcase those qualities at exceptionally high degrees.

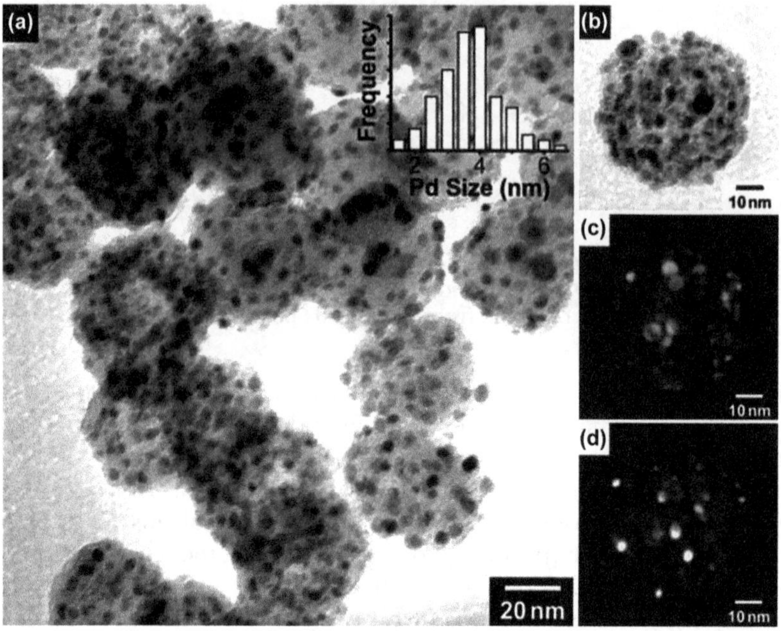

Figure 4.4 TEM images of Pd/N-MCNPs: (a) low-magnification bright-field image (inset: size distribution histogram of the Pd nanoparticles formed), (b) high-magnification bright-field image for a single nanoparticle, and the corresponding dark-field images obtained from (c) Pd(111) and (d) Pd(200) reflections. Reproduced from ref. 57 with permission from The Royal Society of Chemistry.

The efficiency of well-known organic reactions such as Suzuki, Sonogashira and Heck has been enhanced *via* magnetic catalysts. Yoon *et al.* have developed a palladium based magnetic catalyst (Pd/N-MCNPs) where, Pd nanoparticles are impregnated into nitrogen-doped magnetic carbon nanoparticles (Figure 4.4).[57] The latter was made by mixing a carbon precursor with iron-doped polypyrrole nanoparticles at 800 °C. The resulting material consisted of nitrogen atoms included in the carbon matrix due to the presence of pyrrole. The role of the nitrogen atoms within the carbon matrix is to anchor the Pd catalyst particles onto the carbon support by providing strong electronic interactions, thus eliminating the need for further functionalization. The Pd/N-MCNPs was used to catalyze Heck, Suzuki and Sonogashira coupling reactions efficiently. The effective dispersion of Pd onto the magnetic support is said to improve catalytic efficiency by providing a high number of active sites per area. The magnetic support provided robustness together with facile recovery, and recyclability.

Zhang *et al.* reported the development of another palladium catalyst, supported with silica-coated magnetic Fe_3O_4 nanoparticles (Fe_3O_4@SiO_2–Pd), as an efficient heterogeneous catalyst for Suzuki coupling of organoboronic acids with alkynyl bromides.[58] The magnetic nanoparticles, which are protected from oxidation by the silica coating, allowed the catalysts to be easily recovered and were reused up to 16 times without significant loss of activity. The Fe_3O_4@SiO_2–Pd catalyst successfully catalyzed C–C cross coupling of a variety of aromatic and aliphatic boronic acids, potassium aryltrifluoroborates, and sodium tetraphenylborate with 1-bromo-2-substituted acetylenes to produce the corresponding products in ethanol.

Other examples of highly recyclable, core–shell magnetic catalysts include a 2,2,6,6-tetramethylpiperidine-1-oxyl (TEMPO) catalyst, anchored onto silica-coated Fe_3O_4 nanoparticles for aerobic oxidation of alcohols and recyclable for up to 20 cycles,[59] magnetite (Fe_3O_4) and palladium composite nanoparticles, coated with silica for catalytic hydrogenation of 4-nitrophenol[60] and iron oxide-poly(glycidyl methacrylate) (PGMA) core–shell magnetic nanoparticles, functionalized with sulfonic acid groups for the esterification of free fatty acid (FFA) for production of biodiesel.[61]

Hu *et al.* reported ruthenium-based chiral catalysts that are magnetically recoverable for asymmetric hydrogenation of aromatic ketones.[62] The ruthenium(II) complex, [Ru(binap-PO_3H_2)-(dpen)Cl_2](binap = 2,2'-bis(diphenylphosphino)-1,1'-binaphthyl, dpen = 1,2-diphenylethylenediamine) was immobilized on Fe_3O_4 magnetic nanoparticles *via* the phosphonate group, resulting in a heterogenized chiral catalyst that is easily recoverable. The magnetic property not only allowed efficient recovery *via* decantation but also increased the robustness of the catalyst allowing 14 times recyclability with retention of activity as well as enantioselectivity. A recent publication by Yu *et al.* depicts an exceptionally stable magnetic catalyst, that was recyclable up to 8 times for the fast reduction of *p*-nitrophenol to *p*-aminophenol.[63] The catalyst itself is a complex composite, synthesized by first coating magnetic Fe_3O_4 nanoparticles with a thin layer of SiO_2, followed by coupling with poly(ethyleneimine) polymer (PEI) *via* electrostatic self-assembly. Next, glutathione-capped gold nanoparticles again electrostatically self-assembled onto the Fe_3O_4/SiO_2 composite followed by growth of silver on to the surface of the nanocomposite *via* reduction of Ag^+ to Ag^0 *via* dopamine in an alkaline solution. Finally, a 5 nm thick layer of polydopamine (PDA) is formed on the composite surface due to auto-polymerization and forms the complex nanocomposite depicted as Fe_3O_4@SiO_2@PEI-Au/Ag@PDA (Figure 4.5). The authors stress the simplicity of the synthesis due to the self-assembling nature of multiple steps and the use of moderate reaction conditions. The presence of Fe, Au and Ag allows high magnetization and stability to the catalyst which is reflected in its high recyclability of up to 8 times, with no apparent loss of catalytic activity. Due to its high catalytic efficiency, stability and magnetic recoverability, this catalyst has potential applications in multiple areas such as industrial wastewater treatment, production of pharmaceuticals, petrochemicals and food processing as well as other catalytic processes. However, it must be noted that, due to the very presence of multiple and expensive metal constituents, its large scale application and economic viability is questionable.

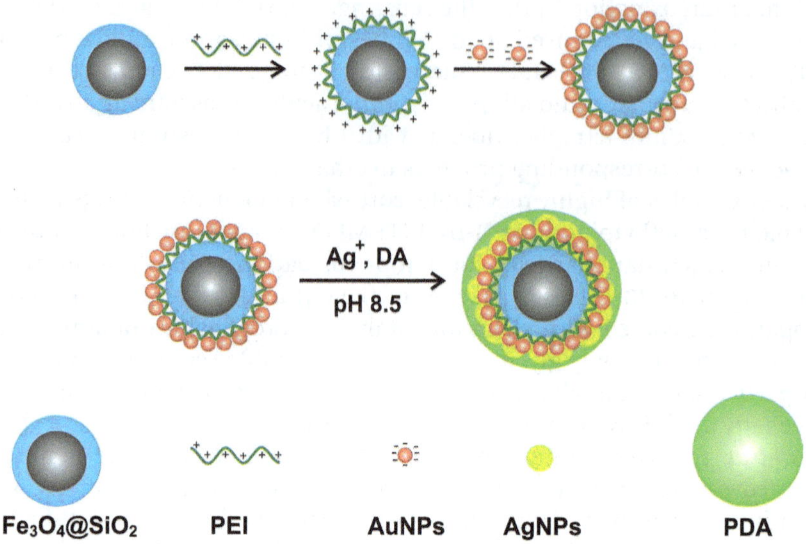

Figure 4.5 Scheme of synthesis of the Fe_3O_4@SiO_2@PEI–Au/Ag@PDA nanocomposite. Reprinted from Macmillan Publishers Ltd: Scientific Reports (ref. 63). Copyright (2016).

4.3.2 Environmentally-Benign Synthesis and Low Toxicity

The synthesis of catalysts often involves either harmful or toxic chemicals or consumption of a large amount of energy due to calcination at high temperatures, *etc.* All these factors have detrimental environmental impacts. In order for magnetic nanocatalysts to be considered sustainable, it is important that their synthesis be achieved according to environmentally benign methods. At the same time, if the catalyst itself has low or no toxicity towards living organisms, it is an added advantage. Recent developments show a growing interest among scientists to achieve the above. Zarghani *et al.* reported an eco-friendly method to synthesize magnetite–chitin (Fe_3O_4@chitin), as a catalyst for synthesis of 5-substituted-1H-tetrazoles under solvent-free conditions.[64] The catalyst was produced *via* hydrothermal synthesis, by mixing in $FeCl_3{\cdot}6H_2O$ into a dispersion of chitin in deionized water, first, at room temperature under argon. Secondly, ammonium hydroxide was added dropwise and the mixture was stirred for 5.0 h at 75 °C, under argon, to produce Fe_3O_4@chitin. This process offers significant advantages such as high product yields, fast reaction rates, and sustainable features such as mild reaction conditions, minimal chemical waste production and facile and environmentally-benign synthesis. In addition, the catalyst is recyclable for up to 6 cycles without any significant loss of activity.

Yang *et al.* reported a series of iron-cobalt mixed oxide nanocatalysts for the decomposition of 2,4-dichlorophenol *via* the heterogeneous activation of peroxymonosulfate (PMS).[65] The activation of PMS leads to the generation of sulfate radicals and the degradation of the 2,4-dichlorophenol, due to their high oxidizing ability. The study targeted the accomplishment of environmentally-favorable synthesis features of the catalyst such as calcination temperature, Fe–Co loading, and use of a TiO_2 support. The catalysts were synthesized by thermal oxidation of Fe and Co salts. It was found that ferromagnetic $CoFe_2O_4$ composites formed *via* a mixed phase of Fe and Co displayed the most efficient and environmentally-friendly activation of PMS. Conventionally, PMS activation is achieved *via* homogeneous catalysis with Co(III) species using Co_3O_4, which shows adverse environmental effects due to the possibility of water pollution by leaching of Co as well as low reaction efficiency of PMS activation by Co(III). However, the cobalt species in $CoFe_2O_4$ are of Co(II), which is highly efficient in PMS activation and the leaching of Co is minimized due to strong Fe–Co linkages of the composite (Figure 4.6). Moreover, the catalyst is conveniently recovered by an external magnet, due to its ferromagnetic nature. The presence of Fe was also found to have the added advantage of increasing the amount of hydroxyl groups on the Fe–Co surface, which enables the formation of Co(II)–OH complexes, which in turn are crucial for the mechanism of heterogeneous PMS activation. It was found that the optimum calcination temperature of 700 °C and thermal oxidation of a mixed phase of Fe and Co renders the most effective catalyst. It was also

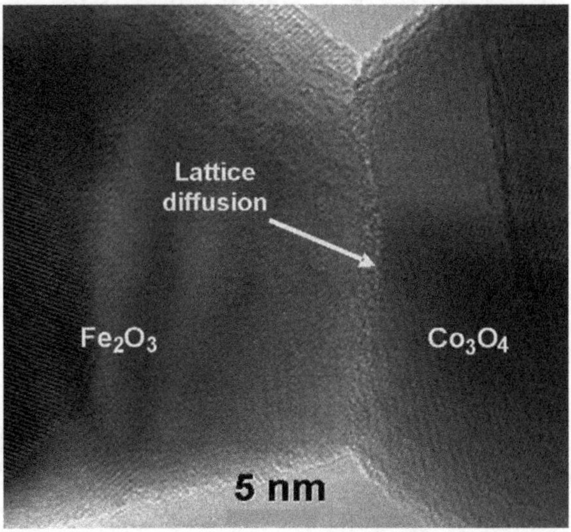

Figure 4.6 HR-TEM images of 1Fe1Co500. Fusion of the Fe_2O_3 to Co_3O_4 lattices at their grain boundaries, suggest the existence of intimate Fe–Co interactions (Fe–Co linkages), which greatly facilitate PMS activation. Reprinted from Applied Catalysis B: Environmental, **88**(3–4), Q. Yang, H. Choi, S. R. Al-Abed, D. D. Dionysiou, Iron-cobalt mixed oxide nanocatalysts: Heterogeneous peroxymonosulfate activation, cobalt leaching, and ferromagnetic properties for environmental applications, 462–469, Copyright (2009) with permission from Elsevier.[65]

found, surprisingly, that use of a TiO_2 support as an attempt to further minimize Co leaching led to very high amounts of leaching, suggesting that TiO_2 may suppress the formation of Fe–Co linkages. Thus, this study represents environmentally-benign features for the synthesis of the magnetic catalyst as well as the nature of the catalytic reaction.

4.3.3 Energy and Cost-Efficiency

Magnetic nanomaterials help reduce the cost of heterogeneous catalysts by replacing expensive ligands with a magnetic component and eliminating extreme synthesis conditions. Cano *et al.* reported a low cost synthesis procedure for a magnetic catalyst for the Suzuki–Miyura cross coupling reaction.[66] Traditionally, C–C coupling *via* the above-mentioned reaction requires the use of expensive ligands such as phosphanes or carbenes. The synthesis and handling of these ligands is complex and cumbersome and they require the addition of additives, such as organic ammonium salts in order for the reaction to occur. Furthermore, Pd metal in the ligands contaminates the final product, thereby reducing its efficiency. Magnetic nanocatalysts are an excellent alternative to eliminate these drawbacks. It was reported that a Fe_3O_4–PdO composite catalyst efficiently catalyzes the Suzuki–Miyura reaction, under mild conditions (Scheme 4.4).

This process was found to be extendable to the coupling of alkenyl boronic reagents with 4-methoxy iodobenzene. The Fe_3O_4–PdO catalyst was easily recoverable by an external magnet, and was reused successfully up to 3 times. The lack of expensive ligands and use of relatively cheap ingredients, together with recyclability makes this catalyst extremely desirable, especially for commonly-used organic reactions such as the Suzuki–Miyura reaction.

Ji *et al.* reported a procedure for the synthesis of reduced graphene oxide–Ni (RGO–Ni) nanocomposites *via* a facile method that used *in situ* reduction

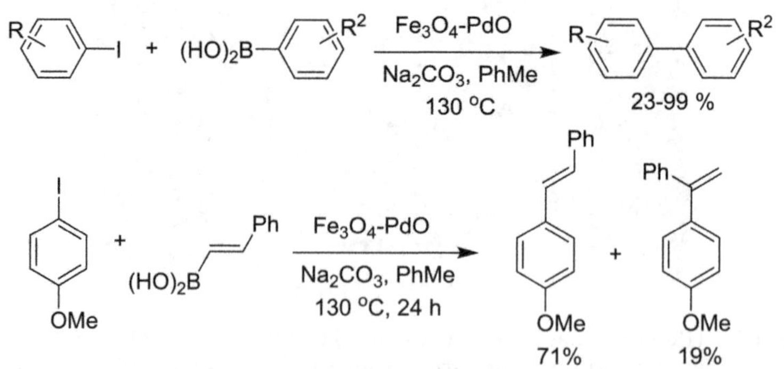

Scheme 4.4 Fe_3O_4–PdO catalyzed Suzuki–Miyaura C–C cross coupling reaction. Adapted from Tetrahedron, 67(30), R. Cano, D. J. Ramón, M. Yus, "Impregnated palladium on magnetite, a new catalyst for the ligand-free cross coupling Suzuki–Miyaura reaction," 5432–5436. Copyright (2011) with permission from Elsevier.[66]

of graphene, as a cost-effective alternative to the RANEY® Ni catalyst.[67] Nanocomposites made from graphene are considered a cheaper alternative to carbon nanotubes for various applications such as supercapacitors, batteries, electronics, photovoltaics and catalysts. The most cost-effective method of graphene production is the chemical reduction of exfoliated graphite oxide, due to its high scalability. Several RGO–Ni catalysts were synthesized by a facile low-temperature solution method, which consisted of first the production of graphene oxide by oxidizing graphite powder with conc. H_2SO_4 at 0 °C and $NaNO_3$ and $KMnO_4$ at 10 °C. Next, an aqueous graphite oxide colloid was mixed with varying amounts of $NiCl_2{\cdot}6H_2O$ and subsequent processing to generate two types of RGO–Ni, based on amount of Ni loading. The catalysts denoted as RGO–Ni-1 and RGO–Ni-2, effectively catalyzed the reduction of *p*-nitrophenol into *p*-aminophenol with $NaBH_4$ (Figure 4.7).

Such aromatic amines have high industrial applicability as precursors for various synthetic products. It was found that the RGO–Ni catalysts had even higher activity than the well-known RANEY® Ni catalyst for reduction of *p*-nitrophenol into the relevant amine. This superior activity is thought to arise due to the synergistic coupling between RGO and the Ni nanoparticles. Due to its magnetic property, this material also has potential as a cost-effective alternative for many other applications apart from catalysis, such as data storage and biomedicine.

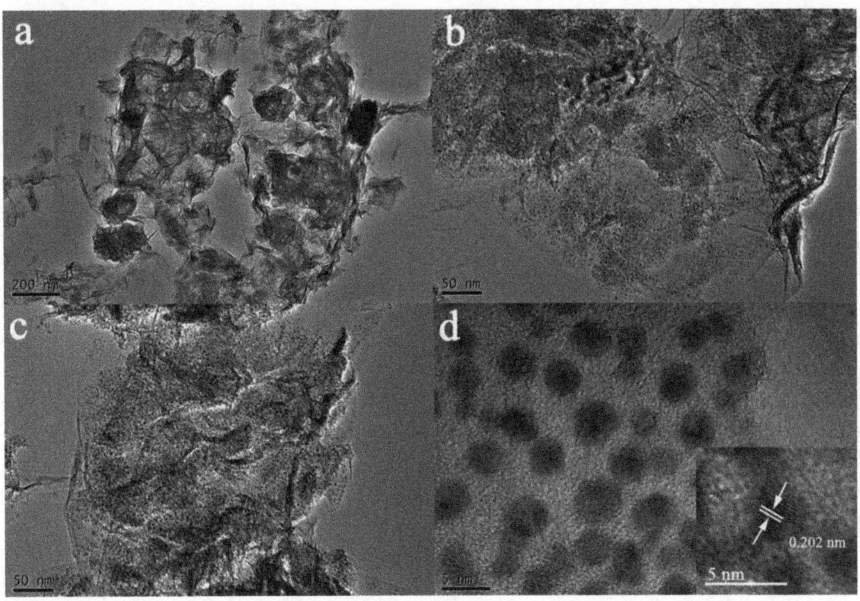

Figure 4.7 (a)–(c) TEM and (d) HRTEM images of the RGO–Ni-1. Reproduced from ref. 67 with permission from The Royal Society of Chemistry.

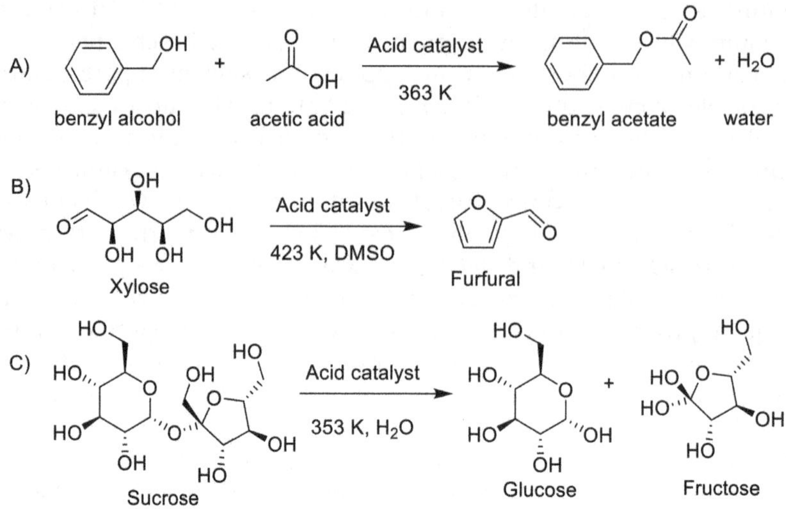

Scheme 4.5 The three reactions catalyzed by the solid acid catalyst: (A) esterification of benzyl alcohol and acetic acid, (B) dehydration of xylose to produce furfural, and (C) hydrolysis of sucrose. Reprinted by permission from Macmillan Publishers Ltd: Scientific Reports (ref. 68), Copyright (2013).

Liu *et al.* reported a highly sustainable process for development of a magnetic solid acid catalyst from biomass waste.[68] The catalyst was produced using sawdust, by integrating Fe and conducting a fast pyrolysis-sulfonation process. This produced a magnetic material which is highly porous, with a high surface area, due to the catalytic effect of Fe. Three typical acid-catalyzed reactions were used to characterize the catalytic ability of the material—esterification, dehydration, and hydrolysis—and it showed excellent catalytic activity for all three reactions (Scheme 4.5). The catalyst was also magnetically collected and recycled up to 5 times without any significant loss of activity. The process represents a sustainable technique for utilization of environmentally-benign biomass waste into highly useful materials.

The preceding sections depict the many sustainable features of magnetic nanocatalysts, and the new and exciting possibilities it imparts on their application in heterogeneous catalysis. The rapidly expanding literature on the subject shows continuous evolution and expansion, suggesting a bright future for sustainable practices in heterogeneous catalysis.

4.4 Summary

This chapter served to highlight the importance of magnetic nanocatalysts as a rapidly evolving sustainable technology. Characteristics of sustainable materials were highlighted while showcasing how magnetic nanomaterials are equipped to transform the field of heterogeneous catalysis, *via* their inherent environmentally-friendly features. The application of magnetic nanomaterials as heterogeneous catalysts and catalyst supports have been discussed using chosen examples. Some of the most prominent sustainable features of magnetic nanocatalysts, which are recovery and recyclability, environmentally-benign synthesis, low toxicity and energy and cost-efficiency, were also discussed.

Acknowledgements

The EPSRC is gratefully acknowledged for support *via* a "Frontier Engineering" award, EP/K038656/1.

References

1. Center for Sustainable Materials at Rutgers University, *What Are Sustainable Materials?* Rutgers, The State University of New Jersey, New Jersey, USA, 2010.
2. F.-M. Belz and K. Peattie, *Sustainability Marketing: A Global Perspective*, Wiley, United Kingdom, 2009.
3. M. Braungart and W. McDonough, *Cradle to Cradle: Remaking the Way We Make Things*, North Point Press, USA, 2002.
4. M. Ioelovich, *ISRN Chem. Eng.*, 2012, **2012**, 7.
5. J. S. Kim, Y. Y. Lee and R. W. Torget, *Appl. Biochem. Biotechnol.*, 2001, **91–93**, 331–340.
6. W. S. Mok, M. J. Antal and G. Varhegyi, *Ind. Eng. Chem. Res.*, 1992, **31**, 94–100.
7. V. Russo, R. Tesser, E. Santacesaria and M. Di Serio, *Ind. Eng. Chem. Res.*, 2013, **52**, 1168–1178.
8. A. S. Perera and M.-O. Coppens, *Catalysis: Volume 28*, The Royal Society of Chemistry, 2016, vol. 28, pp. 119–143.
9. J. M. Campos-Martin, G. Blanco-Brieva and J. L. G. Fierro, *Angew. Chem., Int. Ed.*, 2006, **45**, 6962–6984.
10. C. Samanta, *Appl. Catal., A*, 2008, **350**, 133–149.
11. L. E. Sendelbach, *Toxicology*, 1989, **57**, 227–240.
12. O. G. Mironov, *Helgoländer wissenschaftliche Meeresuntersuchungen*, 1968, **17**, 335–339.
13. P. M. Vitousek, H. A. Mooney, J. Lubchenco and J. M. Melillo, *Science*, 1997, **277**, 494–499.
14. T. K. Tromp, R.-L. Shia, M. Allen, J. M. Eiler and Y. L. Yung, *Science*, 2003, **300**, 1740–1742.
15. S. M. George, *Chem. Rev.*, 1995, **95**, 475–476.
16. J. M. Thomas, J. M. Thomas and W. J. Thomas, *Principles and Practice of Heterogeneous Catalysis*, Wiley, 2015.
17. C. N. Satterfield, *Heterogeneous Catalysis in Industrial Practice*, McGraw Hill Book Co., United States: New York, 2nd edn, 1991.
18. M. Muruganandham, R. P. S. Suri, M. Sillanpää, J. J. Wu, B. Ahmmad, S. Balachandran and M. Swaminathan, *J. Nanosci. Nanotechnol.*, 2014, **14**, 1898–1910.

19. A. T. Bell, *Science*, 2003, **299**, 1688–1691.
20. A. Corma, H. García and F. X. Llabrés i Xamena, *Chem. Rev.*, 2010, **110**, 4606–4655.
21. M. Heitbaum, F. Glorius and I. Escher, *Angew. Chem., Int. Ed.*, 2006, **45**, 4732–4762.
22. M. K. Sabbe, M.-F. Reyniers and K. Reuter, *Catal. Sci. Technol.*, 2012, **2**, 2010–2024.
23. A. K. Gupta and M. Gupta, *Biomaterials*, 2005, **26**, 3995–4021.
24. G. Podaru, S. Ogden, A. Baxter, T. Shrestha, S. Ren, P. Thapa, R. K. Dani, H. Wang, M. T. Basel, P. Prakash, S. H. Bossmann and V. Chikan, *J. Phys. Chem. B*, 2014, **118**, 11715–11722.
25. H. Wang, T. B. Shrestha, M. T. Basel, R. K. Dani, G.-M. Seo, S. Balivada, M. M. Pyle, H. Prock, O. B. Koper, P. S. Thapa, D. Moore, P. Li, V. Chikan, D. L. Troyer and S. H. Bossmann, *Beilstein J. Nanotechnol.*, 2012, **3**, 444–455.
26. L. A. Thomas, L. Dekker, M. Kallumadil, P. Southern, M. Wilson, S. P. Nair, Q. A. Pankhurst and I. P. Parkin, *J. Mater. Chem.*, 2009, **19**, 6529–6535.
27. T. Hyeon, *Chem. Commun.*, 2003, 927–934.
28. J. Zhu, S. Wei, M. Chen, H. Gu, S. B. Rapole, S. Pallavkar, T. C. Ho, J. Hopper and Z. Guo, *Adv. Powder Technol.*, 2013, **24**, 459–467.
29. M. M. Khin, A. S. Nair, V. J. Babu, R. Murugan and S. Ramakrishna, *Energy Environ. Sci.*, 2012, **5**, 8075–8109.
30. S. Mornet, S. Vasseur, F. Grasset, P. Veverka, G. Goglio, A. Demourgues, J. Portier, E. Pollert and E. Duguet, *Prog. Solid State Chem.*, 2006, **34**, 237–247.
31. Z. Li, L. Wei, M. Y. Gao and H. Lei, *Adv. Mater.*, 2005, **17**, 1001–1005.
32. M. Mahmoudi, A. Simchi, M. Imani, A. S. Milani and P. Stroeve, *J. Phys. Chem. B*, 2008, **112**, 14470–14481.
33. R. A. Sheldon and R. S. Downing, *Appl. Catal., A*, 1999, **189**, 163–183.
34. D. J. Cole-Hamilton, *Science*, 2003, **299**, 1702–1706.
35. B. F. G. Johnson, *Coord. Chem. Rev.*, 1999, **190–192**, 1269–1285.
36. L. L. Chng, N. Erathodiyil and J. Y. Ying, *Acc. Chem. Res.*, 2013, **46**, 1825–1837.
37. A. Pourjavadi, M. Tajbakhsh, M. Farhang and S. H. Hosseini, *New J. Chem.*, 2015, **39**, 4591–4600.
38. R. Hudson, C.-J. Li and A. Moores, *Green Chem.*, 2012, **14**, 622–624.
39. R. B. N. Baig and R. S. Varma, *Green Chem.*, 2012, **14**, 625.
40. L. Menini, M. C. Pereira, A. C. Ferreira, J. D. Fabris and E. V. Gusevskaya, *Appl. Catal., A*, 2011, **392**, 151–157.
41. Y. D. Shahamat, M. Farzadkia, S. Nasseri, A. H. Mahvi, M. Gholami and A. Esrafili, *J. Environ. Health Sci. Eng.*, 2014, **12**, 50.
42. M. J. Aliaga, D. J. Ramon and M. Yus, *Org. Biomol. Chem.*, 2010, **8**, 43–46.
43. R. Cano, D. J. Ramón and M. Yus, *J. Org. Chem.*, 2010, **75**, 3458–3460.
44. M. Kotani, T. Koike, K. Yamaguchi and N. Mizuno, *Green Chem.*, 2006, **8**, 735–741.
45. F. Shi, M. K. Tse, S. Zhou, M.-M. Pohl, J. Radnik, S. Hübner, K. Jähnisch, A. Brückner and M. Beller, *J. Am. Chem. Soc.*, 2009, **131**, 1775–1779.

46. R. Cano, M. Yus and D. J. Ramón, *Tetrahedron*, 2012, **68**, 1393–1400.
47. R. Cano, D. J. Ramón and M. Yus, *J. Org. Chem.*, 2011, **76**, 5547–5557.
48. K. H. Lee, B. Lee, K. R. Lee, M. H. Yi and N. H. Hur, *Chem. Commun.*, 2012, **48**, 4414–4416.
49. R. Abu-Reziq, H. Alper, D. Wang and M. L. Post, *J. Am. Chem. Soc.*, 2006, **128**, 5279–5282.
50. M. Z. Kassaee, H. Masrouri and F. Movahedi, *Appl. Catal., A*, 2011, **395**, 28–33.
51. H. Wang, J. Covarrubias, H. Prock, X. Wu, D. Wang and S. H. Bossmann, *J. Phys. Chem. C*, 2015, **119**, 26020–26028.
52. A.-H. Lu, W. Schmidt, N. Matoussevitch, H. Bönnemann, B. Spliethoff, B. Tesche, E. Bill, W. Kiefer and F. Schüth, *Angew. Chem., Int. Ed.*, 2004, **43**, 4303–4306.
53. A.-H. Lu, W.-C. Li, A. Kiefer, W. Schmidt, E. Bill, G. Fink and F. Schüth, *J. Am. Chem. Soc.*, 2004, **126**, 8616–8617.
54. S. Wang, Z. Zhang, B. Liu and J. Li, *Catal. Sci. Technol.*, 2013, **3**, 2104–2112.
55. J. Safari and Z. Zarnegar, *RSC Adv.*, 2015, **5**, 17738–17745.
56. S. Shylesh, V. Schünemann and W. R. Thiel, *Angew. Chem., Int. Ed.*, 2010, **49**, 3428–3459.
57. H. Yoon, S. Ko and J. Jang, *Chem. Commun.*, 2007, 1468–1470.
58. X. Zhang, P. Li, Y. Ji, L. Zhang and L. Wang, *Synthesis*, 2011, **2011**, 2975–2983.
59. B. Karimi and E. Farhangi, *Chem.–Eur. J.*, 2011, **17**, 6056–6060.
60. M. An, J. Cui and L. Wang, *J. Phys. Chem. C*, 2014, **118**, 3062–3068.
61. Zillillah, G. Tan and Z. Li, *Green Chem.*, 2012, **14**, 3077–3086.
62. A. Hu, G. T. Yee and W. Lin, *J. Am. Chem. Soc.*, 2005, **127**, 12486–12487.
63. X. Yu, G. Cheng and S.-Y. Zheng, *Sci. Rep.*, 2016, **6**, 25459.
64. M. Zarghani and B. Akhlaghinia, *RSC Adv.*, 2016, **6**, 31850–31860.
65. H. C. Q. Yang, S. R. Al-Abed and D. D. Dionysiou, *Appl. Catal., B*, 2009, **88**, 462–469.
66. R. Cano, D. J. Ramón and M. Yus, *Tetrahedron*, 2011, **67**, 5432–5436.
67. Z. Ji, X. Shen, G. Zhu, H. Zhou and A. Yuan, *J. Mater. Chem.*, 2012, **22**, 3471–3477.
68. W.-J. Liu, K. Tian, H. Jiang and H.-Q. Yu, *Sci. Rep.*, 2013, **3**, 2419.

CHAPTER 5

Recyclable Magnetic Materials for Biomass Conversion

LEIDY PEÑA[a] AND DONGHAI WANG*[a]

[a]Kansas State University, Biological and Agricultural Engineering, 150 Seaton Hall, Manhattan, KS, 66506, USA
*E-mail: dwang@ksu.edu

5.1 Introduction

Policies aiming at the reduction of CO_2 emissions stimulate the research and development of bio-based materials and chemicals as well as the implementation of environmentally friendly processes. Renewable and/or recyclable materials are sought to replace oil-derived chemicals to reduce the carbon footprint of the chemical industry. Magnetically aided processes are green processes because the magnetic entity can be easily recovered by magnetic means. Industrial applications that use magnetically aided separations are commercially available. Magnetically aided water purification facilitates solid/liquid separation of a magnetic pollutant or a pollutant linked to a magnetic carrier.[1] Magnetic materials that retain their properties even after several process cycles are ideal candidates for green processes because their use minimizes chemical waste. One of the drawbacks of using renewable materials to replace cheap oil-derived chemicals is the non-competitive cost of producing bio-based chemicals. Using recyclable materials with a high capacity, like large surface areas or high selectivity, could make chemical processes environmentally friendly and could make the production of bio-based products economically feasible.

Smart Materials No. 26
Magnetic Nanomaterials: Applications in Catalysis and Life Sciences
Edited by Stefan H. Bossmann and Hongwang Wang

Published by the Royal Society of Chemistry, www.rsc.org

Easy separation of entities with magnetic properties makes them ideal carriers of reaction catalysts. Immobilization of catalysts on the surface of magnetic particles has potential as a solution for the separation and recovery of some expensive catalysts.[2] Recovery and reuse of the catalyst is economically beneficial. Moreover, purification of the final product can be critical when processing chemicals that must meet biocompatibility requirements. In some cases, the magnetic entity can be a catalyst itself. Then, the catalyst can be separated after reaction and fed back into the process as a new reaction catalyst.

Frequently, magnetic particles work as carriers; they don't directly promote the reaction. Magnetic nanoparticles can feature acid groups, heavy metal complexes, organic catalysts, enzymes, and/or biomimetic catalysts. Nanoparticles have very large surface to volume ratios so they can achieve high catalytic activities. Another option is to functionalize the surface of the particles to adjust their properties to meet desired criteria; surface modification can give magnetic materials specific surface chemistry to achieve higher selectivities.

Nanoparticle solutions have the advantage over slurries of macrosized catalysts that they can be dispersed to simulate homogeneous solutions. Using nanoparticles could circumvent mass transfer problems associated with the utilization of solid catalysts in gas and liquid reactions.[3,4] It is of importance that nanoparticles can be easily dispersed in solution, but cannot be easily separated from the solution, which could be problematic in industrial applications. At the lab scale, centrifugation of nanoparticles requires high rotational speeds and prolonged times. Filtration could be an option if membranes with proper size pores are available, but it may also require extended times. Magnetic separation can be less capital- and energy-intensive. Materials difficult to separate out of media, such as proteins or algae, can be recovered using amino-functional groups attached to magnetic flocculants.[5] Magnetic nanoparticles can also be used to mimic homogeneous catalysts, but they can be easily recovered from the reaction media by physical operations. Magnetic nanoclusters can be recovered by applying a permanent magnet.

Using magnetic nanoparticles has two major advantages besides recyclability: high capacity loading of functional groups and easy access of the reactants to active sites because catalytic domains are bound to the surface. The catalysts are usually anchored to the magnetic nanoparticles as a layer of organic coatings using silanes.[6] Magnetic nanoparticles have been used in cross-coupling reactions,[7,8] as a deallylation catalyst,[2] and in the hydroformylation of olefins.[9] Aspartic acid and histidine residues have been bonded to maghemite nanoparticles to hydrolyze phosphoester and carboxylic esters.[10] Extensive reviews on catalytic reactions supported by magnetic nanoparticles are available.[7] Yet, the utilization of nanoparticles for biomass conversion has not been extensively explored. In this chapter, we will focus on the utilization of magnetic nanoparticles in the catalytic hydrolysis of biomass for bio-based product generation. This chapter won't cover magnetic nanoparticles used as a separation aid, such as nanoparticles used in protein purification.

5.2 Magnetic Nanoparticles

5.2.1 Synthesis Methods

The composition, size and morphology of magnetic nanoparticles determine their magnetic properties.[11] Magnetic properties, such as saturation magnetization M_s, coercivity H_c, and remnant magnetization M_R can be tuned by chemical composition and synthesis procedure. For example, magnetic nanoparticles made of elemental iron or cobalt, iron alloys, or iron oxides have different values of M_s as shown in Table 5.1. Nanoparticles that are ferromagnetic particles have high saturation magnetization and are easily magnetically separated from aqueous media but can be difficult to redisperse back into solution due to strong magnetic aggregation. Conversely, magnetic nanoparticles that have superparamagnetic properties show magnetism only in the presence of a magnetic field. Superparamagnetic particles have negligible remnant magnetization and coercivity, and lose magnetism when withdrawn from a magnetic field.[12] For this reason, superparamagnetic particles can be dispersed in a reaction media in the absence of a magnetic field, and when the reaction or process is over, the particles can be gathered and recovered in the presence of a magnetic field. Iron oxide nanoparticles with sizes under 20 nm show superparamagnetic properties; the coercivity decreases with decreasing particle size.[13]

Superparamagnetic iron oxide nanoparticles (SPIONs), such as magnetite Fe_3O_4 and maghemite γ-Fe_2O_3 are favored for biological applications because they are biocompatible.[14] Iron oxide nanoparticles have been used in hyperthermia, drug delivery, biomedical imaging, and diagnostic research studies.[15,16] Due to its higher magnetic properties, magnetite is the preferred iron oxide form but magnetite can be easily oxidized to maghemite γ-Fe_2O_3.[17]

Table 5.1 Magnetization saturation values for various magnetic nanoparticles.

Nanoparticle	Size	Saturation magnetization emu/g	Coercivity kA m^{-1}	Reference
Fe_3O_4	Bulk	90–100		11
Fe_3O_4	6 nm	40.6		29
Fe_3O_4	9 nm	76		19
Fe_3O_4	14 nm	65.1	7.6	18
Fe_3O_4@SiO_2	90–140 nm	0.43–7.62		29
Fe_3O_4	280 nm	84	5	2
Fe_3O_4@SiO_2	2–5 μm	47	5.2	
Fe_3O_4	300 nm	80		14
Fe_3O_4@SiO_2	500 nm	53.3		
$CoFe_2O_4$	Bulk	80		30
$CoFe_2O_4$	3 nm	74		
$CoFe_2O_4$	21 nm	55	16	31
$CoFe_2O_4$ +PVP	22 nm	60	83	
$CoFe_2O_4$ +PEG	22 nm	63	84	
$MnFe_2O_4$	7–18 nm	1–6		12

In addition, maghemite particles exhibit lower saturation magnetization values than magnetite nanoparticles; so a decrease in saturation magnetization in magnetite may indicate oxidation to maghemite. SPIONs can be synthesized by co-precipitation of iron salts,[12,13,18] precipitation in emulsifiers, thermal decomposition,[19] or sonochemical synthesis.[20] Maghemite and magnetite nanoparticles are usually prepared using precipitation methods where iron ions are precipitated in basic solutions.[21] Most procedures use a controlled inert gas atmosphere to avoid oxidation of magnetite to maghemite during synthesis.[22] In a typical experiment, a precursor of ferrous ions Fe^{2+}, such as $FeSO_4 \times 7H_2O$, is precipitated with ammonium hydroxide to produce Fe_3O_4.[23] Silica coated nanoparticles have higher magnetization values (74 emu g^{-1}) than bare iron oxide particles, which is attributed to the reduction of oxidation to maghemite as well as a reduction of dipole–dipole interactions (60 emu g^{-1}).[23] Silica, poly(ethylene glycol), or chitosan can be used to coat the surface of iron oxide nanoparticles to avoid toxic effects or to prevent oxidation of magnetite.

The chemical stability of nanoparticles of Mn, Co, or Ni improves when doping them with iron oxides.[12] Cobalt nanoparticles can reach higher saturated magnetization than iron oxides but are susceptible to oxidation in air.[24] Magnetic nanoparticles containing cobalt, nickel, or manganese are not used in biomedical applications because of the potential toxicity issues of these elements. However, non-biocompatible magnetic nanoparticles can be used in the production of fuels and chemicals. Cobalt iron oxide ($CoFe_2O_4$) nanoparticles are superparamagnetic; so $CoFe_2O_4$ nanoparticles exhibit magnetic properties only when exposed to a magnetic field.[25]

Microemulsions can be used to control the size of the magnetic nanoparticles. Mean particle sizes smaller than 20 nm were obtained using oleic acid.[26,27] Cobalt iron oxide nanoparticles have been synthesized using microemulsion methods.[25,28] In a typical experiment, cobalt(II) chloride and iron(II) chloride solutions are mixed and stirred to oxidize iron(II) to iron(III). Surfactants, such as dodecylsulfate (SDS) are used to generate a microemulsion and keep particles from agglomerating. Methylamine is used to precipitate $CoFe_2O_4$ nanoparticles that can be later separated magnetically from the reaction solution. A scheme of this reaction is shown in Figure 5.1.

To prevent surface oxidation and to be able to tune the surface chemistry of cobalt iron oxide nanoparticles, a silica multilayer is placed on the surface

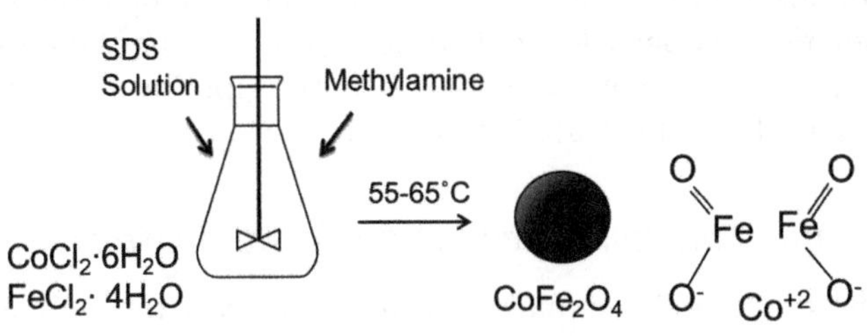

Figure 5.1 Scheme of the synthesis of $CoFe_2O_4$ nanoparticles.

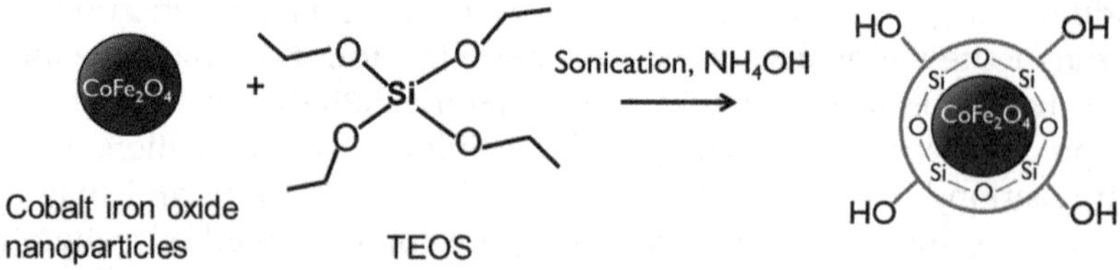

Figure 5.2 Silica coating of cobalt iron oxide nanoparticles.

of the magnetic particles. Coating magnetic nanoparticles with silica generates a hydroxylated surface amenable to modification with silanes.[32,33] Silica coating procedures are described in the literature.[25,28,34] In a typical experiment, ammonium hydroxide and tetraethylorthosilicate (TEOS) are added to an alcohol dispersion of magnetic nanoparticles and the mixture is sonicated and stirred simultaneously for 4 h at room temperature; prior to this, TEOS is diluted in isopropanol and is added dropwise to a nanoparticle solution. The silica coated magnetic nanoparticles are washed with large amounts of water and separated out from the solution using a permanent magnet. A scheme of this coating procedure is depicted in Figure 5.2.

5.2.2 Functionalization of Silica Coated Nanoparticles

Silica coated magnetic nanoparticles are converted into active catalysts by functionalization of the silica surface. The greater the number of silanol groups the greater the number of catalytic sites that can be placed in the surface. There are several pathways to functionalize silica; the simplest way is physisorption where soluble catalysts are immobilized on the silica surface. However, this method is inefficient because the catalytically active molecules can easily become desorbed and solubilized by the solvent during catalysis.[35] Chemisorption is the preferred method because it reduces the amount of catalyst leached from the surface. Through silanization silanol groups react with silylating agents. The catalyst can be anchored to the silane group before or after it is attached to the superficial silanol. Several reports can be found on the modification of porous silicas with mercaptopropyl groups; after grafting thiol groups to a silica surface the –SH functionalities are oxidized to generate sulfonic acid groups.[36–39] In previous work, we reported propylsulfonic acid-functionalization of non-porous silica coated nanoparticles.[40,41] Silica coated magnetic nanoparticles with various levels of functionalization can be synthesized. A scheme of the synthesis of mercapto-propyl acid functionalized nanoparticles is illustrated in Figure 5.3.

5.2.2.1 Amino-Functionalized Magnetic Nanoparticles

Using amino-functionalized magnetic nanoparticles for cellulase immobilization could enable the recovery of these expensive biocatalysts. The carboxylic groups of the protein are cross-linked with the amino-groups

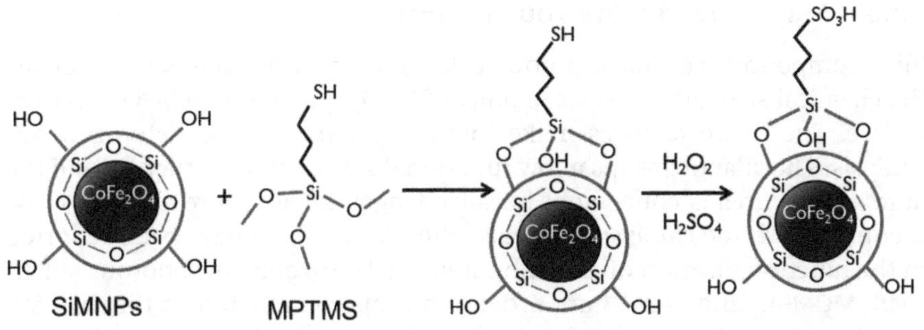

Figure 5.3 Acid-functionalization of silica coated nanoparticles.

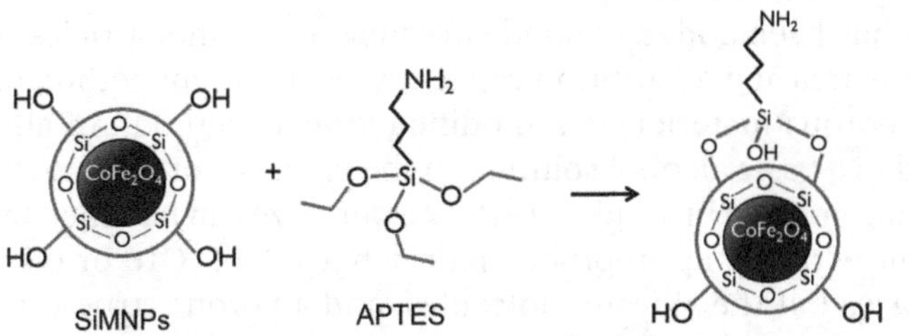

Figure 5.4 Scheme of the amino-functionalization of silica coated iron oxide particles.

of solid supports; in this way, amino-functionalized silicas can be used to support cellulases.[42,43] Mesoporous silica can be synthesized using various organosilane precursors: 3-aminopropyltriethoxysilane (APTES), 3-mercaptopropyltrimethoxysilane (MPTMS), vinyl-trimethoxysilane (VTMS), and phenyltrimethoxysilane (PTMS). The mechanisms used to support the enzyme are chemical linkage to the functional groups and physical adsorption on the pores of the surface.

The most common amine precursors to obtain functionalized silicas are APTES and *N*-(2-aminoethyl)-3-aminopropyltriethoxysilane (AAPTES) or their trimethoxy counterparts APTS and AAPTS.[42–45] Amino-functionalized silica nanoparticles can be synthesized through the co-condensation of tetraethoxysilane (TEOS) with APTS or AAPTS in the presence of a surfactant agent.[44] Amine groups can be grafted to the surface of $CoFe_2O_4$ nanoparticles using AAPTES as the amine precursor.[28] Similarly, silica coated magnetic nanoparticles (SiMNPs) can be functionalized with APTES or AAPTS. In a typical procedure, SiMNPs are placed in a three necked round bottom flask with toluene. The mixture is sonicated for 30 min, and followed by addition of APTES or AAPTS. The nanoparticles are left overnight at 80 °C for 16 h. After the reaction, the nanoparticles are magnetically separated and rinsed with toluene and hexane; then the amino-functionalized nanoparticles are dried at 105 °C under vacuum (Figure 5.4).

5.2.3 Magnetic Mesoporous Materials

Silica compounds can form porous gels which provide large surface areas. The chemical structure consists of a net of Si–O–Si bonds with Si–OH in their surface. These two features make them very attractive for heterogeneous catalysis. The silanol groups allow functionalization of the structure and the large surface area is convenient because reagents have more chances to be in contact with the catalyst and react. Numerous works have been reported on the functionalization of amorphous silica. Hexagonal mesoporous silicas HMS, MCM-41 and SBA-15 are ordered mesoporous materials that are frequently used as the support for catalytic domains.[46] FDU-12 type silica has also been used because it contains larger pore sizes than MCM-41 and SBA-15.[42] Larger pore sizes are desired because they can host big molecules such as enzymes and antibodies.

Porosity has been added to solid core magnetic nanoparticles to increase their surface area and adsorption capacities. A common method for the generation of porous materials is a modified Stöber method. An alkylsilane is hydrolyzed in a basic alcohol solution in the presence of a porogen like cetyltrimethyl ammonium bromide CTAB. The pore size can be tuned by choosing a surfactant with the appropriate chain length: C14, C16 or C18. The surfactant spaces out the silicate molecules, and a porous structure is created when the porogen is burned off the silica material. Magnetic porous materials, such as iron oxide mesoporous silica nanoparticles have been used for protein absorption.[29] The mesoporous silica nanoparticles can have multiple iron oxide cores (Fe_3O_4, 6 nm in diameter), the typical core possesses a magnetic saturation value of 40.6 emu g^{-1}. After coating with mesoporous silica, the magnetization value has ranged from 0.43–7.62 depending on the number of iron oxide nanoparticles. Silica coating enlarges the nanoparticles to 90–140 nm. Besides size, the saturation magnetization of the composite is also controlled by the number of magnetic nanoparticles. Iron oxide nanoparticles are synthesized in hydrophobic media; before silica coating, iron oxide nanoparticles are capped with stearic acid which helps to switch the environment of the nanoparticles from non-polar to polar. The mesoporous layer is generated using a micellar solution of C_XTAB (X = 14, 16, or 18) as a porogen. Higher temperatures of 30–80 °C help the formation of micelles and the creation of ordered pores. More hydrophobic surfactants (with longer carbon chains) produce longer micelles and wider pores. Surfactants with shorter carbon chains may require temperatures where they exhibit more hydrophobic behavior and have a better chance of forming micelles. This may happen at lower temperatures compared to their longer carbon chain counterparts. Synthesis of ordered mesoporous shells at 30 °C can be performed using myristyltrimethylammonium bromide (C_{14}TAB) and at 80 °C when using octadecyltrimethylammonium bromide (C_{18}TAB).[29]

Multicore silica coated magnetic nanoparticles were used to immobilize deallylation catalysts by Kitamura *et al.*[2] The Fe_3O_4 nanoparticles were 280 nm in diameter with a saturation magnetization of 84 emu g^{-1} and a coercivity of 63 Oe (5.0 kA m^{-1}). After these iron oxide nanoparticles were coated with silica using sodium silicate, their size was enlarged to 2–5 μm, their saturation magnetization decreased to 47 emu g^{-1}, and, their coercivity increased to 65 Oe (5.2 kA m^{-1}).

Utilizing mesoporous silicas offers numerous possibilities for chemical modification; however, the following diffusion problems can be encountered in applications, such as enzyme immobilization: (1) the pore size limits the amount of enzyme that the material can host and (2) at some stage in the enzyme catalyzed reaction, mass transport problems of the reagents and/or products can occur.

5.3 Biomass Derivation: Catalysis

Lignocellulosic biomass is an abundant carbon source that is currently being investigated for its potential as feedstock for biofuel production. Processing lignocellulosic biomass is a difficult task because biomass is a very recalcitrant material designed to protect the plant against environmental conditions and microbial degradation. Lignocellulosic biomasses have three major components including cellulose, hemicellulose, and lignin and their ratio varies with the biomass origin. Biomass also contains minor compounds classified as extractives and ashes. Water or alcohol extractives can be fats, proteins, starches, monomeric sugars, gums, resins, essentials oils, *etc.* A scheme of the main components of lignocellulosic biomass is shown in Figure 5.5.

Most sugars in lignocellulosic biomass are in its cellulose fraction. Cellulose is a polymer of D-glucopyranose monomers linked by β-1,4 bonds

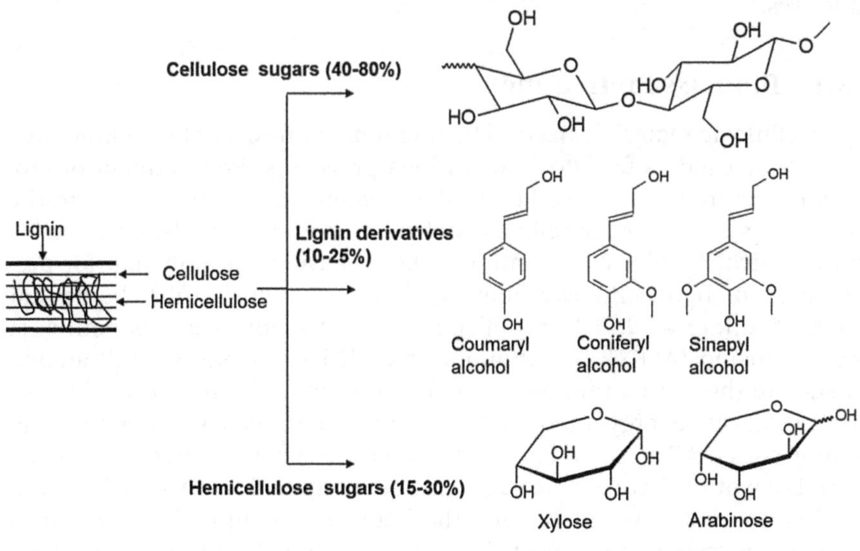

Figure 5.5 Major components of lignocellulosic biomass. Adapted with permission from ref. 47. Copyright (2006) American Chemical Society.

with a degree of polymerization of 8000–15 000.[48,49] Because of its molecular arrangement, cellulose is highly crystalline, water insoluble, and highly resistant to depolymerization.[50] The cellulose polymer has to be depolymerized into single molecules of glucose so the sugars can be fermented to alcohol. The breakdown of cellulose into its monomer components is efficiently done by cellulolytic enzymes. However, these enzymes are high value catalysts that require specific processing conditions such as temperature and pH.

Hemicelluloses are short chain, amorphous polymers of five (pentoses) and six (hexoses) carbon sugars such as xylose, arabinose, glucose, mannose, galactose, and rhamnose. The polymer chain is frequently substituted with sugars, acetyl groups, and/or uronic acids such as glucoronic, mannuronic or galacturonic acid. Since hemicelluloses are amorphous, they can be hydrolyzed at milder conditions compared to celluloses which can be partially crystalline.

The carbohydrate component in biomasses, that is, celluloses and hemicelluloses, is protected by a lignin sheath that must be removed before saccharification.[49] Lignin is a highly branched polyphenolic resin, has an amorphous structure, and is not easy to solubilize.[51] Its chemical composition gives lignin a less hydrophilic character compared to the polysaccharide fraction of the biomass.[51] The removal of the amorphous polymers of lignin enhances the enzymatic digestibility of cellulose[52] because lignin restricts hydrolysis by shielding cellulose surfaces and absorbing and deactivating enzymes.[53] Lignin also prevents swelling of the fiber, thereby affecting enzyme accessibility to the cellulose. As by-products from cellulosic ethanol production, lignin derivatives would be a renewable alternative replacement for current oil-based products such as phenol formaldehyde resins and wood adhesives.[54,55]

5.3.1 Biomass Pretreatment

Lignocellulosic materials need to be fractionated into smaller compounds that can be used as feedstock for various processes. Pretreatment of biomass is required to remove hemicelluloses and lignin, and facilitate the enzymatic attack over the cellulose polymer. Several methods, physical[56–58] and chemical[59] and a combination of both, have been evaluated for pretreatment of lignocellulosic biomass.[60] Steam explosion,[61–63] liquid hot water,[64,65] dilute acid,[66–71] and alkaline pretreatments such as ammonia fiber explosion (AFEX) and ammonia recycled percolation (ARP) among others, are the most common methods reported in the literature.[72] Physical methods are energy intensive and chemical methods have environmental drawbacks. Alkaline and ammonia fiber explosion methods have been proved to work effectively over agricultural waste and hardwood; however, the first one is not recyclable and the latter one requires high pressures and cooling systems to recover the ammonia after the process. Moreover, basic catalysis is not effective for the pretreatment of softwoods. The main goal of pretreatment is the exposition of the cellulose fibrils to increase the available area for enzymatic action. A scheme of the pretreatment of cellulosic biomass is shown in Figure 5.6.

Numerous studies have used mineral acids as the preferred catalyst for pretreatment of lignocellulosic biomass.[66–71] The acid hydrolysis is usually combined with thermal and physical treatments. The principle of these methods is the acid catalysis of the hemicelluloses and the lignin–carbohydrate linkages.[69] Cellulose digestibility increases with hemicellulose removal; but complete hemicellulose hydrolysis with hot water or steam is achieved at a temperature range of 190–240 °C.[73] When acid solutions instead of water are used, effective hemicellulose solubilization is achieved at lower temperatures (140–180 °C).[65,69,74] Low pH and high pretreatment temperatures result in a larger hemicellulose solubilization and improved cellulose digestibility, but the generation of inhibitory products is also larger. Even low concentrations of sulfuric acid can degrade cellulose and hemicellulose sugars to hydroxymethylfurfural (HMF) and furfural among other undesirable compounds. These sugar degradation products are inhibitory of cellulose enzymatic hydrolysis and sugar fermentation.[75,76] At high temperatures the solubilized sugars from hemicelluloses are degraded to furfural and formic acid;[70,77,78] these by-products, besides being inhibitors, lower the sugar recovery.[79] Pretreatment methods that solubilize hemicelluloses at low temperatures and high pH are necessary to maximize the utilization of sugars from both hemicelluloses and celluloses.

The most common pretreatment methods employ mineral acids and high temperatures to remove the lignin component.[62,80,81] Under these conditions, most of the sugars from hemicelluloses are degraded and can't be used for alcoholic fermentation.[75,76,82–84] Some of the common temperatures

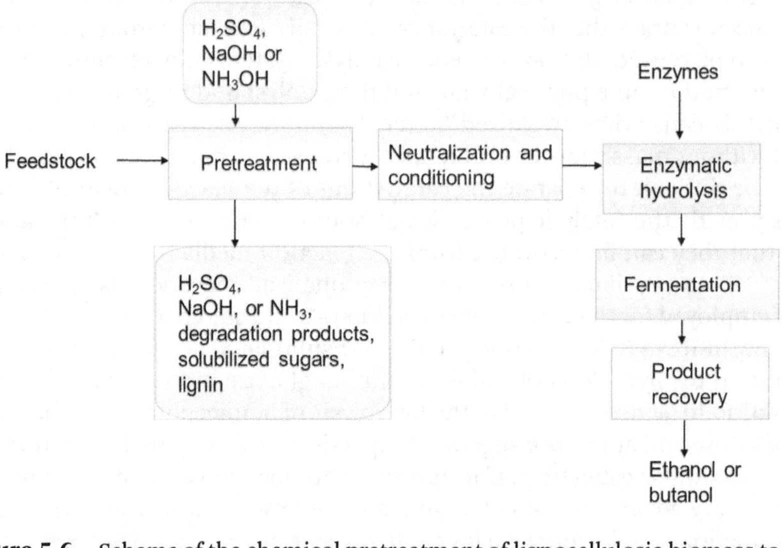

Figure 5.6 Scheme of the chemical pretreatment of lignocellulosic biomass to produce biofuel.

Table 5.2 Temperatures and acid concentrations employed to pretreat corn stover.

Temperature	Sulfuric acid concentration wt. %	Time/rate (min)	Xylan solubilization	Reference
140 °C	0.98	40	32.4%	66
160 °C	0.5	40	93.1%	85
180 °C	0.1	10 ml min^{-1}	95–97%	69
200 °C	0.0	15	53.0%	86

and acid concentrations used in the pretreatment of corn stover are shown in Table 5.2. The lower the acid concentration, the stronger the thermal treatment that the sample must go through is. High temperatures require high energy inputs, which would affect the economic feasibility of the process. Higher acid concentrations allow the use of lower temperatures, but these acid solutions are usually avoided because of the neutralization and corrosion issues associated with their use. A recoverable and reusable catalyst that allows the use of low temperatures is desirable for this process. Magnetic nanoparticles functionalized with –SO_3H groups could be one of the options for applying solid acids to the pretreatment of lignocellulosic biomass.

5.3.2 Biomass Fractionation Using Functionalized Nanoparticles

A catalyst must accelerate a chemical reaction or a physical phenomenon without being modified in the process. In heterogeneous catalysis, the reagents and the catalyst are present in different states, usually the reagents are in the liquid or gaseous phase and the catalyst is solid. These reactions have the advantage that the catalyst can be easily separated from the reaction media to be reused. In homogeneous catalysis systems, the reagents and catalyst are in the same physical state and the catalyst and reagents are in close contact, because they are mixed better. Therefore, the reaction can be catalyzed without mass transfer problems.[87] However, in homogeneous catalysis, it can be difficult to separate the catalyst out of the reaction mixture.[88] Solid acids provide the catalytic properties of homogeneous acids with the advantage that they can be recovered from the reaction media by physical separation.[38,89,90] In previous studies, acid-functionalized mesoporous silicas have been employed for the conversion of cellobiose and starch to glucose.[91,92] Zeolites, AMBERLYST-15, heteropolyacids, and sulfonated activated-carbon have been used for hydrolysis of cellulose into its glucose monomers.[93,94] Similar types of solid acids were used in the hydrolysis of hemicellulose and the recovery of xylose and arabinose sugars.[95] Dispersions of functionalized nanoparticles have similar catalytic performances to homogeneous catalysts as well as the recovery advantages of heterogeneous catalysts.[6,96] Because of their size, acid-functionalized nanoparticles could provide effective catalysis over a solid substrate such as biomass, and because they are magnetic, the nanoparticles

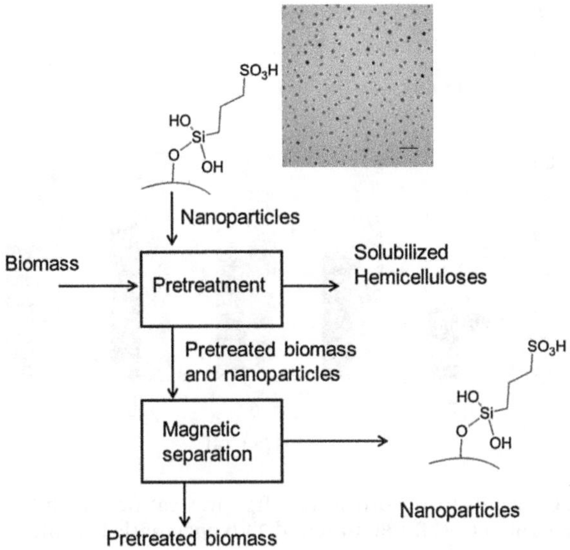

Figure 5.7 Scheme of biomass pretreatment using acid-functionalized nanoparticles. Modified from Peña *et al.*, ref. 41.

can be recovered from the reaction media using a strong magnetic field.[8,33,34] Figure 5.7 shows a scheme of the biomass pretreatment with acid-functionalized nanoparticles. The insert shows the TEM image of propyl-sulfonic acid-functionalized nanoparticles, the scale bar represents 100 nm.

In an attempt to solubilize hemicellulose sugars from wheat straw at low temperatures, alkylsulfonic and perflurosulfonic acid nanoparticles were used to catalyze xylan and arabinan at 80 °C for 24 h. The hemicellulose conversion was 24% and 9% compared to 7% of the control without catalyst. A subsequent experiment used 160 °C for 2 h; the fraction of hemicelluloses recovered increased to 45–46% for 0.25% catalysts compared to 35% for the control with no catalyst load. Biomass hydrolysis using acid-functionalized nanoparticles at these low temperatures solubilized hemicelluloses to oligomers of C5 sugars and not to the monomeric form of the sugars.[40]

The effect of degree of functionalization (measured as %S) of the propyl-sulfonic acid-functionalized nanoparticles (PS) on corn stover pretreatment and their ability to increase biomass susceptibility to an enzymatic hydrolysis is shown in Figure 5.8. Pretreatment was carried out at 160 °C for 60 min and 0.5% catalyst. After pretreatment, approximately 2.5 g of wet biomass was incubated in acetate buffer at pH 4.8 with a cellulose cocktail at 50 °C for 24 h. After enzymatic hydrolysis, the solutions were analyzed by HPLC to account for glucose and xylose. The control experiments had no catalyst and were carried out under the same conditions as experiments using PS nanoparticles. The results showed that the sugar yield after enzymatic hydrolysis increased as the degree of functionalization (measured as %S) of the

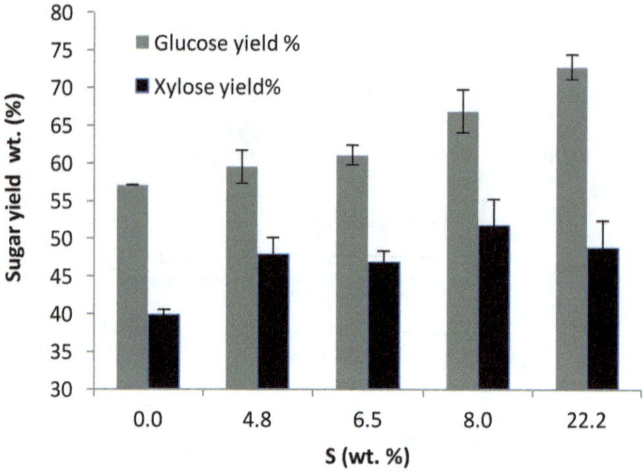

Figure 5.8 Sugar yields from corn stover after pretreatment with 1 g of PS nanoparticles at 160 °C for 60 min and 24 h enzymatic hydrolysis.

nanoparticles increased. Biomass pretreatment at 160 °C with nanoparticles that had 22%S gave sugar yields of 73% glucose and 52% xylose after enzymatic hydrolysis, which is 16% higher in glucose yield and 12% higher in xylose yield than that from the control. The effect of degree of functionalization (measured as %S) of the propylsulfonic acid-functionalized nanoparticles on the hydrolysis of cellobiose was also significant; the hydrolysis reached a maximum of 96% with nanoparticles with 6–8% sulfur content which was equivalent to 0.06% sulfuric acid solutions.[97]

In another set of experiments with lower catalyst concentrations, a catalyst load of 0.2 g nps g^{-1} biomass, which was equivalent to acid solutions of 0.05 mM H^+, and higher temperatures of 160, 180 and 200 °C were used for biomass hydrolysis.[41] The sum of the sugars released during both pretreatment and enzymatic hydrolysis was reported as total glucose and total xylose yields. These experiments showed that there is a stronger effect from the pretreatment temperature compared to that of the catalyst on the total glucose yield after enzymatic hydrolysis (Figure 5.9A). The total glucose yield was modeled as a function of pretreatment temperature and load of PS-nanoparticles per gram of biomass. According to the model 80% glucose yield can be obtained between 176 and 184 °C using 30 to 0 mg S g^{-1} biomass. Maximum glucose yields can be obtained at 200 °C at any catalyst level but the total xylose yields are minimized at this temperature. Hemicelluloses are degraded to hydroxymethylfurfural, furfural and formic acid when biomass was pretreated at 200 °C. A maximum xylose yield of 50% can be achieved at between 160 and 186 °C at any catalyst level (Figure 5.9B). Therefore, a pretreatment temperature set at 184 °C together with a catalyst load of 30 mg S per gram of biomass, which is equivalent to 24 mmol H^+ per liter of slurry, are required to maximize both total glucose and xylose yields.

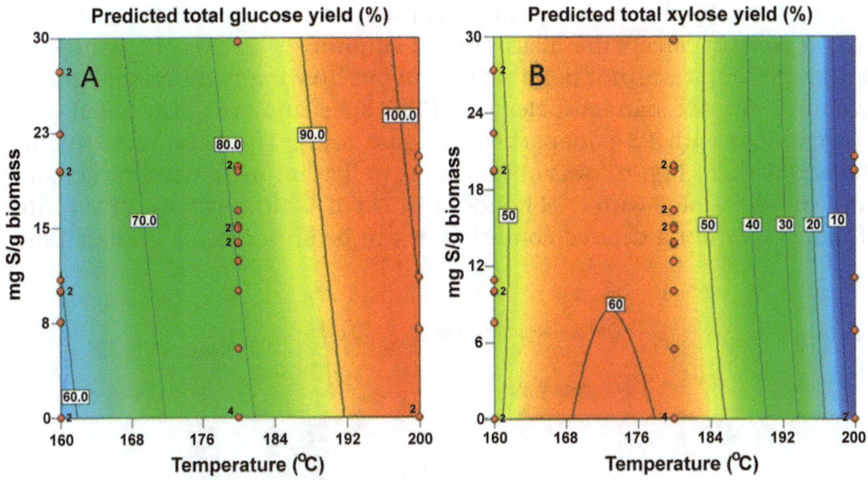

Figure 5.9 Glucose yield (A) and xylose yield (B) after pretreatment with PS nanoparticles and 24 h enzymatic hydrolysis with cellulose.

Beside pretreatment experiments, PS nanoparticles were also evaluated for their ability to hydrolyze the β-1-4 glycosidic bond of cellobiose.[97] The performance of PS nanoparticles in the hydrolysis of cellobiose was tested at 175 °C for 30 min and 2.5 g PS nps g^{-1} cellobiose. PS nanoparticles with an average sulfur content of 6% hydrolyzed 96% of the initial amount of cellobiose, which was significantly higher than the conversion of the control without catalyst (32.8%) (Figure 5.10). Cellobiose conversion is not specific to glucose; HMF and possibly levulinic and formic acids were also generated during the hydrolysis. Nanoparticles with a small number of sulfonic acids could not convert cellobiose probably due to a catalytic site deficiency. On the other hand, nanoparticles with a very large degree of functionalization (18.6% S) did not perform as well as the nanoparticles with a medium number of sulfonic groups (6.1%); this may be due to the hydrophobicity of the nanoparticles with a large propylsulfonic acid content which are hard to wet in aqueous sugar solutions. PS nanoparticles with 6.1% S were used over a total of 10 cycles for the hydrolysis of cellobiose at 175 °C. The sulfur content which is an indicator of the number of sulfonic groups remained the same at 10.6%.

5.3.3 Recyclability of Acid-Functionalized Magnetic Nanoparticles

Elemental analysis was also performed on PS nanoparticles after biomass hydrolysis to confirm that the functional groups did not leach out during pretreatment. Despite physical separation, which was used to remove the biomass left after pretreatment, biomass remained bound to the nanoparticles as evidenced by the higher organic content of the nanoparticles. Higher carbon content affects the mass ratio of sulfonic groups on the nanoparticles; therefore, it was not possible to find out how many functional groups remained on the nanoparticles. An FTIR spectrum was also taken after pretreatment with PS nanoparticles (Figure 5.11). The band at 1510 cm^{-1} characteristic of lignin[98] was observed, as well as an increment in the band intensity associated with C–H bonds at 1457 cm^{-1}. Although, the absorption of material is not a desired condition, it can be an indicator of the affinity

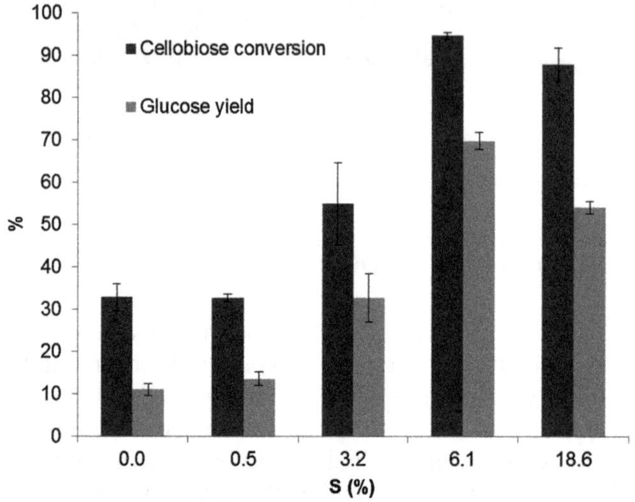

Figure 5.10 Cellobiose conversion and glucose yield during hydrolysis using propylsulfonic acid nanoparticles.

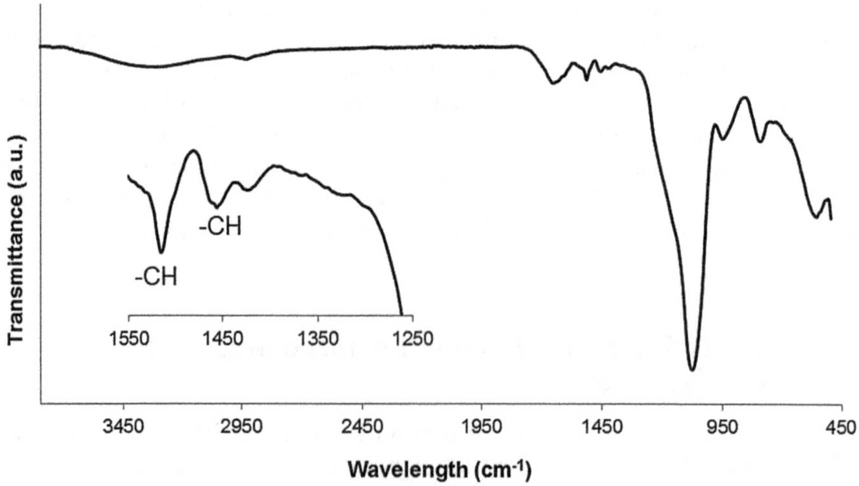

Figure 5.11 FT-IR spectra of PS nanoparticles after pretreatment of corn stover. With permission from ref. 41.

of these nanoparticles towards cellulose-derived materials, necessary for effective catalysis. Reports also showed that propylsulfonic acid-functionalized nanoparticles were recycled after one cycle reaction in non-aqueous solvents.[34] In other studies, sulfonic acid groups linked to the surface of iron oxide nanoparticles retained their catalytic activity after 17 cycles of sucrose conversion to glucose.[99]

5.3.4 Recyclable Enzymes for Biomass Hydrolysis

The use of enzymes as industrial catalysts is attractive because of their specificity and effectiveness. A hundred percent of the reagents involved in the reaction are converted to the final product, and without side-product generation. Enzymes are very complex molecules that decrease the activation energy for a specific reaction. Enzymes are produced *via* biotechnological processes; industrial scale production often requires highly developed processes. The use of top technologies for the industrial production of enzymes increases their price ostensibly. In many cases, catalysis of industrial reactions using enzymes is non-feasible. Therefore less efficient, synthetic catalysts are more commonly used because they are more affordable. Immobilization of enzymes has been proposed to confer these biocatalysts with recoverability attributes.

A great variety of materials has been used for immobilization of enzymes. Supports from natural and synthetic sources are available. Agarose, dextran, and succinates are examples of the natural supports popular because of their biocompatibility.[100,101] A large number of reports are available on the immobilization of enzymes on synthetic supports such as polymers, mesoporous silica, silicon oxide nanoparticles.[102–105] Moreover, supports with hydrophilic or hydrophobic surface functional groups have been developed to host enzyme systems;[106] amine or carboxylic acid groups are typically the functional groups employed to attach enzymes covalently.[107]

Alternative supports for enzyme immobilization are nanoparticles. Magnetic nanoparticles could be good enzyme carriers because, if well dispersed, nanoparticles slurries behave like a homogeneous solution.[108] Dispersability is required during the reaction, while post-reaction the enzyme–nanoparticle complex needs to be separated out of the solution by decantation or applying a magnetic field. Stability of the bio-nanocatalyst is important to keep the enzyme attached to the solid carrier during their recovery from the reaction media.

Cellulases are catalysts used in the derivatization of cellulose; they are a group of synergistic enzymes including endo-1,4-β-D-glucanases, exo-1,4-β-glucosidase and β-D-glucosidase. These enzymes break down cellulose following different mechanisms. Endoglucanases and exoglucosidases convert the cellulose polymer to shorter chains of glucose. Continuous action of these two enzymes would generate cellobiose molecules which are finally converted to glucose by β-D-glucosidase. Unfortunately, their cost limits the feasibility of cellulose-derived bioproducts. Because of current cellulase prices, enzymatic hydrolysis is the most expensive step in obtaining alcohol from biomasses. Consequently, some efforts have been placed on cellulase immobilization aiming at its recoverability in order to reduce the cost of employing enzymes.[43,100,101,109]

Siliceous materials have been used to support cellulases.[42,43] The enzymes are supported on the silica surface by chemical linkage to functional groups or by physical adsorption on the silica pores.[106] Mesoporous silica has been synthesized with various organosilane precursors: 3-aminopropyltrimethoxysilane (APTMS), 3-aminopropyltriethoxysilane (APTES), 3-mercaptopropyltrimethoxysilane (MPTMS), vinyltrimethoxysilane (VTMS), and phenyltrimethoxysilane (PTMS). Typically, immobilization procedures are carried out at either room temperature[105,110] or low temperatures.[43,111] After incubation, the immobilized enzymes can be rinsed with a buffer and separated magnetically to remove free cellulase. The amount of enzyme retained in the support is calculated from the difference in enzyme concentration before and after incubation: 19.1 and 17.6 mg cellulose g^{-1} on large-pore ordered-mesoporous silica functionalized with APTES and VTMS groups, respectively;[42] while 17–250 mg cellulose g^{-1} chitosan were reported using a chitosan support[112] and 23.4 mg cellulose g^{-1} on large-pore mesoporous silica functionalized with APTMS.[113] Chitosan coated iron oxide nanoparticles supported 112 mg cellulose g^{-1} and retained up to 50% activity after ten hydrolysis cycles.[114] Although cellulase loses, in most cases, close to 50% of their activity, immobilization improves their thermal and pH stability.[114] Amino-functionalized silica coated iron oxide particles were used to immobilize cross-linked cellulose aggregates. After incubation for 24, 48, and 72 h at 65 °C, the immobilized enzyme retained 74%, 68% and 61% of its initial activity while the free enzyme only kept 43%, 31% and 26%. Moreover, free cellulase lost 100% of its relative enzymatic activity at pH over 4.8 while the immobilized enzyme retained 45% of its initial activity.[115]

5.4 Conclusions

Acid-functionalized nanoparticles are still in an early stage of development. There are stability issues that must be resolved in order to be recycled for a considerable number of cycles; these nanoparticles leach sulfonic acid moieties during reaction. There are two problems with leaching: (1) The nanoparticles cannot be recycled for a considerable number of cycles, and (2) The catalysis may be present, but in the homogeneous phase, which defeats the purpose of using acid-functionalized nanoparticles instead of traditional mineral acids.

Acid-functionalized nanoparticles have a small density of acid groups because most of their weight is in the magnetic core (0.1 mmol H^+ equivalents g^{-1} catalyst) while sulfuric acid has a density of 0.02 mol H^+ equivalents g^{-1} H_2SO_4. Therefore, the mass of nanoparticles is 200-fold that of sulfuric acid to provide a solution with the same hydronium concentration. Increasing the number of H^+ equivalents per unit mass is necessary for this technology to outperform homogeneous mineral acids.

Recyclability of cellulases could advance the processability of lignocellulosic biomass. However, with the currently available immobilization technologies, the enzyme activity is largely affected. Finally, synergistic effects of nanoscale supports and enzymes could enhance the enzyme activity on solid substrates, such as biomass.

References

1. R. D. Ambashta and M. Sillanpaa, *J. Hazard. Mater.*, 2010, **180**, 38–49.
2. T. Hirakawa, S. Tanaka, N. Usuki, H. Kanzaki, M. Kishimoto and M. Kitamura, *Eur. J. Org. Chem.*, 2009, 789–792.
3. G. Rothenberg, *Nat. Chem.*, 2010, **2**, 9–10.
4. C. A. Witham, W. Y. Huang, C. K. Tsung, J. N. Kuhn, G. A. Somorjai and F. D. Toste, *Nat. Chem.*, 2010, **2**, 36–41.
5. K. Lee, J. G. Na, J. Y. Seo, T. S. Shim, B. Kim, R. Praveenkumar, J. Y. Park, Y. K. Oh and S. G. Jeon, *ACS Appl. Mater. Interfaces*, 2015, **7**, 18336–18343.
6. V. Polshettiwar and R. Varma, *Green Chem.*, 2010, **12**, 743–754.
7. J. Fan and Y. Gao, *Section Title: Catalysis, Reaction Kinetics, and Inorganic Reaction Mechanisms*, 2006, vol. 1, pp. 457–475.
8. P. D. Stevens, J. Fan, H. M. R. Gardimalla, M. Yen and Y. Gao, *Section Title: Benzene, Its Derivatives, and Condensed Benzenoid Compounds*, 2005, vol. 7, pp. 2085–2088.
9. T.-J. Yoon, W. Lee, Y.-S. Oh and J.-K. Lee, *Section Title: Industrial Organic Chemicals, Leather, Fats, and Waxes*, 2003, vol. 27, pp. 227–229.
10. Y. Zheng, C. Duanmu and Y. Gao, *Section Title: General Organic Chemistry*, 2006, vol. 8, pp. 3215–3217.
11. A. S. Teja and P. Y. Koh, *Prog. Cryst. Growth Charact. Mater.*, 2009, **55**, 22–45.
12. Q. Chen and Z. J. Zhang, *Appl. Phys. Lett.*, 1998, **73**, 3156.
13. J. Chatterjee, Y. Haik and C.-j. Chen, *J. Magn. Magn. Mater.*, 2003, **257**, 113–118.
14. Y. Deng, D. Qi, C. Deng, X. Zhang and D. Zhao, *J. Am. Chem. Soc.*, 2008, **130**, 28–29.
15. A. Kumar and M. Gupta, *Biomaterials*, 2005, **26**, 3995–4021.
16. B. Issa, I. M. Obaidat, B. A. Albiss and Y. Haik, *Int. J. Mol. Sci.*, 2013, **14**, 21266–21305.
17. A. Kabelitz, A. Guilherme, M. Joester, U. Reinholz, M. Radtke, R. Bienert, K. Schulz, R. Schmack, R. Kraehnert and F. Emmerling, *CrystEngComm*, 2015, **17**, 8463–8470.
18. T. Hosono, H. Takahashi, A. Fujita, R. J. Joseyphus, K. Tohji and B. Jeyadevan, *J. Magn. Magn. Mater.*, 2009, **321**, 3019–3023.
19. D. Maity, S.-g. Choo, J. Yi, D. Jun and M. X. Jun, *J. Magn. Magn. Mater.*, 2009, **321**, 1256–1259.
20. W. Wu, Q. G. He and C. Z. Jiang, *Nanoscale Res. Lett.*, 2008, **3**, 397–415.
21. Y. Mizukoshi, T. Shuto, N. Masahashi and S. Tanabe, *Ultrason. Sonochem.*, 2009, **16**, 525–531.

22. S. Alibeigi and M. R. Vaezi, *Chem. Eng. Technol.*, 2008, **31**, 1591–1596.
23. N. Mahmed, O. Heczko, A. Lancok and S. P. Hannula, *J. Magn. Magn. Mater.*, 2014, **353**, 15–22.
24. J. Connolly, T. G. St Pierre, M. Rutnakornpituk and J. S. Riffle, *J. Phys. D: Appl. Phys.*, 2004, **37**, 2475–2482.
25. A. J. Rondinone, A. C. S. Samia and Z. J. Zhang, *J. Phys. Chem. B*, 1999, **103**, 6876–6880.
26. P. Guardia, B. Batlle-Brugal, A. G. Roca, O. Iglesias, M. P. Morales, C. J. Serna, A. Labarta and X. Batlle, *J. Magn. Magn. Mater.*, 2007, **316**, E756–E759.
27. J. Salado, M. Insausti, I. G. de Muro, L. Lezama and T. Rojo, *J. Non-Cryst. Solids*, 2008, **354**, 5207–5209.
28. N. T. S. Phan and C. W. Jones, *J. Mol. Catal. A: Chem.*, 2006, **253**, 123–131.
29. L. Zhang, S. Qiao, Y. Jin, H. Yang, S. Budihartono, F. Stahr, Z. Yan, X. Wang, Z. Hao and G. Q. Lu, *Adv. Funct. Mater.*, 2008, **18**, 3203–3212.
30. K. Mooney, J. Nelson and M. Wagner, *Chem. Mater.*, 2004, **16**, 3155–3161.
31. C. I. Covaliu, I. Jitaru, G. Paraschiv, E. Vasile, S. S. Biris, L. Diamandescu, V. Ionita and H. Iovu, *Powder Technol.*, 2013, **237**, 415–426.
32. P. S. Haddad, E. L. Duarte, M. S. Baptista, G. F. Goya, C. A. P. Leite and R. Itri, *Prog. Colloid Polym. Sci.*, 2004, **128**, 232–238.
33. M. Zhang, B. L. Cushing and C. J. O'Connor, *Nanotechnology*, 2008, **19**, 1–5.
34. C. S. Gill, B. A. Price and C. W. Jones, *Section Title: Catalysis, Reaction Kinetics, and Inorganic Reaction Mechanisms*, 2007, vol. 251, pp. 145–152.
35. A. Corma and P. Serna, *Section Title: Industrial Organic Chemicals, Leather, Fats, and Waxes*, 2006, vol. 313, pp. 332–334.
36. T. Asefa and Z. M. Tao, *Can. J. Chem.*, 2012, **90**, 1015–1031.
37. E. Cano Serrano, J. M. Campos Martin and J. L. G. Fierro, *Chem. Commun.*, 2003, 246–247.
38. A. Corma and H. Garcia, *Section Title: General Organic Chemistry*, 2006, vol. 348, pp. 1391–1412.
39. I. Diaz, C. Marquez-Alvarez, F. Mohino, J. Perez-Pariente and E. Sastre, *J. Catal.*, 2000, **193**, 283–294.
40. L. Peña, M. Ikenberry, K. L. Hohn and D. Wang, *J. Biomater. Nanobiotechnol.*, 2012, **3**, 342–352.
41. L. Peña, F. Xu, K. Hohn and D. Wang, Propyl-Sulfonic Acid Functionalized Nanoparticles as Catalyst for Pretreatment of Corn Stover, *J. Biomater. Nanobiotechnol.*, 2014, **5**, 8–16.
42. S. Hartono, S. Qiao, J. Liu, K. Jack, B. Ladewig and Z. Hao, *J. Phys. Chem. C*, 2010, **114**, 8353–8362.
43. M. Kitaoka, H. Taniguchi and T. Sasaki, *J. Ferment. Bioeng.*, 1989, **67**, 182–185.
44. S. Shylesh, A. Wagner, A. Seifert, S. Ernst and W. Thiel, *Chem.–Eur. J.*, 2009, **15**, 7052.

45. V. Zelenak, M. Badanicova, D. Halamova, J. Cejka, A. Zukal and N. Murafa, *Chem. Eng. J.*, 2008, **144**, 336–342.
46. B. Rac, A. Molnar, P. Forgo, M. Mohai and I. Bertoti, *J. Mol. Catal. A: Chem.*, 2006, **244**, 46.
47. S. Iborra, A. Corma and G. Huber, *Chem. Rev.*, 2006, **106**, 4044–4098.
48. R. M. Brown, *J. Polym. Sci., Part A: Polym. Chem.*, 2003, **42**, 487–495.
49. M. M. Chang, T. Y. C. Chou, G. T. Tsao and A. Fiechter, in *Bioenergy*, Springer Berlin Heidelberg, 1981, vol. 20, pp. 15–42.
50. A. C. O'Sullivan, *Cellulose*, 1997, **4**, 173–207.
51. D. Fengel, *J. Polym. Sci., Part C*, 1971, 383.
52. B. Yang and C. E. Wyman, *Section Title: Fermentation and Bioindustrial Chemistry*, 2004, vol. 86, pp. 88–95.
53. H. Ooshima, D. S. Burns and A. O. Converse, *Biotechnol. Bioeng.*, 1990, **36**, 446–452.
54. J. Salvado and N. El Mansouri, *Ind. Crops Prod.*, 2006, **24**, 8.
55. A. Effendi, H. Gerhauser and A. Bridgwater, *Renewable Sustainable Energy Rev.*, 2008, **12**, 2092.
56. S.-H. Lee, Y. Teramoto and T. Endo, *Bioresour. Technol.*, 2010, **101**, 769–774.
57. S.-H. Lee, S. Inoue, Y. Teramoto and T. Endo, *Bioresour. Technol.*, 2010, **101**, 9645–9649.
58. S. Banerjee, R. Sen, R. A. Pandey, T. Chakrabarti and D. Satpute, *Biomass Bioenergy*, 2009, **33**, 1680–1686.
59. A. T. W. M. Hendriks and G. Zeeman, *Bioresour. Technol.*, 2009, **100**, 10–18.
60. N. Mosier, C. Wyman, B. Dale, R. Elander, Y. Y. Lee, M. Holtzapple and M. Ladisch, *Section Title: Cellulose, Lignin, Paper, and Other Wood Products*, 2005, vol. 96, pp. 673–686.
61. J. Kristensen, L. Thygesen, C. Felby, H. Jorgensen and T. Elder, *Biotechnol. Biofuels*, 2008, **1**, 5.
62. D. Y. Corredor, X. S. Sun, J. M. Salazar, K. L. Hohn and D. Wang, *J. Biobased Mater. Bioenergy*, 2008, **2**, 43–50.
63. L. Ramos, *Quim. Nova*, 2003, **26**, 863.
64. C. G. Liu and C. E. Wyman, *Bioresour. Technol.*, 2005, **96**, 1978.
65. S. G. Allen, D. Schulman, J. Lichwa, M. J. Antal and E. Jennings, *Ind. Eng. Chem. Res.*, 2001, **40**, 2352–2361.
66. T. Lloyd and C. Wyman, *Bioresour. Technol.*, 2005, **96**, 1967–1977.
67. B. Dien, H.-J. Jung, K. Vogel and M. Casler, *Biomass Bioenergy*, 2006, **30**, 880–891.
68. C. Sievers, T. Marzialetti, T. J. C. Hoskins, M. B. Valenzuela Olarte, P. Agrawal and C. Jones, *Bioresour. Technol.*, 2009, **100**, 4758–4765.
69. C. Liu and C. E. Wyman, *Section Title: Food and Feed Chemistry*, 2004, vol. 43, pp. 2781–2788.
70. T. Marzialetti, C. Sievers and P. Agrawal, *Ind. Eng. Chem. Res.*, 2008, **47**, 7131–7140.

71. R. Torget and T. Hsu, in *Applied Biochemistry and Biotechnology*, ed. N. J. Clifton, Humana Press, 1994, vol. 45–46, pp. 5–21.
72. J. Kristensen, J. Borjesson, M. Bruun, F. Tjerneld and H. Jorgensen, *Enzyme Microb. Technol.*, 2007, **40**, 888.
73. J. F. Saeman, *Ind. Eng. Chem.*, 1945, **37**, 43–52.
74. J. D. McMillan, M. E. Himmel, J. O. Baker and R. P. Overend, *Enzymatic Conversion of Biomass for Fuels Production*, American Chemical Society, 1994, vol. 566, pp. 292–294.
75. O. Bobleter, *Prog. Polym. Sci.*, 1994, **19**, 797–841.
76. N. S. Mosier, C. M. Ladisch and M. R. Ladisch, *Biotechnol. Bioeng.*, 2002, **79**, 610–618.
77. M. J. Antal, T. Leesomboon, W. S. Mok and G. N. Richards, *Carbohydr. Res.*, 1991, **217**, 71–85.
78. D. F. Root, J. F. Saeman and J. F. Harris, *Forest Prod. J.*, 1959, **9**, 158–165.
79. E. Palmqvist and B. Hahn Hagerdal, *Bioresour. Technol.*, 2000, **74**, 25–33.
80. Y. Lee and S. Kim, *Bioresour. Technol.*, 2002, **83**, 165–171.
81. E. Viola, F. Nanna, E. Larocca, M. Cardinale, D. Barisano and F. Zimbardi, *Ind. Crops Prod.*, 2007, **26**, 195.
82. K. Cejpek, J. Velisek and O. Novotny, *Czech J. Food Sci.*, 2008, **26**, 113–131.
83. J. Delgenes, *Enzyme Microb. Technol.*, 1996, **19**, 220–225.
84. J. M. Oliva, M. J. Negro and F. Saez, *Process Biochem.*, 2006, **41**, 1223–1228.
85. J. Shen and C. Wyman, *Bioresour. Technol.*, 2011, **102**, 9111–9120.
86. S. L. Stuhler, *Effects of Solids Concentration, Acetylation, and Transient Heat Transfer on Uncatalyzed Batch Pretreatment of Corn Stover*, Dartmouth College, Hanover, NH, 2002.
87. B. Cornils and W. A. Herrmann, *J. Catal.*, 2003, **216**, 23–31.
88. D. J. Cole-Hamilton, *Science*, 2003, **299**, 1702–1706.
89. M. A. Harmer, Q. Sun, A. J. Vega, W. E. Farneth, A. Heidekum and W. F. Hoelderich, *Green Chem.*, 2000, **2**, 7–14.
90. M. Yurdakoc, M. Akcay, Y. Tonbul and K. Yurdakoc, *Turk. J. Chem.*, 1999, **23**, 319–327.
91. J. A. Bootsma and B. H. Shanks, *Appl. Catal., A*, 2007, **327**, 44–51.
92. P. L. Dhepe, M. Ohashi, S. Inagaki, M. Ichikawa and A. Fukuoka, *Catal. Lett.*, 2005, **102**, 163.
93. A. Onda, T. Ochi and K. Yanagisawa, *Green Chem.*, 2008, **10**, 1033–1037.
94. K. Shimizu, H. Furukawa, N. Kobayashi, Y. Itaya and A. Satsuma, *Green Chem.*, 2009, **11**, 1627–1632.
95. P. Dhepe and R. Sahu, *Green Chem.*, 2010, **12**, 2153–2156.
96. A. Bell, *Science*, 2003, **299**, 1688–1691.
97. L. Peña, K. L. Hohn, J. Li, X. S. Sun and D. Wang, *J. Biomater. Nanobiotechnol.*, 2014, **5**, 241–253.
98. R. J. A. Gosselink, J. E. G. van Dam, C. Boeriu and D. Bravo, *Ind. Crops Prod.*, 2004, **20**, 205–218.
99. M. Ikenberry, L. Pena, D. M. Wei, H. W. Wang, S. H. Bossmann, T. Wilke, D. H. Wang, V. R. Komreddy, D. P. Rillema and K. L. Hohn, *Green Chem.*, 2014, **16**, 836–843.

100. J. Zhou, *J. Agric. Food Chem.*, 2010, **58**, 6741–6746.
101. C. Li, M. Yoshimoto, K. Fukunaga and K. Nakao, *Bioresour. Technol.*, 2007, **98**, 1366–1372.
102. M. S. M. Eldin, M. R. El-Aassar, A. A. E. Zatahry and M. M. B. Al-Sabah, *Adv. Polym. Technol.*, 2014, **33**, 11.
103. K. Ho, X. Mao, L. Gu and P. Li, *Langmuir*, 2008, **24**, 11036–11042.
104. R. Singh, Y. Zhang, N. Nguyen, M. Jeya and J. Lee, *Appl. Microbiol. Biotechnol.*, 2011, **89**, 337.
105. L. Bautista, G. Morales, R. Sanz and F. Bautista, *Bioresour. Technol.*, 2010, **101**, 8541–8548.
106. S. Hirsh, M. Bilek, N. Nosworthy, A. Kondyurin, C. D. Remedios and D. McKenzie, *Langmuir*, 2010, **26**, 14380.
107. S. Kumar, A. K. Jana, M. Maiti and I. Dhamija, *J. Nanopart. Res.*, 2014, **16**, 1–23.
108. L. Matuszewski, *Pol. Marit. Res.*, 2008, **15**, 49.
109. C. I. Simionescu, V. Popa, M. Popa and S. Maxim, *J. Appl. Polym. Sci.*, 1990, **39**, 1837–1846.
110. N. Brun, A. B. Garcia, H. Deleuze and C. Sanchez, *Chem. Mater.*, 2010, **22**, 4555–4562.
111. M. Kumakura and I. Kaetsu, *Process Biochem.*, 1988, **23**, 51–54.
112. X. Mao, G. Guo, J. Huang, Z. Du, Z. Huang and L. Ma, *J. Chem. Technol. Biotechnol.*, 2006, **81**, 189–195.
113. R. Chang, J. Jang and K. Wu, *Green Chem.*, 2011, **13**, 2844–2850.
114. L. Zang, J. Qiu, W. Zhang, E. Sakai and Y. Wei, *Ind. Eng. Chem. Res.*, 2014, **53**, 3448–3454.
115. K. J. Khorshidi, H. Lenjannezhadian, M. Jamalan and M. Zeinali, *J. Chem. Technol. Biotechnol.*, 2016, **91**, 539–546.

CHAPTER 6

Catalytic Enhancement of NMR Sensitivity for Advanced Spectroscopic and Imaging Studies in Catalysis and Life Sciences

K. V. KOVTUNOV[a,b], V. V. ZHIVONITKO[a,b], I. V. SKOVPIN[a,b], O. G. SALNIKOV[a,b] AND I. V. KOPTYUG*[a,b]

[a]International Tomography Center, SB RAS, Laboratory of Magnetic Resonance Microimaging, 3A Institutskaya St., Novosibirsk 630090, Russian Federation; [b]Novosibirsk State University, 2 Pirogova St., Novosibirsk 630090, Russian Federation
*E-mail: koptyug@tomo.nsc.ru

6.1 Introduction

NMR is a powerful and versatile instrument for scientific research and practical applications. An even wider applicability is feasible, provided that NMR sensitivity can be further improved. One of the major factors that limits sensitivity of NMR, and consequently one of the major opportunities for improving it, is related to the fact that in a conventional NMR experiment the polarization of nuclear spins amounts to not more than 100 ppm even in

Smart Materials No. 26
Magnetic Nanomaterials: Applications in Catalysis and Life Sciences
Edited by Stefan H. Bossmann and Hongwang Wang

Published by the Royal Society of Chemistry, www.rsc.org

the strongest static magnetic fields available today for researchers. Polarizing nuclear spins in a sample far beyond the thermal equilibrium level could therefore improve NMR sensitivity by up to 10^4-fold. This, however, will not be achievable in any foreseeable future by means of increasing the magnetic fields of conventional NMR spectrometers. In addition, recent years have witnessed a new trend in NMR applications, namely the introduction of commercial low-field NMR systems, which have proven to be quite useful in a number of applications and allow one to perform NMR analysis virtually anywhere, be it a fume hood in a chemistry lab or a chemical reactor in an industrial environment. For the relatively low magnetic fields of such benchtop/portable/mobile NMR systems which are often based on the use of permanent magnets, nuclear spin polarization levels are even lower (*e.g.*, $\sim 5 \times 10^{-6}$ at 1.4 T for 60 MHz ^{1}H NMR frequency), so that achieving the nuclear polarization on the order of unity would provide an even higher (*e.g.*, $\sim 2 \times 10^5$-fold) sensitivity boost than in the high-field NMR/MRI systems equipped with modern superconducting magnets. It is clear then that the ultimate nuclear spin polarization levels can be achieved only by going beyond the polarizations that thermal equilibrium (Boltzmann statistics) can provide.

It is not surprising, therefore, that in recent years significant attention was directed toward the development and use of nuclear spin hyperpolarization (HP) techniques.[1–6] This family of techniques encompasses several approaches to polarize nuclear spins to the levels that exceed significantly the levels attained at thermal equilibrium when a sample is placed in a strong static magnetic field. Among the members of this HP family, arguably the most advanced and the most versatile is the dynamic nuclear polarization (DNP) technique which can be used to produce HP solids, liquids and even gases. However, DNP is also the most instrumentally sophisticated and demanding technique in the HP family, which makes it the most expensive one in terms of equipment and operational costs. In addition, each of the existing HP techniques has its advantages and limitations, and none is universal in terms of the types of substances that can be hyperpolarized, sample purity and biocompatibility, experimental conditions required, *etc.* These considerations are among the main reasons why alternative techniques are being considered by the broad HP NMR/MRI community.

One of the viable alternatives to DNP is based on the physical and chemical properties of molecular hydrogen (H_2), or, more specifically, on the use of one of its nuclear spin isomers known as parahydrogen. In the latter, the nuclear spins $I = 1/2$ of the two hydrogen atoms are combined in a state with the total nuclear spin $I = 0$. Being spinless, parahydrogen is, in a sense, the "dark matter" of NMR as it does not interact with radiofrequency radiation or produce an observable NMR signal. But when a parahydrogen molecule is activated in certain chemical processes, its $I = 0$ state can be converted to a major polarization of nuclear spins and thus can provide a dramatic enhancement of the NMR signals of the species involved in such processes. Since this signal enhancement is produced by a chemical (*e.g.*, catalytic) reaction, this parahydrogen-induced polarization (PHIP) phenomenon can be a powerful yet unique tool for mechanistic studies in catalysis and other subfields of chemistry. PHIP can also produce HP substances for other applications of NMR/MRI including biomedical and/or *in vivo* applications, but this is where it has to compete hard with its rivals, such as DNP.

6.2 Parahydrogen *vs.* Nuclear Spin Isomers of Other Symmetric Molecules

As a consequence of the Pauli exclusion principle, molecular hydrogen (H_2) and other symmetric molecules with equivalent atoms (*e.g.*, H_2O, H_2CO, CH_4, NH_3, *etc.*) exist as several different species known as nuclear spin isomers of molecules (NSIM) which differ in the value of the total nuclear spin of the equivalent nuclei. In particular, for H_2 the Pauli principle requires that its molecular wave function should be antisymmetric with respect to the permutation of its two identical nuclei since the latter are fermions. For the ground vibronic state, which is the common situation in most practical cases, this implies that molecular states of H_2 allowed by the Pauli principle can only combine a symmetric (resp., antisymmetric) rotational state with an antisymmetric (resp., symmetric) nuclear spin state. Parahydrogen (p-H_2) with its total nuclear spin $I = 0$ has the spin state $(1/2)^{0.5}(|\alpha\beta\rangle - |\beta\alpha\rangle)$ (here, $|\alpha\rangle = |1/2, 1/2\rangle$ and $|\beta\rangle = |1/2, -1/2\rangle$ are the states of the two spins) which is antisymmetric with respect to the permutation of the two hydrogens: $(1/2)^{0.5}(|\beta\alpha\rangle - |\alpha\beta\rangle) = -(1/2)^{0.5}(|\alpha\beta\rangle - |\beta\alpha\rangle)$. Therefore, only the symmetric rotational states (with even rotational quantum numbers $J = 0, 2, \ldots$) are allowed for parahydrogen. The rotational states with odd J values then belong to orthohydrogen (o-H_2) with $I = 1$. Other symmetric molecules can have two or more NSIM; for instance, H_2O, CH_3F and acetylene (C_2H_2) also have two (ortho and para) NSIM, whereas ethylene (C_2H_4) has four different NSIM. The number of NSIM of a symmetric molecule depends on its molecular symmetry group.

Since NSIM differ in the allowed wave functions, they are also different in terms of molecular energies. In particular, the lowest ortho and para states have different rotational states and thus have different rotational (total) energies. This difference may not be apparent at high temperatures where multiple rotational states are significantly populated. For instance, at room temperature normal hydrogen (n-H_2) in thermal equilibrium is a 75 : 25 mixture of its ortho and para forms because the $I = 1$ state of orthohydrogen is triply degenerate. However, at low temperatures the equilibrium will be shifted toward the para modification since parahydrogen can exist in the lowest rotational state with $J = 0$ while orthohydrogen cannot. In this respect, H_2 is unique because the energy gap between the lowest ortho and para states is huge (Figure 6.1a) compared to other symmetric molecules. As a result, the equilibrium o : p ratio in H_2 can be shifted significantly toward its para modification by cooling H_2. Already at liquid nitrogen temperature (77 K) the equilibrium o: p ratio is shifted from 75 : 25 to *ca.* 50 : 50 (Figure 6.2), providing the enrichment which is sufficient for many applications in NMR and

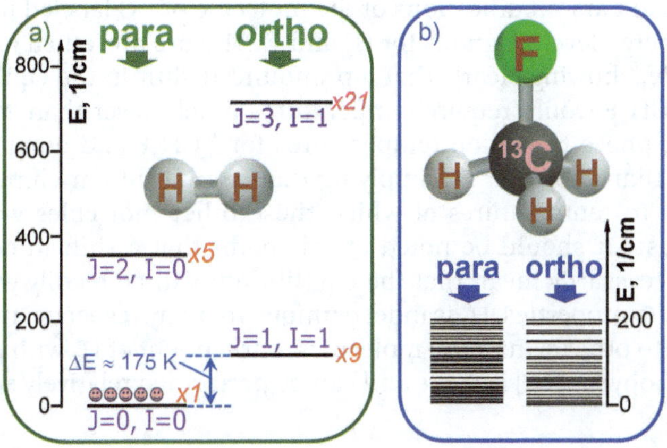

Figure 6.1 (a) Energy level diagram for orthohydrogen and parahydrogen (0–800 cm^{-1} range). The values of nuclear spin (*I*), rotational quantum number (*J*) and degeneracy (*xN*) are shown for each level. At ~20.3 K, the lowest state of p-H_2 is populated almost exclusively (pink circles). (b) Energy diagram for $^{13}CH_3F$ (0–200 cm^{-1} range) shown on the same vertical scale for comparison.

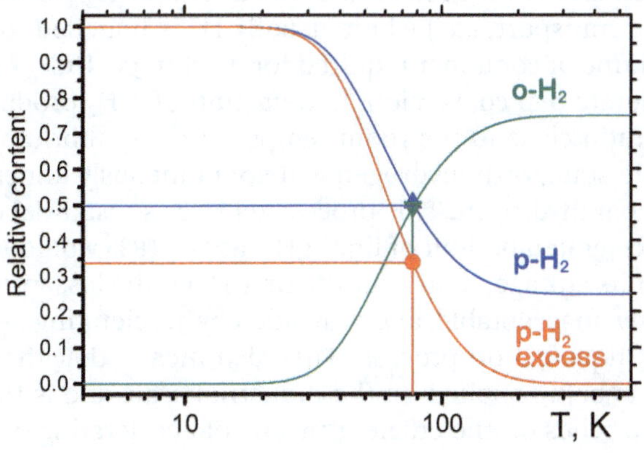

Figure 6.2 The temperature dependence of the relative content of orthohydrogen and parahydrogen in equilibrium, and the corresponding excess of p-H_2 (equal to (4p-1)/3, where p is the relative content of p-H_2). At 77 K (circles), the relative contents of o-H_2 and p-H_2 are very close, and the p-H_2 excess is *ca.* 33%, so that p-H_2 enriched at 77 K gives *ca.* three times lower PHIP effects compared to pure p-H_2 provided that other experimental conditions are kept the same.

catalysis. By cooling H_2 down to its boiling temperature (*ca.* 20.3 K) the equilibrium is shifted further to *ca.* 99.8 : 0.2 o : p ratio, *i.e.*, an equilibrium mixture contains almost pure p-H_2. In contrast, the energy differences between different NSIM of other molecules are dramatically smaller. As an example, Figure 6.1b shows the energy level diagram for the 0–200 cm^{-1} interval of the ortho and para modifications of the molecule of ^{13}C-labeled fluoromethane. The energy level diagrams for H_2 and $^{13}CH_3F$ are presented on the same energy scale, showing clearly that a pronounced shift in the equilibrium of NSIM of $^{13}CH_3F$ would require temperatures much lower than 20 K. At the same time, phase transition temperatures for $^{13}CH_3F$ and other molecules are higher than those for H_2, implying that significant enrichments would correspond to temperatures at which these other molecules would be in a solid phase. It should be noted at this point that a shift in equilibrium does not necessarily mean that the equilibrium will be readily established. Because of its properties, H_2 is indeed unique in many respects. In particular, in contrast to other symmetric molecules, enrichment of H_2 with parahydrogen (or orthohydrogen) is facile and can be performed relatively easily.

6.3 Producing Parahydrogen, and Orthohydrogen–Parahydrogen Interconversion

As stated in the preceding section, thermodynamics provides an opportunity to shift the o–p equilibrium for H_2 toward its para modification by simply cooling H_2, but the enrichment process can be severely limited by its kinetics. This is well-known in industrial applications of H_2 (rocket fuels, fuels for automotive transport, *etc.*) where usually H_2 is liquefied to significantly reduce the volume of container required for its storage. Unless measures are taken to accelerate o–p conversion, liquefaction of n-H_2 produces liquid H_2 with the o:p ratio close to the room temperature equilibrium of 75:25. In this metastable state, orthohydrogen will spontaneously undergo a gradual conversion to parahydrogen. This process releases substantial energy which leads to an accelerated boil-off of liquid H_2: about 18% would be lost during the first day of its storage,[7] and after the first 100 h the loss would be ~40%.[8] This is certainly unacceptable, and is avoided by accelerating o–p conversion during the H_2 liquefaction process. This also means that the p-H_2 enrichment process is being applied on the industrial scale and is thus fairly well developed, and relies on the efficient means of accelerating o–p conversion of H_2 at cryogenic temperatures.

Acceleration of hydrogen o–p conversion for parahydrogen enrichment is where magnetic materials such as magnetic nanoparticles (MNPs) come into play for the first time in the context of parahydrogen applications in catalysis and NMR/MRI. NSIM interconversion is often slow because of the relatively large differences in energy between different NSIM of a molecule in comparison to the strength of the magnetic interactions (*e.g.*, dipolar, spin-rotational, hyperfine, *etc.*) which can interconvert these NSIM. This is particularly true for the interconversion of the ortho and para modifications of H_2, for which the o–p gap is particularly large while the intramolecular dipolar interaction does not have the right symmetry to induce the o–p conversion. Thus, to accelerate the equilibration of H_2 at reduced temperatures, one has to supply efficient magnetic interactions able to flip one of the nuclear spins relative to another along with changing the rotational quantum number of an H_2 molecule. The simple way of looking at this is as follows: for the conversion to be efficient it is generally required that the two H atoms in the H_2 molecule should be briefly made inequivalent; that is, that the two hydrogens of an H_2 molecule should experience different magnetic environments (possibly averaged by the molecular motion). While a uniform magnetic field (local or external) cannot induce o–p conversion, a substantial gradient of a local magnetic interaction provided by the surface of a magnetic material can be quite efficient in this respect. In general, o–p conversion can be induced by any magnetic material, be it ferromagnetic, antiferromagnetic, ferrimagnetic or paramagnetic, with a lot of research reported on comparing the efficiency of different materials and their types. For instance, Svadlenak and Scott compared γ-Fe_2O_3, α-Fe_2O_3, ZnO and a series of iron oxide–zinc oxide mixtures in the o–p conversion of H_2 at 77 K and concluded that ferromagnetic γ-Fe_2O_3 was the most efficient material for o–p conversion while paramagnetic ZnO and precursors of Zn ferrite (paramagnetic with some ferromagnetic character) were the least efficient.[9] The exceptionally high efficiency of ferromagnetic hydrous ferric oxide reported by Weitzel *et al.*[7] disagreed with the previously developed theories of paramagnetic o–p conversion and provided evidence that ferromagnetic materials may indeed be much more efficient in o–p conversion of H_2 compared to paramagnetic ones. Similarly and independently, Buyanov has found that metal hydroxide gels with Cr^{3+}, Mn^{4+}, Fe^{3+}, Co^{3+} and Ni^{2+} ions are very efficient in o–p conversion of H_2.[10]

However, the notion of a magnetic particle inducing an o–p conversion by producing a magnetic field gradient is certainly and drastically oversimplified. The actual physics behind the o–p conversion process is a lot more sophisticated, and a realistic description should account for the energy and momentum exchange required for the o–p conversion, and should be based on a detailed consideration of the combined H_2-surface system, including the states of the electrons and nuclei of both H_2 and the surface, and the dynamics of their interactions with each other and with the lattice of the solid material.[8] Some indications that point to the process being intrinsically more sophisticated come from the observations of fairly efficient o–p conversions on certain diamagnetic materials including pretreated Al_2O_3 and various diamagnetic metals, *e.g.*, Cu and Ag surfaces.

While o–p interconversion may still take place on essentially diamagnetic materials, the absence of magnetic interactions comparable in energy to the o–p energy gap often makes this process extremely inefficient. As a result, other processes start to dominate, and interaction of H_2 with the surfaces of diamagnetic particles often leads to very different consequences. In particular, because of the different molecular properties of o-H_2 and p-H_2, the strength of adsorption of ortho and para modifications of H_2 on solid surfaces is different, with o-H_2 adsorbing more strongly on materials such as Al_2O_3, SiO_2, *etc.* This makes it possible to separate o-H_2 and p-H_2 chromatographically. At the same time, such separation only works when the residence time of molecules adsorbed on the surface is shorter compared to the rate of o–p conversion on this surface. The latter can be induced by the paramagnetic impurities in the oxide material, paramagnetic centers existing on the oxide surface (*e.g.*, surface defects and oxygen vacancies) and possibly even by the nuclear spins (*e.g.*, ^{27}Al for Al_2O_3[11]), and depends crucially on the metal oxide preparation and pretreatment. Therefore, in certain cases the paramagnetic o–p conversion and chromatographic o–p separation can compete with each other. Nevertheless, chromatographic separation of o-H_2 and p-H_2 is a well-established phenomenon, which has been used to produce H_2 mixtures enriched with o-H_2. This possibility was important in particular in the early days of exploration of the PHIP effect.[12,13] Indeed, after the theoretical prediction[14] and experimental demonstration[15] of the PHIP effect by Bowers and Weitekamp, it was recognized that it had in fact been observed earlier than its official discovery but misinterpreted as originating from a different (CIDNP) hyperpolarization mechanism.[16] When o-H_2 is used instead of p-H_2 in PHIP experiments, the observed enhanced signals are expected to change their sign, which was indeed observed experimentally and helped to distinguish it from other known signal enhancement mechanisms. However, production of p-H_2 is more robust and provides a three-fold stronger signal enhancement compared to o-H_2.

There is another mechanism of o–p conversion of H_2, with particular relevance to catalysis. Molecular hydrogen is readily chemisorbed on the surfaces of many catalytic solids. In particular, on metals H_2 is often activated dissociatively.[17–19] Combined with a high mobility of hydrogens on many metal surfaces (the H atom jumps between neighboring metal atoms once every 10^{-9} s or so[20,21]), this leads to a rapid randomization of hydrogens on the surface so that recombination of two random H atoms and the subsequent desorption of the newly formed H_2 molecule leads to an efficient o–p interconversion of H_2 mixtures when they are brought into contact with such materials. In addition, such chemical dissociation of H_2 can induce rapid o–p interconversion even in the absence of chemical exchange, *i.e.*, if the surface diffusion of H atoms is comparatively slow (*e.g.*, on metal oxides as compared to metals) and the two nascent H atoms recombine to form the original H_2 molecule again. This is because nuclear spin relaxation of surface H atoms can be relatively rapid once the H_2 molecule dissociates. Rapid o–p interconversion of molecular hydrogen upon its chemisorption was used already in the 1930s to gain mechanistic information on catalytic processes on metal surfaces,[22] *i.e.*, long before the NMR was even discovered. This chemical o–p conversion mechanism is important to take into account when using p-H_2 in the PHIP experiments as it operates in parallel with the PHIP production and reduces the actual p-H_2 fraction in the enriched H_2 gas used in the experiments, and thus reduces the levels of the observed PHIP effects.

On nominally diamagnetic materials such as alumina,[23] it is possible to have both chemical and magnetic o–p conversion mechanisms operating simultaneously, with their relative contributions depending on experimental conditions such as temperature, material pretreatment, *etc.* Chemical exchange processes are clearly operative on alumina surfaces, especially at higher temperatures (*e.g.*, above 90 K[23]). Evacuation or gamma-irradiation of alumina creates centers that are active in o–p conversion and may affect the efficiency of both mechanisms since the centers created can be both paramagnetic and catalytically active. The involvement of the chemical conversion mechanism is usually demonstrated in the H–D equilibration experiments using H_2 + D_2 mixtures and observing the formation of HD which can only be formed chemically *via* a number of mechanisms. Both o–p conversion of H_2 and the production of HD in H_2 + D_2 mixtures was reported by Matthes *et al.*[24] for Vaska's complex $Ir(CO)Cl(PPh_3)_2$, both as a frozen solution in C_6D_6 and as a solid powder, even at 77 K. The authors concluded that the chemical mechanism of o–p conversion was more important than the magnetic one even at such a low temperature. The efficient contact of H_2 with the Ir complex in frozen solution was explained as a result of complex segregation and formation of gas-filled pores upon sample freezing. These results shed more light on the observation of PHIP effects in early studies where they were not recognized as such, because frozen samples containing metal complexes and H_2 were stored at liquid N_2 temperatures prior to performing the NMR experiments.[16,25]

The equilibration of H_2 at reduced temperatures may be additionally accelerated if, in addition to efficient magnetic interactions, the surface interacting with the H_2 molecule is able to provide the means to reduce the o–p energy gap. The reduction of the o–p energy separation can be achieved if the rotation of the molecule is hindered due to its interaction with a solid surface. In particular, physisorption and chemisorption of H_2 on surfaces of solid materials can switch the regime from free to hindered rotation or even to replace it with quantum tunneling with a dramatically reduced o–p energy gap. In addition, interaction with a surface of a magnetic particle can provide a magnetic interaction which is strong enough to induce rapid o–p interconversion under conditions of the reduced energy gap for H_2 molecule physisorbed or chemisorbed on this surface.[26,27] At the same time, it should be stressed that the reduction of the o–p energy separation for a surface H_2 molecule implies that surface o–p equilibrium can be rather different from that in the gas phase at the same temperature. For instance, Limbach *et al.* suggested that even a relatively weak interaction of an H_2 molecule with a surface upon physisorption can reduce the energy gap from *ca.* 10^{12} to *ca.* 10^9 Hz.[26] This is expected to shift the o–p equilibrium dramatically, so that an appreciable enrichment of p-H_2 would require temperatures well below 4 K, while for higher temperatures the equilibration on the surface would produce a 75:25 o–p mixture, *i.e.*, the o:p ratio corresponding to normal hydrogen. A more conservative estimate of an o–p energy gap reduction upon physisorption of H_2 on alumina was provided by Hall[11] who described a model in which an H_2 molecule adsorbed on alumina interacts with the local electric field provided by the positively charged Al and negatively charged O atoms of the surface. This was shown to create a barrier to rotation and to reduce the o–p gap from *ca.* 175 to 29 cal mol^{-1}. This six-fold o–p gap reduction leads to essentially the same conclusion: that a pronounced p-H_2 enrichment on the surface cannot be achieved at 77 K or even at 20 K. At the same time, in practice when H_2 is brought into contact with a magnetic material at a reduced temperature, p-H_2 enrichment corresponding to the gas phase o–p equilibrium does take place. This implies that the enrichment in the gas phase is accomplished in the final step of the process, namely upon desorption of H_2 from the solid surface. The probability of desorption is governed by the Boltzmann factor and should be proportional to the probability for an H_2 molecule to gather the energy required to desorb into a state allowed by the Pauli principle. Thus, the spin statistics can be very different on the surface and in the gas phase which is in direct contact with this surface. The role of the magnetic surface is then not to establish the o:p ratio that corresponds to a gas phase equilibrium at a given temperature, but rather to convert H_2 to a statistical 75:25 o–p mixture, which then equilibrates with the gas phase conditions upon desorption. If this picture is indeed valid, it could then suggest ways to manipulate the o:p ratio in the gas phase, such as affecting the desorption probability by supplying external energy for the desorption process (*e.g.*, laser ablation) or manipulating spin statistics (*e.g.*, by polarizing nuclear spins of the surface species). For developing a complete and detailed model of o–p interconversion, it is likely that one has to take into account the differences in adsorption of o-H_2 and p-H_2 in addition to their differential desorption from a solid surface.

In any case, when H_2 is brought into contact with a magnetic material at a reduced temperature, the enrichment of H_2 with p-H_2 proceeds quite rapidly. The materials conventionally used for the practical o–p conversion of H_2 include hydrous ferric oxide [FeO(OH)], chromia on alumina, activated charcoal, *etc.* The process can be implemented either in a batch mode or even by continuously flowing H_2 through the o–p converter. Depending on the required degree of p-H_2 enrichment, the apparatus may be a simple coiled metal tubing containing a packed bed of paramagnetic particles and immersed in liquid nitrogen, or a more sophisticated commercial or home-built device assembled around a closed-circuit cryostat that can go down to *ca.* 20 K (or 15 K if conversion is to be performed on liquid H_2). While the latter approach produces an almost pure p-H_2, the enrichment at 77 K yields a mixture of H_2 NSIM comprising *ca.* 77% normal H_2 (with o:p = 75:25) and *ca.* 33% of pure p-H_2. This means that the fraction of p-H_2 can be increased further only by a factor of three by going from this mixture obtained at 77 K to pure p-H_2 obtained near 20 K. Consequently, this will provide only a factor of three gain in the PHIP signal enhancements. We have also observed that a noticeable p-H_2 enrichment can be obtained if an H_2 tank is kept outdoors in the conditions of a Siberian winter (*e.g.*, 240 K). The enrichment in this case likely takes place upon interaction of H_2 with the walls of the steel container.

Interestingly, the efficiency of o–p conversion over magnetic particles is significantly higher at reduced temperatures than at, *e.g.*, room temperature. It was reported that when parahydrogen was passed over charcoal at room temperature, no conversion took place.[28] We observed similar behavior for FeO(OH) when using a liquid nitrogen converter containing a granular bed of this catalyst for a continuous production of hydrogen gas enriched with p-H_2. For the loaded amount of FeO(OH), a gas flow rate at which *ca.* 50% of p-H_2 was produced reliably, as confirmed by ^{1}H NMR of the flowing H_2 gas, was established. After that, an insert with the same amount of the FeO(OH) catalyst was placed in the H_2 gas flow path downstream of the converter, and the H_2 flow was reestablished with the same flow rate. This insert was maintained at room temperature, and the original presumption was that it would convert the enriched p-H_2 back to n-H_2 as the two volumes of the catalyst (one at 77 K and another at RT) were the same, and so were the gas flow rates of H_2 through these two catalyst beds. It was found, however, that the presence of the catalyst bed at RT had no effect whatsoever, and the modified setup was still producing the gas with *ca.* 50% p-H_2. Assuming that the magnetic properties of the catalysts are the same at different temperatures, this drastic reduction of o–p conversion efficiency is possibly explained by the difference in residence times of H_2 molecules in the packed beds and the durations of the contacts of H_2 with the magnetic surface: at RT both events should be much shorter in time compared to cryogenic temperatures, which would reduce the efficiency of o–p conversion dramatically at RT.

The slow spontaneous o–p conversion of H_2 may be a bit of a nuisance for the enrichment of p-H_2, but it becomes a major advantage once the enriched p-H_2 is produced, because its back conversion into n-H_2 is slow as long as it is no longer in contact with a magnetic material (magnetic particles, container walls, oxygen, *etc.*). As a result, liquid and gaseous hydrogen enriched in p-H_2 can survive for days and even months in the enriched form. Therefore, it is easy to handle enriched p-H_2 while preparing and performing PHIP experiments. In particular, back conversion of p-H_2 in a commercial aluminum cylinder was reported to be an exponential process with a characteristic time constant of *ca.* 64 days.[29] Back conversion can also be accelerated while p-H_2 is being used in the PHIP experiments, for instance in the gas phase due to the presence of traces of O_2,[30] or in the liquid phase in the presence of paramagnetic species (metal complexes,[31] nitroxides,[32] *etc.*). Back conversion of p-H_2 can proceed upon its chemical activation of transition metal complexes in solution since oxidative addition of H_2 to the metal center is reversible, while conversion can be initiated as soon as the chemical bond in the H_2 molecule is broken.[33] At the same time, back conversion of p-H_2 is often significantly retarded on metal complexes and supported metals in the presence of an unsaturated compound which can undergo hydrogenation on these catalysts.

It would be very useful to employ NSIM of molecules other than H_2 for signal enhancement in chemical reactions. Unfortunately, as mentioned above (see Figure 6.1), thermodynamics of those molecules makes it very difficult or impossible to achieve NSIM enrichment by reducing temperature. So far, the cryogenic and chromatographic enrichment processes described above could be applied successfully only to H_2 and its isotopologues. Therefore, applications utilizing NSIM of other molecules have to await efficient methods for NSIM enrichment of such molecules. Two approaches of relevance to

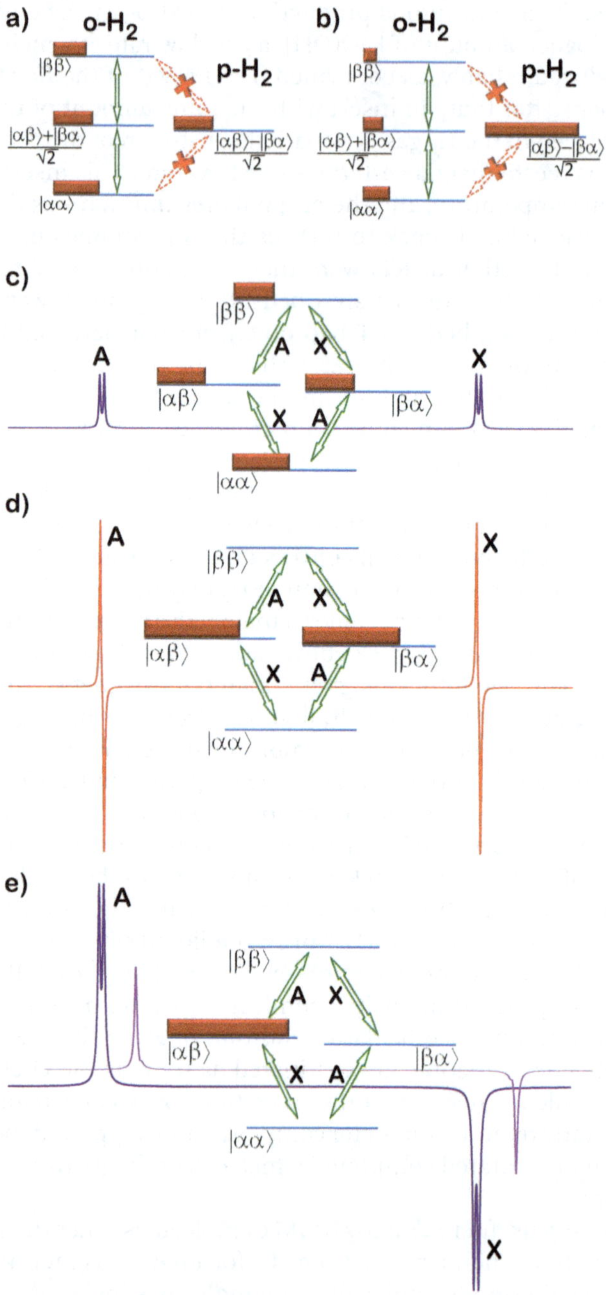

Figure 6.3 (a) Energy level diagram of the ^{1}H nuclear spins of H_2 and the spin wave functions for the strongly coupled A_2 spin system. The length of each bar symbolizes the relative population of the respective spin state under thermal equilibrium. Vertical arrows show the allowed NMR transitions for o-H_2 which coincide in frequency and thus result in a single line in the ^{1}H NMR spectrum of H_2. NMR transitions between o-H_2 and p-H_2

this work have been demonstrated recently. One is based on the use of parahydrogen to produce enriched NSIM of ethylene in the catalytic reaction of parahydrogen with acetylene.[34] Another uses DNP to polarize nuclear spins of a molecule (*e.g.*, ethylene) to alter the relative amounts of NSIM of ethylene.[35] Both approaches have been demonstrated at the proof-of-principle levels, but practical enrichment processes are yet to be developed based on these observations.

6.4 Parahydrogen-Induced Polarization: Enhancing NMR Signals Using Parahydrogen

As mentioned in the Introduction, parahydrogen can be used in certain chemical reactions to dramatically enhance NMR signals of the species involved.[15,36–41] This is usually achieved in catalytic transformations using catalysts that can chemically activate molecular hydrogen, in particular in H_2 activation by transition metal complexes[36,37,42] and metal-free frustrated Lewis pairs,[43–45] and in catalytic hydrogenations.[36,37,41,46]

There are several conditions that need to be met to observe PHIP effects: (i) upon activation of H_2, both hydrogen atoms of one H_2 molecule should end up in the same reaction intermediate and/or product molecule, *i.e.*, the addition of H_2 should be pairwise; (ii) in the reaction product or intermediate, these two hydrogens should not be identical any longer, either chemically, or at least magnetically; (iii) the two H atoms in the product/intermediate should be close to each other so that they retain a non-zero mutual spin-spin coupling that can be resolved in the NMR spectrum under experimental conditions; (iv) the processes involving H_2 activation and transformation should be faster than the nuclear spin relaxation processes that act to bring the nuclear spins to thermal equilibrium once the H_2 molecule is broken.

In the NMR terminology, H_2 represents an A_2 system—a strongly coupled spin system of two chemically and magnetically equivalent nuclei. The energy level diagram of the nuclear spin states of H_2 is shown in Figure 6.3a. Only the transitions belonging to o-H_2 (with $I = 1$) give a conventional (thermal

---

are forbidden so that p-H_2 is NMR silent. (b) Same as (a), but after partial enrichment of H_2 with p-H_2. (c) Energy level diagram and the spin wave functions of a weakly coupled AX spin system under thermal equilibrium. Also shown is the corresponding ^{1}H NMR spectrum. (d) Energy level diagram, the wave functions and the ^{1}H NMR spectrum of an AX spin system formed from p-H_2 right after its incorporation into a reaction product in a high magnetic field (PASADENA). The NMR signals are maximized if a 45°-pulse is used and are close to zero for a 90°-pulse. (e) Same as (d), but for an ALTADENA experiment (hydrogenation in the Earth's field with subsequent adiabatic sample transfer to NMR magnet for spectrum acquisition). The bottom spectrum corresponds to a 90°-pulse; the top one corresponds to a short (<<90°) rf detection pulse. The spectra in (c)–(e) are not shown on the same vertical scale.

equilibrium) NMR signal, while p-H_2 is NMR silent. Enrichment of an H_2 mixture with p-H_2 does not change this situation but only decreases the thermal equilibrium signal of o-H_2 as its concentration in the sample becomes smaller (Figure 6.3b). Imagine now that a chemical reaction converts the A_2 spin system of H_2 to a weakly coupled AX spin system in a reaction product, *i.e.*, the two H atoms now have different chemical shifts, with this difference in frequency units significantly exceeding their non-zero spin–spin coupling in the product molecule. The energy level diagram for this AX spin system is shown in Figure 6.3d. Nuclear spin states, being associated with good quantum numbers,[47] are often quite rigorously conserved in rapid chemical processes. As the spin wave function of p-H_2 is $(1/2)^{0.5}(|\alpha\beta\rangle-|\beta\alpha\rangle)$, *i.e.*, is composed of the $|\alpha\beta\rangle$ and $|\beta\alpha\rangle$ states, only these states of the AX spin system will be populated after reaction, whereas the other two will be empty (see Figure 6.3d). Such spin level population distribution corresponds to a dramatic enhancement of the NMR signals of this AX spin system because the population differences for the allowed transitions are now on the order of the total number of nuclear spins in the sample, whereas under thermal equilibrium (Figure 6.3c) these differences are a Boltzmann factor smaller (*e.g.*, 10^{-4}–10^{-5} for ^{1}H nuclei in a magnetic field of an NMR spectrometer). Figure 6.3d also shows how the spectrum of this non-equilibrium AX spin system is expected to look: in addition to a large signal enhancement it is also characterized by an unusual emission/absorption pattern of each multiplet which is readily distinguishable from the conventional NMR spectra. Such antiphase multiplets are normally observed when the reaction with parahydrogen is performed inside an NMR probe in the high magnetic field of an NMR spectrometer (PASADENA[14,15]). Alternatively, if the reaction is carried out in a low (*e.g.*, Earth's) magnetic field and the sample is then adiabatically transferred to an NMR instrument for observation, the populations of nuclear spin sublevels and the resulting spectra are quite different (ALTADENA,[48] Figure 6.3e), but the observed signals are still readily distinguishable from the multiplet patterns in conventional NMR spectra. While the use of p-H_2 naturally produces the hyperpolarization on the ^{1}H nuclei of the reaction products, it can be also transferred to other types of magnetic nuclei either spontaneously or by manipulating the magnetic fields experienced by the nuclei involved.

Because of the unusual shapes of NMR lines and particularly because of a major signal enhancement, PHIP effect provides a highly sensitive NMR-based tool which can be very useful in studying the mechanisms and kinetics of reactions that involve catalytic activation of H_2, including homogeneous hydrogenation reactions catalyzed by transition metal complexes in solution.[36,49,50] In addition to the detection of reaction products formed upon the pairwise addition of p-H_2 to a substrate, the high sensitivity provided by PHIP allows one to detect the short-lived reaction intermediates that are present at very low concentrations and thus are normally beyond the NMR detection limit. Their detection becomes possible because the symmetry of an H_2 molecule entering a catalytic cycle is often broken once reaction intermediates such as metal dihydride complexes are formed, accompanied by a major enhancement of their NMR signals (Figure 6.4). PHIP has already proven to be a powerful tool in the studies of hydrogenations and related processes in homogeneous solutions, but an extension of applications of this hypersensitive mechanistic tool to heterogeneous catalytic processes is clearly highly desirable and can be very informative in such studies. Furthermore, combination of PHIP with MRI makes it possible to study processes in model catalytic reactors and microreactors,[51,52] with one example of such applications shown in Figure 6.5.

However, potential applications of PHIP are certainly not limited to catalysis. In particular, combining the major signal enhancement provided by PHIP with biomedical MRI and MRS research currently attracts major attention.[1,53–60] The very first study in which PHIP was used to enhance MR images *in vivo* exploited the homogeneous hydrogenation of acetylenedicarboxylic acid dimethyl ester with parahydrogen catalyzed by a cationic rhodium complex in acetone.[53] The reaction produced maleic acid dimethyl ester with PHIP of the hydrogen nuclei. However, the hyperpolarization usually decays exponentially in time with a characteristic time constant T_1, the nuclear spin relaxation time of the hyperpolarized nuclei. After PHIP is generated, for ^{1}H nuclei this provides an observation window of no more than 10 seconds or even less, which is not sufficient for *in vivo* studies. This is because preparation and handling of an injectable sample and the delivery of the hyperpolarized substance to the region of interest in a lab animal takes longer than a

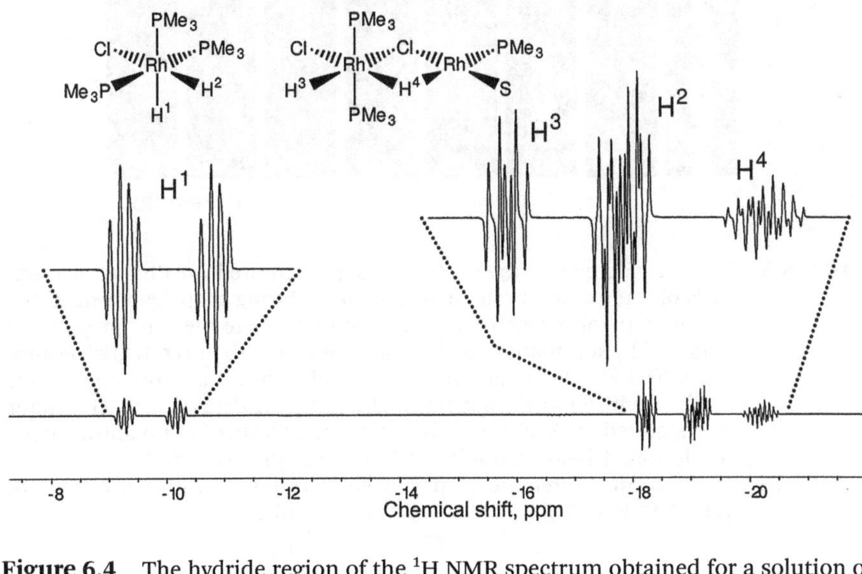

Figure 6.4 The hydride region of the ^{1}H NMR spectrum obtained for a solution of $[Rh(NBD)Cl]_2$ and PMe_3 in acetone-d6 after bubbling with p-H_2 at 320 K. The spectrum shows the hydride resonances of the complexes Rh(H)-$_2$Cl(PMe_3)$_3$ and (H)(Cl)Rh(PMe_3)$_2$(μ-Cl) (μ-H)Rh(PMe_3)S (S is acetone-d6, NBD is norborna-2,5-diene). Reproduced with permission from ref. 46 Copyright 2001 American Chemical Society.

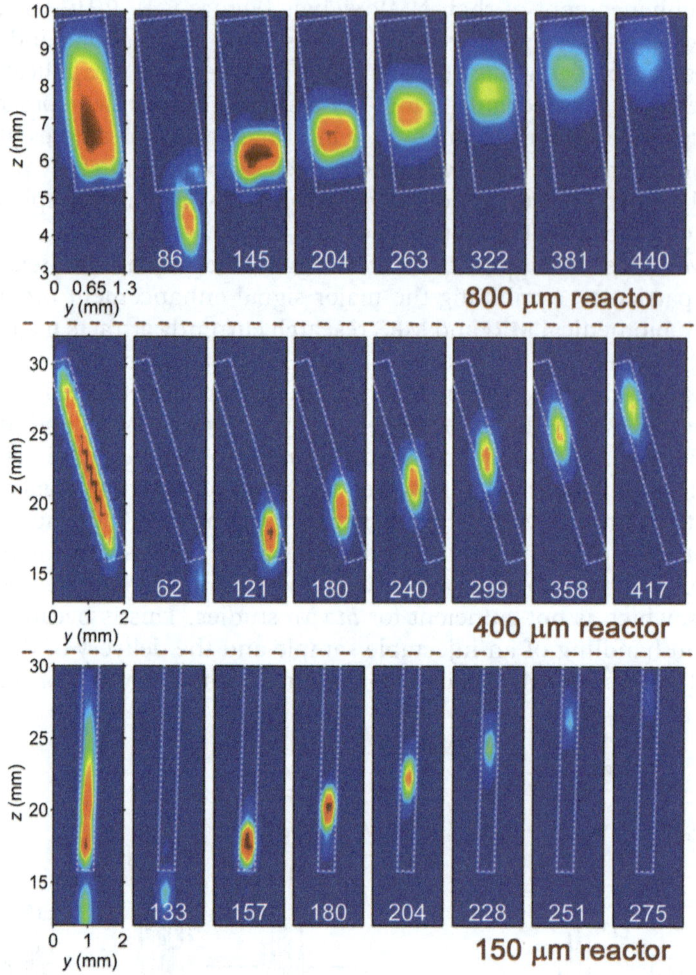

Figure 6.5 Remotely detected time-of-flight images acquired for Rh/SiO_2 catalyst beds of three different diameters at 60 °C during propylene hydrogenation with parahydrogen. The leftmost images are the time projections obtained by summation of all images shown to their right. Travel times are indicated in the panels in milliseconds. The catalyst bed regions are outlined with white dashed lines. The complete data set for each reactor was acquired in 13 minutes with a time resolution of 12 ms and a spatial resolution of 160–250 μm in the horizontal direction and 0.62–2.2 mm in the vertical direction. Reproduced from ref. 52 with permission of WILEY-VCH Verlag GmbH & Co. KGaA, Weinheim.

few seconds, resulting in a complete loss of hyperpolarization of ^1H nuclei by the time the measurement can be performed. For this reason, the polarization of ^1H nuclei in maleic acid dimethyl ester was transferred to ^{13}C nuclei using magnetic field cycling. After that, the reaction solution containing maleic acid dimethyl ester with hyperpolarized ^{13}C nuclei was injected into the tail vein of a rat, and an angiographic ^{13}C MR image of the blood vessels was acquired in less than a second.[53] This and other ^{13}C-labeled hyperpolarized substances were used in later *in vivo* angiographic and perfusion MRI studies.[1,54,58]

An even more promising and challenging avenue for *in vivo* applications of hyperpolarized magnetic resonance is that of the *in vivo* studies of metabolism.[4,56,61–64] The idea behind these applications is that if a natural metabolite is hyperpolarized and then injected into the blood stream, it will reach the target organ or tissue, get transported into the cells with a normal or pathological metabolism, and will enter the metabolic (*e.g.*, TCA) cycle. Provided that the hyperpolarization lives long enough (*e.g.*, for ^{13}C or other heteronuclei which have T_1 times much longer than those of ^1H), the administered metabolite can get converted to other metabolites before nuclear spins return to the thermal equilibrium, and thus the product metabolites will retain some of the original hyperpolarization. This not only boosts the detection sensitivity for *in vivo* MRI/MRS of various metabolites, but also makes it possible to study fluxes of metabolites through various enzymes *in vivo*, which is normally not possible under the steady-state conditions maintained by a living system. As a result, an early diagnosis of various diseases including, but not limited to, cancer, becomes possible, and so does the detection of an early response to treatment of various pathologies.

In the context of molecular imaging of cellular metabolism based on the use of PHIP, several hyperpolarized metabolites have been considered.[56,58,59,65] For instance, conversion of hyperpolarized [1-^{13}C]succinate into hyperpolarized [1-^{13}C]malate, [4-^{13}C]malate, [1-^{13}C]fumarate, [1-^{13}C]glutamate and [6-^{13}C]citrate has been detected in cancer models in mice.[58] The spatial localization of hyperpolarized molecules in the animal after injection is visualized by performing spatially resolved ^{13}C MRS experiments. Hyperpolarized diethyl [2,3-D_2, 1-^{13}C]succinate was used for real-time *in vivo* studies of several steps of the TCA cycle.[59] After its injection into normal mice, a significant fraction of the hyperpolarization of ^{13}C nuclei was shown to be retained through three or more enzymatic reactions, leading to the *in vivo* detection of hyperpolarized malate, succinate, fumarate, and aspartate. These and other proof-of-principle studies demonstrate the significant potential of *in vivo* MRI/MRS based on the utilization of PHIP to produce hyperpolarized metabolites.

Currently, the instrumental implementation of PHIP is based on a homogeneous hydrogenation of unsaturated precursors using transition metal complexes as catalysts.[54,66–70] The entire process of hydrogenation followed by ^1H to ^{13}C polarization transfer can be done in under five seconds, providing polarization levels of up to 15–50% for ^{13}C NMR, corresponding to signal enhancements of more than 10^5 at 1.5 T. However, any further viable extension of PHIP-based hyperpolarization to biomedical research poses additional major challenges. The solution of a HP agent for *in vivo* injection needs to be biocompatible and non-toxic. This most likely means that hydrogenation of the precursor molecule with p-H_2 needs to be performed in an aqueous medium. This is because the lifetime of hyperpolarization (a few to a few tens of seconds depending on conditions and the type of polarized nucleus) is too short to attempt an extraction of an HP substance from an organic reaction medium with subsequent dissolution in water/saline. Another important challenge that one faces is that of rapidly removing the dissolved metal complex catalyst from the solution after the reaction is complete. While it is possible in principle to remove certain metal complexes using, *e.g.*, functionalized ion-exchange resins, the lack of a reliable procedure for bringing the metal complex catalyst concentration down within seconds to the acceptable levels of toxicity is still a major obstacle for initiating in-human trials.[71]

6.5 Parahydrogen-Induced Polarization with Heterogeneous Catalysts

The problem of rapid and efficient catalyst removal after the reaction that was raised in the previous section is well-known in industrial catalysis, and numerous solutions have been developed over the decades of research to achieve this. Most such approaches are based on combining in various ways the advantages of homogeneous and heterogeneous catalysis, using in particular a variety of strategies to turn a suitable homogeneous catalyst into its heterogeneous counterpart.[72–74] However, for a long time the PHIP effects were associated with homogeneous catalysis only, and were even considered as evidence for the homogeneous nature of the underlying chemistry. Notably, even an observation of PHIP effects with colloidal hydrogenation catalysts[75] did not alter the common perception that PHIP is a prerequisite of homogeneous catalysis.

The situation started to change rapidly after the first observation of PHIP effects using heterogeneous catalysts was reported in 2007.[76] This study relied on the use of Wilkinson's catalyst, $RhCl(PPh_3)_3$, immobilized on silica gel or a polymer. Hydrogenation of unsaturated hydrocarbons with p-H_2 using these catalysts produced the PHIP effects in the NMR spectra of their products, both for liquid (styrene to ethylbenzene) and gas phase (propylene to propane) hydrogenations. The experiments with propylene hydrogenation in the gas phase were key to proving beyond doubt that the PHIP effects were indeed observed in a heterogeneous process, because in liquid phase hydrogenations it is notoriously difficult to completely exclude any metal complex leaching into solution. At the same time, it had not been fully realized at that point that in long-term gas phase hydrogenations the immobilized metal complex, while being kept at an elevated temperature in an atmosphere rich in H_2, may be transforming into a supported metal catalyst comprising metal nanoparticles on a solid support. Therefore, some of those experiments may have been the very first examples of the use of nanoparticles for PHIP production. This possibility was not considered by the authors at that time, largely because metal nanoparticles that may form in the reduction of supported metal complexes were not expected to produce any PHIP effects. The observations were in line with an apparently straightforward (but in the end not so obvious) presumption that the pairwise oxidative addition–reductive elimination mechanism of the hydrogenation with metal complexes would remain unchanged upon metal complex immobilization. It should be stressed that even if reduction of metal complexes did take place in some cases, this does not alter the main conclusion made by the authors[76] that the PHIP effects were for the first time observed in heterogeneous hydrogenations.

The demonstration of PHIP with authentic heterogeneous catalysts comprising metal (Pt and Pd) nanoparticles supported on γ-Al_2O_3 followed just a year later.[77] This was quite unexpected, because in contrast to metal complexes, hydrogenation on metal surfaces is expected to proceed *via* dissociative chemisorption of H_2 (the Horiuti–Polanyi mechanism[17,78]). The subsequent random diffusion of surface H atoms is very rapid, with a time between jumps from one surface site to the neighboring one of, *e.g.*, 10^{-9} s for Pd surfaces.[20,21] This was expected to be incompatible with the main requirement for PHIP observation, namely the pairwise addition of H_2 to an unsaturated substrate. However, experimental demonstrations of PHIP effects with supported metal nanoparticle catalysts clearly show that the reality once again is proven to be a lot more complex than originally anticipated.

In subsequent studies, observation of PHIP effects was successfully extended to the hydrogenation of various unsaturated hydrocarbons over many other supported metal catalysts, which now include Pt, Pd, Rh, Ir, Ru, Au and Ag nanoparticles supported on various supports (SiO_2, ZrO_2, Al_2O_3, TiO_2, CeO_2, activated carbon, multiwalled carbon nanotubes, aluminosilicate fiberglass, *etc.*).[39,40,79–85] In the gas–solid hydrogenation of propylene over Pt/γ-Al_2O_3 catalysts with different Pt nanoparticle sizes,[86] the largest PHIP effects were observed for the catalyst with the smallest (<1 nm) Pt nanoparticles; in increasing the Pt particle size the PHIP effects almost disappeared for 2–4 nm Pt nanoparticles and then increased again for sizes up to 12 nm, but did not reach the level observed for the smallest Pt nanoparticles (Figure 6.6). Hydrogenation of propylene over Rh/TiO_2 catalysts with different Rh nanoparticle sizes demonstrated essentially the same behavior.[87] The Rh/TiO_2 catalyst with 1 wt% metal loading was shown to produce stronger PHIP effects than that with 5 wt% metal loading. HRTEM was used to establish that for the 5 wt% sample the average Rh nanoparticle size was *ca.* 1 nm, whereas for the 1 wt% sample the particles were too small to see them. Yet, the combination of EDX and XPS methods confirmed the presence of metallic Rh on the surface of both catalysts. In contrast, for propylene hydrogenation over Pd/TiO_2 the PHIP effects increased with increasing nanoparticle size from 0.7 nm to 1.5 nm to 11 nm,[88] while propyne hydrogenation over Pd/TiO_2 demonstrated an even more complex behavior, with the two products, propylene and propane, showing opposing trends with the variation of nanoparticle sizes.

The observed PHIP effects also depend on the metal and support type. For instance, hydrogenation of propylene over Pd nanoparticles supported on

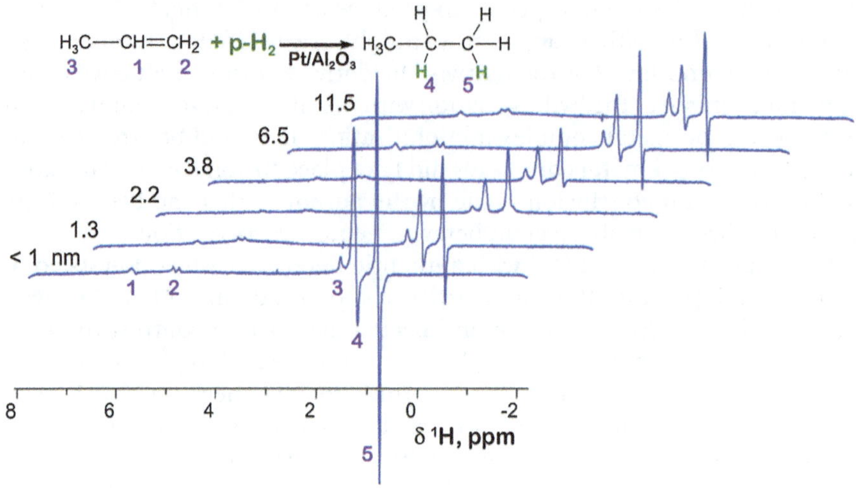

Figure 6.6 ^{1}H NMR spectra acquired during hydrogenation of propylene with parahydrogen over the Pt/γ-Al_2O_3 catalysts with different metal particle sizes. The signals of propane labeled 4 and 5 exhibit pronounced PASADENA PHIP effects in most of the spectra. Adapted with permission from ref. 87 Copyright 2011 American Chemical Society.

ZrO_2, SiO_2, and γ-Al_2O_3 produced propane without any PHIP. Hydrogenation of propyne also gave some propane without any PHIP, but propylene produced in the same reaction was hyperpolarized.[88] A different behavior was observed for Pd/TiO_2 catalyst which produced PHIP effects for all products, both in propyne and propylene hydrogenations. Similar trends were observed in the hydrogenation of 1,3-butadiene over supported Pd catalysts.[39] The fact that PHIP effects are only observed for the products of partial hydrogenation of alkynes and dienes for most supported Pd catalysts may be related to the different reaction mechanisms, *e.g.*, the participation of the surface and subsurface hydrogens in the hydrogenations. In contrast, for supported Pt nanoparticles PHIP effects are usually observed for all products in the hydrogenation of both propyne and propylene independent of the oxide support used. Supported metal catalysts were also shown to produce PHIP effects in liquid–solid hydrogenations of numerous alkynes and alkenes, including hydrogenation of dissolved propylene,[39,89] styrene,[89] methyl propiolate[90] in organic solvents and acrylamide[89] and allyl methyl ether in D_2O. At the same time, PHIP effects were also successfully observed for bulk unsupported Pt black and Pd black catalysts as well as their oxides[91] and unsupported mercapto-succinic acid-capped Pt nanoparticles.[92]

Metal nanoparticles supported on TiO_2 often produce larger PHIP effects compared to the same metals on other oxide supports.[39,86,88,93] For instance, Figure 6.7 shows that for Pt nanoparticles supported on four different oxide supports, Pt supported on TiO_2 gives the largest PHIP effects in the hydrogenation of propylene. This may be explained in terms of the strong

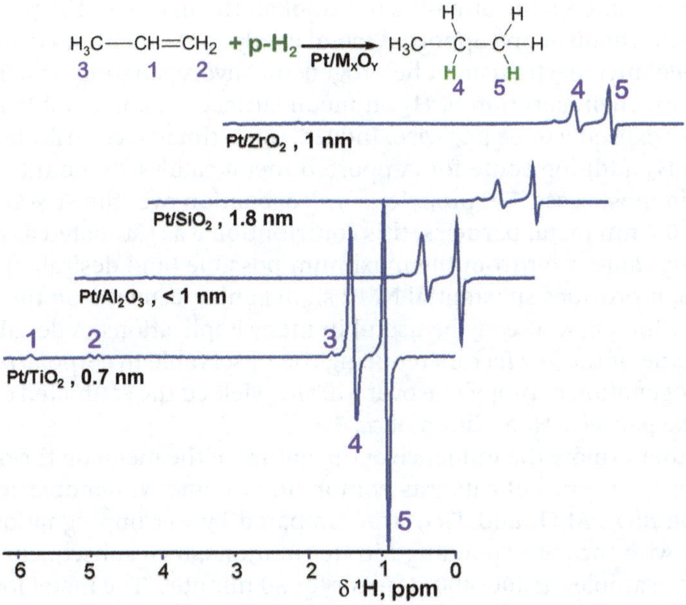

Figure 6.7 ^{1}H NMR spectra obtained during hydrogenation of propylene with parahydrogen for Pt metal catalysts supported on different oxides. The reaction was performed at 345–350 K. The PASADENA polarization signals correspond to the protons of the CH_2 (4) and CH_3 (5) groups of propane. The signals from the CH_3, CH_2 and CH groups of propylene are labeled as 3, 2 and 1, respectively; these signals show no hyperpolarization. The sizes of Pt nanoparticles are indicated in the figure and correspond to the maximum PHIP effect for the respective oxide.

metal-support interaction (SMSI) effect, which was recently studied in the context of PHIP effects for Pd/TiO_2 catalysts.[94] In particular, it was found that PHIP effects in the hydrogenation of 1,3-butadiene, while pronounced for catalysts reduced at 100 °C in H_2, almost completely disappeared for the catalysts reduced at 500 °C. At the same time, the associated hydrogenation activity of the catalysts dropped by 20–30% only for the same reduction temperatures. Interestingly, *in situ* XPS measurements revealed formation of $Pd^{\delta+}$ state after reduction at 500 °C. SMSI effects were also studied for metal-oxide-supported Ir nanoparticles by Zhao *et al.*[95] who reported a ~20-fold increase in the pairwise selectivity following a reduction treatment of the Ir/TiO_2 catalyst at 500 °C.

Overall, the results reported in the literature demonstrate that PHIP effects show a very complex and not easily explicable dependence on numerous factors such as the type of metal and metal particle size, the type of support, the nature of substrate, the reaction conditions, *etc.* Despite the availability of a large body of experimental data, it has so far proven impossible to find unifying empirical correlations that could reveal the nature of active centers responsible for PHIP production with supported metal catalysts.

This further hinders the attempts to establish the nature of the pairwise H_2 addition mechanism on supported metal catalysts. As mentioned above, the Horiuti–Polanyi mechanism of heterogeneous hydrogenation which implies dissociative chemisorption of H_2 on metal surfaces is not suitable for PHIP as it is not expected to be pairwise. Indeed, the estimated contribution of the pairwise H_2 addition route for supported metal catalysts amounts to a few percent in most cases. For propylene hydrogenation over the Pt/γ-Al_2O_3 catalyst with 0.6 nm metal particles, this contribution was estimated as *ca.* 3%.[77] While this value is far from the maximum possible (and desirable) value of *ca.* 100%, it provides substantial NMR signal enhancements on the order of 100-fold which may already be useful in many applications. A detailed analysis of some of the key factors reducing the observable hyperpolarization in the hydrogenation of propylene over Rh/TiO_2 yielded the estimated contribution of the pairwise H_2 addition of *ca.* 7%.[93]

To further explore the influence of the nature of the metal on the observed PHIP effects, a series of catalysts comprising Co and Ni nanoparticles supported on SiO_2, Al_2O_3 and TiO_2 were prepared by wet impregnation of the supports with the corresponding nitrate solutions, with subsequent calcination of the samples at 300–400 °C in air for 30 minutes. The metal loading in these catalysts was *ca.* 5 wt%. In the hydrogenation of propylene at 500 °C, a Co/SiO_2 catalyst showed low activity and the PHIP effects were barely observable. The Co/Al_2O_3 catalyst was more active at the same conditions, and PHIP effects, while not very large, were clearly observed (Figure 6.8a). Interestingly, the Co/TiO_2 catalyst was active even at a lower temperature of 400 °C, and also demonstrated many more substantial PHIP effects for propane (Figure 6.8b,c). The pairwise selectivity for propylene hydrogenation over Co/TiO_2 at 400–500 °C can be estimated as 2–3% even without introducing corrections for polarization losses caused by spin relaxation and incomplete adiabaticity of the sample transfer from the Earth's field to the probe on the NMR spectrometer in the ALTADENA experiment; applying this correction[93] brings the pairwise selectivity in this case to the remarkable level of 19–25%. This high pairwise selectivity may be tentatively attributed to the very small size of Co nanoparticles in the Co/TiO_2 catalyst, which were barely detectable by TEM and thus are smaller than 0.5 nm in diameter. In contrast, the supported Ni catalysts in general demonstrated substantially higher conversions in the hydrogenation of propylene to propane, which decreased significantly when the temperature was increased from 400 to 500 °C. Ni catalysts also exhibited the tendency to catalyze hydrocarbon cracking as a side reaction, leading to the formation of the pronounced amounts of methane and ethane. At the same time, while PHIP effects were observed with Ni/Al_2O_3 and Ni/TiO_2 catalysts, the pairwise selectivity was much lower compared to the Co catalysts, and for the Ni/SiO_2 catalyst no PHIP effects could be observed.

The observation of PHIP effects with magnetic nanoparticles is quite remarkable in itself. While no attempts have been made so far to characterize the properties of metal nanoparticles used in PHIP experiments, some tentative conclusions can be made at this stage. Nanoparticles of Pt and Pd

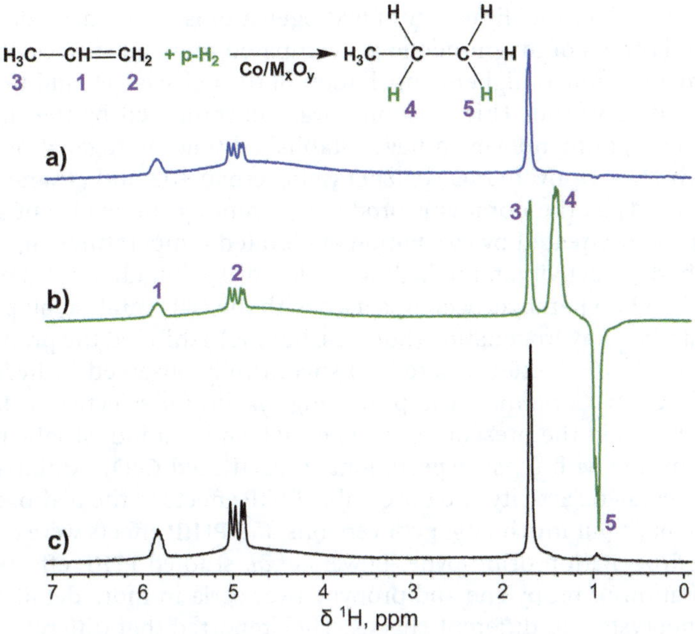

Figure 6.8 (a) ^{1}H NMR spectrum acquired during hydrogenation of propylene with parahydrogen over the Co/Al_2O_3 catalyst at 500 °C. (b) ^{1}H NMR spectrum acquired during hydrogenation of propylene with parahydrogen over Co/TiO_2 catalyst at 400 °C, and (c) the spectrum for the same reaction acquired after the gas flow was interrupted and nuclear spins relaxed to thermal equilibrium. The ^{1}H NMR signals of propane labeled 4 and 5 exhibit pronounced ALTADENA PHIP effects in spectrum b (compare with spectrum c).

are expected to be either diamagnetic or weakly paramagnetic, especially when their sizes are very small.[96] In contrast, Ni and Co MNPs used in the experiments described above are most likely superparamagnetic, given their small size and the relatively high temperatures used in the catalytic hydrogenations over these catalysts. Paramagnetic and superparamagnetic nanoparticles possess a high proton relaxivity and are thus employed as efficient contrast agents for MRI. Accelerated relaxation of nuclear spins is expected to diminish or even completely destroy the PHIP effects. The substantial PHIP effects observed over the Co/TiO_2 catalyst, however, demonstrate that the use of MNPs may be quite feasible in the context of PHIP.

While supported metal nanoparticle catalysts do produce PHIP effects, it is hard to expect them to be able to produce maximum NMR signal enhancements (corresponding to 100% pairwise selectivity) as a result of the dominance of the non-pairwise Horiuti–Polanyi mechanism on many metals, in particular metals of the platinum group. In search of more efficient catalysts for PHIP production, the attention was turned to metal oxides which are known to activate H_2 and hydrogenate unsaturated hydrocarbons.[97] It was expected that since the diffusivity of hydrogen atoms on metal oxides is usually several orders of magnitude lower compared to metal surfaces,[98,99] they could provide a much higher contribution of the pairwise H_2 addition to an unsaturated substrate. This reasoning was substantiated by the results of some early experiments which have established that hydrogenation of ethylene with H_2 + D_2 over Cr_2O_3 or ZnO yields ethane-d2 and ethane-d0 (but not ethane-d1) as the dominant products.[100] Among the oxides of alkaline earth metals pretreated by evacuation at elevated temperatures (up to 1000 °C), the highest activity in the hydrogenation of 1,3-butadiene was reported for CaO.[101] When parahydrogen was used with the CaO catalyst, all products of 1,3-butadiene hydrogenation (butenes, butane) exhibited the pronounced PHIP effects.[91] PHIP effects were also successfully observed in hydrogenations over CeO_2. Ceria may be a promising system for selective hydrogenation of alkynes in the presence of alkenes, which is an industrially relevant process. In line with these expectations, preactivated CeO_2 exhibited some moderate catalytic activity and observable PHIP effects in the hydrogenation of propylene,[91] but much larger conversions and PHIP effects were observed in the hydrogenation of propyne. Bowers *et al.* studied PHIP effects in the hydrogenation of propylene and propyne over ceria in more detail[102] using CeO_2 nanocrystals of different shapes. They reported that different crystalline facets of CeO_2 differ in conversion but produce similar PHIP effects, and estimated the contribution of the pairwise H_2 addition as <3%. However, in a later study[103] the facet dependence of PHIP effects was found in the hydrogenation of propyne over CeO_2, with ceria nanorods producing the largest PHIP effects with the contribution of pairwise H_2 addition of up to 8%. The authors related the shape dependence of PHIP effects to the different surface concentrations of oxygen vacancies at different facets [100,110,111] of ceria nanocrystals.

Catalytic activity and PHIP effects were also successfully observed in the hydrogenation of unsaturated hydrocarbons over several transition metal oxides, including PtO_2, PdO, ZrO_2, Cr_2O_3 [91] and Rh_2O_3.[104] In particular, hydrogenation of 1,3-butadiene over Cr_2O_3 performed under ALTADENA conditions demonstrated the PHIP effects which were increasing with the reaction temperature increasing from 200 to 400 °C, while the overall reaction conversion demonstrated the opposite trend.[87] In contrast to ALTADENA experiments, for hydrogenations performed under PASADENA conditions, *i.e.*, inside the NMR probe, the reaction temperatures are limited to much lower values. Despite that, hydrogenation of 1,3-butadiene over Cr_2O_3 produced quite pronounced PHIP effects even at 130 °C. According to the literature, an appropriate activation can provide a chromia catalyst which is even more active in catalytic hydrogenations of unsaturated hydrocarbons,[100] but this is yet to be achieved in the context of PHIP research. The results obtained for Cr_2O_3 are quite remarkable, as it is most likely paramagnetic at reaction temperatures used in the PHIP experiments.[105] Paramagnetic centers are known to dramatically accelerate the rates of nuclear spin relaxation of molecular species in their vicinity, and thus paramagnetic catalysts could be expected to be unsuitable for producing PHIP effects. The results obtained with chromia, however, demonstrate that perhaps paramagnetic and possibly other types of magnetic catalysts should not be *a priori* excluded from the search for a better catalyst for PHIP production.

6.6 Summary and Outlook

The possibility to increase magnetization of nuclear spins by four to five orders of magnitude compared to the thermal equilibrium value provides an opportunity to dramatically enhance observable signal in NMR spectroscopy and imaging. This not only translates into enormous time savings while performing NMR/MRI experiments, but also opens entirely new possibilities for applications that are usually impractical or even impossible. In catalytic applications of NMR, such a major signal enhancement provides novel opportunities for advanced mechanistic studies by allowing one to directly detect short-lived reaction intermediates at concentrations that are insufficient for detection in conventional experiments. In life sciences, hyperpolarization has already proven to be a useful tool for advanced MRI and MRS studies *in vivo*, enabling the ultrafast imaging of blood flow and perfusion as well as spatially resolved real-time MRS studies of cell metabolism in normal and pathological tissues and organs. Further development of hyperpolarization technology holds promise for rapidly advancing these directions of research and practice.

Parahydrogen-induced polarization (PHIP) is a promising member of the family of hyperpolarization techniques of modern magnetic resonance. It utilizes the correlated nuclear spin state of parahydrogen to produce strong NMR signal enhancements upon its chemical activation. Compared to other hyperpolarization techniques, PHIP may be less versatile at present, but at the same time it is inexpensive and relatively simple to implement in practice. It is not surprising, therefore, that it continues to attract significant attention from researchers working in the field of hyperpolarization in NMR.

A practical implementation of a PHIP experiment comprises two stages: (i) the production of parahydrogen, and (ii) its subsequent use to produce hyperpolarized species. The first stage heavily relies on the use of magnetic materials (including MNPs). Indeed, conversion of normal hydrogen gas to parahydrogen is performed in the presence of magnetic particles to accelerate the equilibration of the ortho–para composition of H_2, which at cryogenic temperatures results in its enrichment with parahydrogen. These processes are well-developed currently, and PHIP applications are not limited by the efficiency of p-H_2 enrichment. Nevertheless, more efficient ortho–para conversion catalysts that may result from the rapid developments of the field of MNPs synthesis and applications would be of interest for industrial uses of liquid H_2. In contrast, the second stage of implementing PHIP may still benefit greatly from the progress in the science and technology of MNPs. The advances are feasible along several lines. First, one of the key challenges at the moment is the rapid and reliable removal of a homogeneous catalyst from reaction solution after hydrogenating an unsaturated substrate with parahydrogen to yield the hyperpolarized reaction product. It has been demonstrated already that immobilization of transition metal complexes on suitable solid supports is a viable approach to observing PHIP in heterogeneous hydrogenations of unsaturated substrates.[39,76,106–109] However, this has not so far been attempted with the use of MNPs as catalyst supports. Immobilization of a metal complex directly on a MNP is one possibility;[110] furthermore, the use of MNPs encapsulated in a carbon, polymer or silica shell makes their functionalization with catalytically active species such as transition metal complexes relatively straightforward.[111–113] Second, the described above relatively recent extension of PHIP to heterogeneous hydrogenations over supported and bulk metal nanoparticle catalysts and metal oxide nanoparticles[39,40,82,83,86,88,91,106] opens the way to producing PHIP effects using bare[114] or decorated MNPs[112] as well. Importantly, while one can usually expect an accelerated decay of nuclear spin polarization in the presence of (ferro-, para-, superpara-, *etc.*) magnetic materials and surfaces, the results presented previously (particularly for Cr_2O_3 and Co/TiO_2 hydrogenation catalysts) provide a hint that the magnetic nature of catalytic nanoparticles may not be entirely detrimental for PHIP observation. This may provide an opportunity to combine an efficient production of hyperpolarized molecules in solution with magnetic recovery of the catalyst to produce biocompatible hyperpolarized contrast agents for advanced *in vivo* MRI/MRS. In catalytic research, extension of the PHIP technique beyond hydrogenation reactions is desirable. So far, PHIP has been demonstrated in the studies of heterogeneous hydrodesulfurization[115] and oligomerization[116] processes. Demonstration of substantial PHIP effects in hydrogenations over supported Co nanoparticles may pave the way to the challenging extension of PHIP technique toward the Fischer–Tropsch synthesis.

Acknowledgements

The authors acknowledge the Russian Science Foundation (grant 14-13-00445) for financial support of this work. IVK thanks Prof. D. Budker (Mainz/Berkeley) for useful discussions.

References

1. S. Mansson, E. Johansson, P. Magnusson, C.-M. Chai, G. Hansson, J. S. Petersson, F. Stahlberg and K. Golman, *Eur. Radiol.*, 2006, **16**, 57–67.
2. *Hyperpolarization Methods in NMR Spectroscopy*, ed. L. T. Kuhn, Springer, 2013.
3. J. Kurhanewicz, D. B. Vigneron, K. Brindle, E. Y. Chekmenev, A. Comment, C. H. Cunningham, R. J. De Berardinis, G. G. Green, M. O. Leach and S. S. Rajan, *et al.*, *Neoplasia (Ann Arbor, MI, U. S.)*, 2011, **13**, 81–97.
4. I. V. Koptyug, *Mendeleev Commun.*, 2013, **23**, 299–312.

5. A. V. Atsarkin and W. Kockenberger, *Appl. Magn. Reson.*, 2012, **43**, 1–221.
6. D. A. Barskiy, A. M. Coffey, P. Nikolaou, B. M. Goodson, R. T. Branca, G. J. Lu, M. G. Shapiro, M. S. Rosen, D. M. Mikhaylov and V.-V. Telkki, *et al.*, *Chem. – Eur. J.*, 2017, **23**, 725–751.
7. D. H. Weitzel, W. V. Loebenstein, J. W. Draper and O. E. Park, *J. Res. Natl. Bur. Stand. (U. S.)*, 1958, **60**, 221–227, Research Paper 2840.
8. E. Ilisca, *Prog. Surf. Sci.*, 1992, **41**, 217–335.
9. R. E. Svadlenak and A. B. Scott, *J. Am. Chem. Soc.*, 1957, **79**, 5385–5388.
10. R. A. Buyanov, *Kinet. Katal.*, 1960, **1**, 617–619.
11. W. K. Hall, *Acc. Chem. Res.*, 1975, **8**, 257–263.
12. J. Bargon, J. Kandels and K. Woelk, *Angew. Chem.*, 1990, **102**, 70–71.
13. J. Bargon, J. Kandels and K. Woelk, *Z. Phys. Chem.*, 1993, **180**, 65–93.
14. C. R. Bowers and D. P. Weitekamp, *Phys. Rev. Lett.*, 1986, **57**, 2645–2648.
15. C. R. Bowers and D. P. Weitekamp, *J. Am. Chem. Soc.*, 1987, **109**, 5541–5542.
16. P. F. Seidler, H. E. Bryndza, J. E. Frommer, L. S. Stuhl and R. G. Bergman, *Organometallics*, 1983, **2**, 1701–1704.
17. J. Horiuchi and M. Polanyi, *Trans. Faraday Soc.*, 1934, **30**, 1164–1172.
18. G. C. Bond, *Metal-Catalysed Reactions of Hydrocarbons*, Springer, 2005.
19. A. Farkas and L. Farkas, *J. Am. Chem. Soc.*, 1942, **64**, 1594–1599.
20. L. H. Rayhel, R. L. Corey, D. T. Shane, D. F. Cowgill and M. S. Conradi, *J. Phys. Chem. C*, 2011, **115**, 4966–4970.
21. A. Renouprez, P. Fouilloux, R. Stockmeyer, H. M. Conrad and G. Goeltz, *Ber. Bunsenges. Phys. Chem.*, 1977, **81**, 429–432.
22. A. Farkas and L. Farkas, *J. Am. Chem. Soc.*, 1939, **61**, 3396–3401.
23. S. Ohno and I. Yasumori, *Bull. Chem. Soc. Jpn.*, 1968, **41**, 2227–2233.
24. J. Matthes, S. Gruendemann, G. Buntkowsky, B. Chaudret and H. H. Limbach, *Appl. Magn. Reson.*, 2013, **44**, 247–265.
25. J. Bargon, *eMagRes*, John Wiley & Sons, Ltd, 2007.
26. H.-H. Limbach, G. Buntkowsky, J. Matthes, S. Gruendemann, T. Pery, B. Walaszek and B. Chaudret, *ChemPhysChem*, 2006, **7**, 551–554.
27. S. Paris and E. Ilisca, *J. Phys. Chem. A*, 1999, **103**, 4964–4968.
28. A. Farkas, *Orthohydrogen, Parahydrogen and Heavy Hydrogen*, Cambridge University Press, Cambridge, 1935.
29. B. Feng, A. M. Coffey, R. D. Colon, E. Y. Chekmenev and K. W. Waddell, *J. Magn. Reson.*, 2012, **214**, 258–262.
30. S. Wagner, *Magn. Reson. Mater. Phys., Biol. Med.*, 2014, **27**, 195–199.
31. M. E. Tadros and L. Vaska, *J. Colloid Interface Sci.*, 1982, **85**, 389–410.
32. E. Sartori, M. Ruzzi, R. G. Lawler and N. J. Turro, *J. Am. Chem. Soc.*, 2008, **130**, 12752–12756.
33. J. M. Brown, L. R. Canning, A. J. Downs and A. M. Forster, *J. Organomet. Chem.*, 1983, **255**, 103–111.
34. V. V. Zhivonitko, K. V. Kovtunov, P. L. Chapovsky and I. V. Koptyug, *Angew. Chem., Int. Ed.*, 2013, **52**, 13251–13255.

35. B. Vuichoud, E. Canet, J. Milani, A. Bornet, D. Baudouin, L. Veyre, D. Gajan, L. Emsley, A. Lesage and C. Coperet, *et al.*, *J. Phys. Chem. Lett.*, 2016, 7, 3235–3239.
36. S. B. Duckett and N. J. Wood, *Coord. Chem. Rev.*, 2008, **252**, 2278–2291.
37. J. Nattereer and J. Bargon, *Prog. Nucl. Magn. Reson. Spectrosc.*, 1997, **31**, 293–315.
38. D. Canet, C. Aroulanda, P. Mutzenhardt, S. Aime, R. Gobetto and F. Reineri, *Concepts Magn. Reson., Part A*, 2006, **28A**, 321–330.
39. K. V. Kovtunov, V. V. Zhivonitko, I. V. Skovpin, D. A. Barskiy and I. V. Koptyug, *Top. Curr. Chem.*, 2013, **338**, 123–180.
40. V. P. Ananikov, L. L. Khemchyan, Y. V. Ivanova, V. I. Bukhtiyarov, A. M. Sorokin, I. P. Prosvirin, S. Z. Vatsadze, A. V. Medved'ko, V. N. Nuriev and A. D. Dilman, *et al.*, *Russ. Chem. Rev.*, 2014, **83**, 885–985, 101 pp.
41. R. A. Green, R. W. Adams, S. B. Duckett, R. E. Mewis, D. C. Williamson and G. G. R. Green, *Prog. Nucl. Magn. Reson. Spectrosc.*, 2012, **67**, 1–48.
42. S. B. Duckett and C. J. Sleigh, *Prog. Nucl. Magn. Reson. Spectrosc.*, 1999, **34**, 71–92.
43. V. V. Zhivonitko, V.-V. Telkki, K. Chernichenko, T. Repo, M. Leskela, V. Sumerin and I. V. Koptyug, *J. Am. Chem. Soc.*, 2014, **136**, 598–601.
44. L. E. Longobardi, C. A. Russell, M. Green, N. S. Townsend, K. Wang, A. J. Holmes, S. B. Duckett, J. E. McGrady and D. W. Stephan, *J. Am. Chem. Soc.*, 2014, **136**, 13453–13457.
45. V. V. Zhivonitko, K. Sorochkina, K. Chernichenko, B. Kotai, T. Foldes, I. Papai, V.-V. Telkki, T. Repo and I. Koptyug, *Phys. Chem. Chem. Phys.*, 2016, **18**, 27784–27795.
46. A. Koch and J. Bargon, *Inorg. Chem.*, 2001, **40**, 533–539.
47. M. Quack, *Mol. Phys.*, 1977, **34**, 477–504.
48. M. G. Pravica and D. P. Weitekamp, *Chem. Phys. Lett.*, 1988, **145**, 255–258.
49. H. G. Niessen, D. Schleyer, S. Wiemann, J. Bargon, S. Steines and B. Driessen-Hoelscher, *Magn. Reson. Chem.*, 2000, **38**, 747–750.
50. S. B. Duckett and R. E. Mewis, *Acc. Chem. Res.*, 2012, **45**, 1247–1257.
51. L.-S. Bouchard, S. R. Burt, M. S. Anwar, K. V. Kovtunov, I. V. Koptyug and A. Pines, *Science (Washington, DC, U. S.)*, 2008, **319**, 442–445.
52. V. V. Zhivonitko, V.-V. Telkki and I. V. Koptyug, *Angew. Chem., Int. Ed.*, 2012, **51**, 8054–8058, S8054/8051–S8054/8021.
53. K. Golman, O. Axelsson, H. Johannesson, S. Mansson, C. Olofsson and J. S. Petersson, *Magn. Reson. Med.*, 2001, **46**, 1–5.
54. H. Johannesson, O. Axelsson and M. Karlsson, *C. R. Phys.*, 2004, **5**, 315–324.
55. P. Bhattacharya, K. Harris, A. P. Lin, M. Mansson, V. A. Norton, W. H. Perman, D. P. Weitekamp and B. D. Ross, *MAGMA*, 2005, **18**, 245–256.
56. P. Bhattacharya, B. D. Ross and R. Bunger, *Exp. Biol. Med. (Maywood, NJ, U. S.)*, 2009, **234**, 1395–1416.
57. P. Bhattacharya, E. Y. Chekmenev, W. H. Perman, K. C. Harris, A. P. Lin, V. A. Norton, C. T. Tan, B. D. Ross and D. P. Weitekamp, *J. Magn. Reson.*, 2007, **186**, 150–155.

58. B. D. Ross, P. Bhattacharya, S. Wagner, T. Tran and N. Sailasuta, *AJNR Am. J. Neuroradiol.*, 2010, **31**, 24–33.
59. N. M. Zacharias, H. R. Chan, N. Sailasuta, B. D. Ross and P. Bhattacharya, *J. Am. Chem. Soc.*, 2012, **134**, 934–943.
60. S. Kadlecek, V. Vahdat, T. Nakayama, D. Ng, K. Emami and R. Rizi, *NMR Biomed.*, 2011, **24**, 933–942.
61. K. Golman, L. E. Olsson, O. Axelsson, S. Mansson, M. Karlsson and J. S. Petersson, *Br. J. Radiol.*, 2003, **76**, S118–S127.
62. E. Terreno, D. Delli Castelli, A. Viale and S. Aime, *Chem. Rev. (Washington, DC, U. S.)*, 2010, **110**, 3019–3042.
63. P. Nikolaou, B. M. Goodson and E. Y. Chekmenev, *Chem.–Eur. J.*, 2015, **21**, 3156–3166.
64. S. J. Nelson, J. Kurhanewicz, D. B. Vigneron, P. E. Z. Larson, A. L. Harzstark, M. Ferrone, M. van Criekinge, J. W. Chang, R. Bok and I. Park, *et al.*, *Sci. Transl. Med.*, 2013, **5**, 198ra108.
65. W. H. Perman, P. Bhattacharya, J. Leupold, A. P. Lin, K. C. Harris, V. A. Norton, J.-B. Hoevener and B. D. Ross, *Magn. Reson. Imaging*, 2010, **28**, 459–465.
66. M. Goldman, H. Johannesson, O. Axelsson and M. Karlsson, *Magn. Reson. Imaging*, 2005, **23**, 153–157.
67. M. Goldman, H. Johannesson, O. Axelsson and M. Karlsson, *C. R. Chim.*, 2006, **9**, 357–363.
68. M. Goldman and H. Johannesson, *C. R. Phys.*, 2005, **6**, 575–581.
69. J.-B. Hovener, E. Y. Chekmenev, K. C. Harris, W. H. Perman, T. T. Tran, B. D. Ross and P. Bhattacharya, *MAGMA*, 2009, **22**, 123–134.
70. J.-B. Hovener, E. Y. Chekmenev, K. C. Harris, W. H. Perman, L. W. Robertson, B. D. Ross and P. Bhattacharya, *MAGMA*, 2009, **22**, 111–121.
71. F. Reineri, A. Viale, W. Dastru, R. Gobetto and S. Aime, *Contrast Media Mol. Imaging*, 2011, **6**, 77–84.
72. A. Corma, *Catal. Rev.: Sci. Eng.*, 2004, **46**, 369–417.
73. N. End and K.-U. Schoening, *Top. Curr. Chem.*, 2004, **242**, 241–271.
74. A. Malkov, *Appl. Organomet. Chem.*, 2015, **29**, 338.
75. A. Eichhorn, A. Koch and J. Bargon, *J. Mol. Catal. A: Chem.*, 2001, **174**, 293–295.
76. I. V. Koptyug, K. V. Kovtunov, S. R. Burt, M. S. Anwar, C. Hilty, S.-I. Han, A. Pines and R. Z. Sagdeev, *J. Am. Chem. Soc.*, 2007, **129**, 5580–5586.
77. K. V. Kovtunov, I. E. Beck, V. I. Bukhtiyarov and I. V. Koptyug, *Angew. Chem., Int. Ed.*, 2008, **47**, 1492–1495.
78. P. S. Cremer, X. Su, Y. R. Shen and G. A. Somorjai, *J. Phys. Chem.*, 1996, **100**, 16302–16309.
79. V. V. Zhivonitko, K. V. Kovtunov, I. V. Skovpin, D. A. Barskiy, O. G. Salnikov and I. V. Koptyug, Catalytically Enhanced NMR of Heterogeneously Catalyzed Hydrogenations, in *Understanding Organometallic Reaction Mechanisms and Catalysis: Computational and Experimental Tools*, ed. V. P. Ananikov, Wiley-VCH, Weinheim, Germany, 2014, pp. 145–186.
80. R. Zhou, E. W. Zhao, W. Cheng, L. M. Neal, H. Zheng, R. E. Quinones, H. E. Hagelin-Weaver and C. R. Bowers, *J. Am. Chem. Soc.*, 2015, **137**, 1938–1946.

81. R. Zhou, W. Cheng, L. M. Neal, E. W. Zhao, K. Ludden, H. E. Hagelin-Weaver and C. R. Bowers, *Phys. Chem. Chem. Phys.*, 2015, **17**, 26121–26129.
82. K. V. Kovtunov, V. V. Zhivonitko, L. Kiwi-Minsker and I. V. Koptyug, *Chem. Commun.*, 2010, **46**, 5764–5766.
83. D. A. Barskiy, K. V. Kovtunov, A. Primo, A. Corma, R. Kaptein and I. V. Koptyug, *ChemCatChem*, 2012, **4**, 2031–2035.
84. O. G. Salnikov, D. A. Barskiy, D. B. Burueva, Y. K. Gulyaeva, B. S. Balzhinimaev, K. V. Kovtunov and I. V. Koptyug, *Appl. Magn. Reson.*, 2014, **45**, 1051–1061.
85. A. Corma, O. G. Salnikov, D. A. Barskiy, K. V. Kovtunov and I. V. Koptyug, *Chem. – Eur. J.*, 2015, **21**, 7012–7015.
86. V. V. Zhivonitko, K. V. Kovtunov, I. E. Beck, A. B. Ayupov, V. I. Bukhtiyarov and I. V. Koptyug, *J. Phys. Chem. C*, 2011, **115**, 13386–13391.
87. K. V. Kovtunov, O. G. Salnikov, V. V. Zhivonitko, I. V. Skovpin, V. I. Bukhtiyarov and I. V. Koptyug, *Top. Catal.*, 2016, **59**, 1686–1699.
88. K. V. Kovtunov, I. E. Beck, V. V. Zhivonitko, D. A. Barskiy, V. I. Bukhtiyarov and I. V. Koptyug, *Phys. Chem. Chem. Phys.*, 2012, **14**, 11008–11014.
89. I. V. Koptyug, V. V. Zhivonitko and K. V. Kovtunov, *ChemPhysChem*, 2010, **11**, 3086–3088.
90. A. M. Balu, S. B. Duckett and R. Luque, *Dalton Trans.*, 2009, 5074–5076.
91. K. V. Kovtunov, D. A. Barskiy, O. G. Salnikov, A. K. Khudorozhkov, V. I. Bukhtiyarov, I. P. Prosvirin and I. V. Koptyug, *Chem. Commun.*, 2014, **50**, 875–878.
92. M. Irfan, N. Eshuis, P. Spannring, M. Tessari, M. C. Feiters and F. P. J. T. Rutjes, *J. Phys. Chem. C*, 2014, **118**, 13313–13319.
93. D. A. Barskiy, O. G. Salnikov, K. V. Kovtunov and I. V. Koptyug, *J. Phys. Chem. A*, 2015, **119**, 996–1006.
94. K. V. Kovtunov, D. A. Barskiy, O. G. Salnikov, D. B. Burueva, A. K. Khudorozhkov, A. V. Bukhtiyarov, I. P. Prosvirin, E. Y. Gerasimov, V. I. Bukhtiyarov and I. V. Koptyug, *ChemCatChem*, 2015, **7**, 2581–2584.
95. E. W. Zhao, H. Zheng, K. Ludden, Y. Xin, H. E. Hagelin-Weaver and C. R. Bowers, *ACS Catal.*, 2016, **6**, 974–978.
96. S. K. Ignatov, A. I. Okhapkin, O. B. Gadzhiev, A. G. Razuvaev, S. Kunz and M. Baumer, *J. Phys. Chem. C*, 2016, **120**, 18570–18587.
97. D. L. Harrison, D. Nicholls and H. Steiner, *J. Catal.*, 1967, **7**, 359–364.
98. V. Sharma, P. A. Crozier, R. Sharma and J. B. Adams, *Catal. Today*, 2012, **180**, 2–8.
99. Y. Lykhach, T. Staudt, M. Vorokhta, T. Skala, V. Johanek, K. C. Prince, V. Matolin and J. Libuda, *J. Catal.*, 2012, **285**, 6–9.
100. W. C. Conner and R. J. Kokes, *J. Phys. Chem.*, 1969, **73**, 2436–2438.
101. Y. Tanaka, H. Hattori and K. Tanabe, *Chem. Lett.*, 1976, 37–38.
102. E. W. Zhao, H. Zheng, R. Zhou, H. E. Hagelin-Weaver and C. R. Bowers, *Angew. Chem., Int. Ed.*, 2015, **54**, 14270–14275.
103. E. W. Zhao, Y. Xin, H. E. Hagelin-Weaver and C. R. Bowers, *ChemCatChem*, 2016, **8**, 2197–2201.

104. O. G. Salnikov, K. V. Kovtunov, D. A. Barskiy, A. K. Khudorozhkov, E. A. Inozemtseva, I. P. Prosvirin, V. I. Bukhtiyarov and I. V. Koptyug, *ACS Catal.*, 2014, **4**, 2022–2028.
105. G. Vile, B. Bridier, J. Wichert and J. Perez-Ramirez, *Angew. Chem., Int. Ed.*, 2012, **51**, 8620–8623, S8620/8621–S8620/8624.
106. K. V. Kovtunov, V. V. Zhivonitko, A. Corma and I. V. Koptyug, *J. Phys. Chem. Lett.*, 2010, **1**, 1705–1708.
107. I. V. Skovpin, V. V. Zhivonitko and I. V. Koptyug, *Appl. Magn. Reson.*, 2011, **41**, 393–410.
108. I. V. Skovpin, V. V. Zhivonitko, R. Kaptein and I. V. Koptyug, *Appl. Magn. Reson.*, 2013, **44**, 289–300.
109. V. Skovpin Ivan, V. Zhivonitko Vladimir, P. Prosvirin Igor, F. Khabibulin Dzhalil and V. Koptyug Igor, *Z. Phys. Chem.*, 2016, **231**, 575–592.
110. U. Laska, C. G. Frost, P. K. Plucinski and G. J. Price, *Catal. Lett.*, 2008, **122**, 68–75.
111. J. Govan and Y. K. Gun'ko, *Nanomaterials*, 2014, **4**, 222–241.
112. L. Wu, A. Mendoza-Garcia, Q. Li and S. Sun, *Chem. Rev.*, 2016, **116**, 10473–10512.
113. A. H. Lu, E. L. Salabas and F. Schueth, *Angew. Chem., Int. Ed.*, 2007, **46**, 1222–1244.
114. R. Hudson, Y. Feng, R. S. Varma and A. Moores, *Green Chem.*, 2014, **16**, 4493–4505.
115. O. G. Salnikov, D. B. Burueva, D. A. Barskiy, G. A. Bukhtiyarova, K. V. Kovtunov and I. V. Koptyug, *ChemCatChem*, 2015, **7**, 3508–3512.
116. V. V. Zhivonitko, I. V. Skovpin, M. Crespo-Quesada, L. Kiwi-Minsker and I. V. Koptyug, *J. Phys. Chem. C*, 2016, **120**, 4945–4953.

CHAPTER 7

Development of Magnetic Theranostic Agents

ASANKA S. YAPA[a] AND STEFAN H. BOSSMANN*[a]

[a]Department of Chemistry and Johnson Cancer Center, Kansas State University, Manhattan, KS, 66506 USA
*E-mail: sbossman@ksu.edu

7.1 Introduction

According to Xie *et al.* at the Laboratory of Molecular Imaging and Nanomedicine, National Institute of Biomedical Imaging and Bioengineering, National Institutes of Health, USA, "the term "theranostics" was coined to define ongoing efforts in clinics to develop more specific, individualized therapies for various diseases, and to combine diagnostic and therapeutic capabilities into a single agent. The rationale arose from the fact that diseases, such as cancers, are immensely heterogeneous, and all existing treatments are effective for only limited patient subpopulations and at selective stages of disease development".[1] The promise of theranostics is that through the "marriage" of diagnosis and therapeutics, better therapeutic protocols can be developed. The latter comprise mostly drug delivery, such as chemotherapy of cancer (see Figure 7.1). However, photodynamic therapy (PDT),[2] photothermal therapy (PTT)[2] and hyperthermia (HT)[3] have been reported as well. The last decade has shown that both great opportunities and great challenges have arisen from the pursuit of theranostic nanosystems. Many of these challenges have been overcome by the progress of materials chemistry,

Smart Materials No. 26
Magnetic Nanomaterials: Applications in Catalysis and Life Sciences
Edited by Stefan H. Bossmann and Hongwang Wang

Published by the Royal Society of Chemistry, www.rsc.org

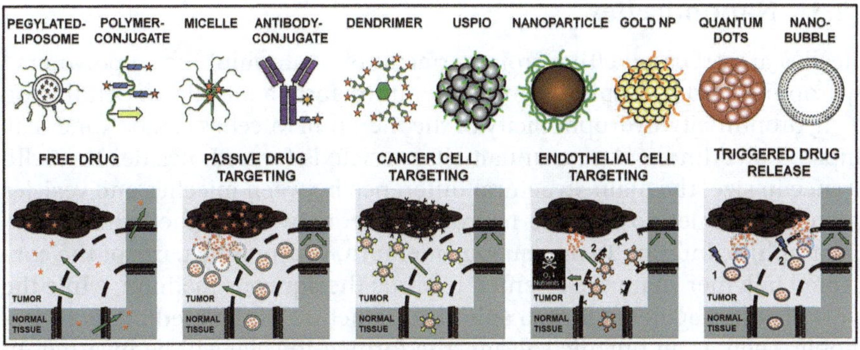

Figure 7.1 Examples of routinely used drug delivery systems and drug targeting strategies. (USPIO: ultra-small superparamagnetic iron oxide.) Reproduced with permission from ref. 8. Copyright (2015) American Chemical Society.

but some general design challenges remain. Multiple materials that are, at least principally, compatible with *in vivo* applications have been developed, such as magnetic nanoparticles,[4] second generation quantum dots and carbon dots,[4] upconversion nanoparticles,[4] mesoporous silica nanoparticles,[5,6] carbon-based nanomaterials,[7] and organic dye-based nanoparticles.[5] In this chapter, we are focusing on theranostic systems that are, in the opinion of the authors, easily translatable to the clinic, and that are designed for using MRI (magnetic resonance imaging) as imaging modality. We anticipate that the clinical research of the next decade will be driven by realistic considerations for clinically translatable nanomaterials. Furthermore, there will be intense competition between various theranostic designs to provide the best disease prognoses and treatment efficacies.[8]

7.2 Theranostic Platforms Derived from Supramolecular Structures

All theranostic applications discussed in this chapter have the utilization of supramolecular self-assembling processes to construct "intelligent" materials that respond to external stimuli in common. Through spontaneous, thermodynamically driven self-assembly, small organic and inorganic building blocks form nanoscopic or mesoscopic systems. These materials are either molecules or small clusters/nanoparticles that can be reproducibly synthesized and easily modified. Structural modifications of the building components and variations of their absolute and relative concentrations result in changes of size and shape of the supramolecular assemblies. This permits the fine-tuning of the supramolecular structures that are being used to design theranostic platforms. One of the major advantages of supramolecular systems is that they are adaptive to changes in external conditions, which makes them well-suited for applications in bioimaging and controlled drug release.[9]

7.2.1 Nanomicelles

Micelles are prepared either from surfactants or amphiphilic co-polymers.[9] Hydrophilic and hydrophobic monomer units form a micelle, depending on the hydrophilicity/hydrophobicity of the medium. Micelles can be spherical, ellipsoid or cylindrical, or a unilamellar vesicle is formed. Besides their difference in size, the main structural difference between micelles and vesicles is that the former possesses a hydrophobic core, whereas the latter feature a hydrophilic interior. In an aqueous medium, hydrophilic parts of the conjugated polymer chain are oriented towards the aqueous medium, while the hydrophobic fragments form a spherical structure. If the medium is hydrophobic, such as an organic solvent, the formed micelle will be inverted. In these structures, the hydrophilic groups are oriented towards the center of the micelle, which is hydrophilic. In either arrangement, the hydrophobic or hydrophilic cores of the micelles have enough free space to entrap small molecules. Therefore, micelles and micelle-derived hybrid structures can be effective drug carriers.

7.2.1.1 Hybrid Nanomicelles for Cancer Theranostics

Thorat *et al.* have developed polymeric micelles (PM) that feature oleic acid and polyethylene glycol (PEG) units.[10] In the core of the polymeric micelle, superparamagnetic $La_{0.7}Sr_{0.3}MnO_3$ nanoparticles (SPMNPs) are stabilized against coagulation. SPMNPs serve a dual function. They enable magnetic fluid hyperthermia (MFH) guided by magnetic resonance imaging (MRI). The widely used anticancer drug Doxorubicin, which acts as a topoisomerase II blocker,[11] is loaded into the polymeric micelles with high loading capacity (>60%). The design of this theranostic platform permits the combination of chemotherapy with hyperthermia (thermochemotherapy[12]), which is known to delay the onset of drug-resistance during chemotherapy. This theranostic platform proved to be very effective against cancer cell cultures *in vitro* (MCF7 breast cancer cell line). The *in vivo* biodistribution of micellar SPMNPs, performed in Wistar rats, did not result in significant cytotoxicity. It should be noted that tumor targeting by micellar SPMNPs proceeds *via* a passive targeting mechanism (extravasation). This is an excellent example of a theranostic platform designed around polymeric micelle-stabilized SPMNPs (see Figure 7.2). It serves as a promising candidate for effective multimodal cancer theranostics by combining MR imaging and magnetic thermochemotherapy.

Tong *et al.* developed a novel strategy to deliver an anticancer drug and an imaging agent simultaneously to tumor sites using nanomicelles that were designed from poly(ethylene glycol)-block-poly(acrylamide-*co*-acrylonitrile) copolymers (see Figure 7.3).[13] The topoisomerase blocker Doxorubicin (DOX)[14] and gadolinium ions, as (Gd^{3+}-DOTA (1,4,7,10-tetraazacyclododecane-1,4,7,10-tetraacetic + Gd^{3+})[15]) were loaded into the hydrophobic core and chelated within the shell of the self-assembled micelles. The resulting theranostic micelles feature magnetic resonance imaging capabilities and significantly increased DOX release with elevated temperature, thus enabling the

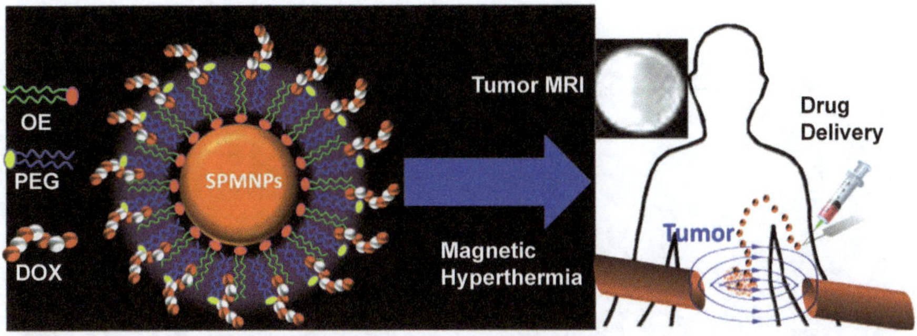

Figure 7.2 Theranostic platform based on polymeric micelle-stabilized superparamagnetic nanoparticles (SPMNPs). This platform passively targets tumor tissue and is able to provide thermochemotherapy with high (MRI-guided) spatiotemporal resolution. PEG: polyethylene glycol, OE: oleic acid, DOX: doxorubicin. Reproduced with permission from ref. 10 with permission. Copyright (2016) American Chemical Society.

combined treatment of solid tumors with a theranostic agent and hyperthermia. Furthermore, high proton relaxivity r_1 (25.88 mM^{-1} s^{-1}) was observed, permitting the imaging of solid tumors and metastases by means of T_1-weighted MRI.[13] On the cellular level, the Gd^{3+}/DOX loaded micelles facilitated enhanced cellular uptake of DOX (more than two times greater when compared to free DOX) by human hepatic carcinoma (BEL-7402) cells.[13,16]

7.2.2 Nanoemulsions

Nanoemulsions are another popular type of soft matter that can be utilized in biomedical applications, especially to treat infections in the reticuloendothelial system (RES), as well as for enzyme replacement therapy in the liver, cancer treatments and vaccination.[17] Nanoemulsions are liquid-in-liquid dispersions (oil-in-water or water-in-oil), in which spherical particles are dispersed in a continuous phase. These formulations are kinetically stable and can be prepared by high-energy methods, such as high pressure homogenization and ultrasonication. Alternatively, low-energy methods, such as phase inversion temperature or the emulsion inversion point method are also possible.[18] In virtually all nanoemulsions, emulsifying agents have to be added to gain thermodynamic stability.[17]

Ganta *et al.* prepared a nanoemulsion for ovarian cancer therapy. It is fabricated from the following components: endothelial growth factor receptor (EGFR) targeting ligand, myrisplatin: platinum-based first line anticancer drug for ovarian cancer, pro-apoptotic agent: C6-ceramide, and chelated Gd^{3+} for MRI-guided delivery. As Figure 7.4 indicates, both the EGFR-targeting ligand and chelated Gd^{3+} reside on the outside of the nanodroplets, whereas myrisplatin and ceramide are confined to their lipid cores.[19,20] This geometrical arrangement permits contact between Gd^{3+} and water molecules, which is mandatory for MR imaging. At the same time, both cytotoxic agents are shielded from interaction with healthy tissue, thus minimizing collateral

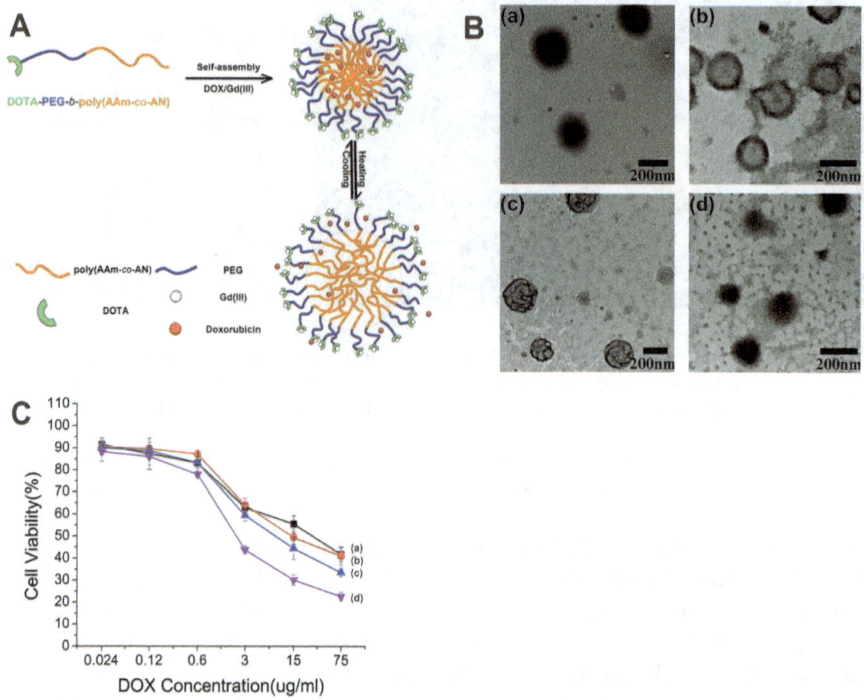

Figure 7.3 A: Schematic illustration of poly(ethylene glycol)-block-poly(acrylamide-*co*-acrylonitrile) [poly(AAm-*co*-AN)] micelles with conjugated DOTA chelators (1,4,7,10-tetra-azacyclododecane-1,4,7,10-tetraacetic), chelated Gd^{3+}, and adsorbed Doxorubicin (DOX). B: TEM images of micelles. (a) PEG_{2k}-b-poly(AAm-*co*-AN)$_{8k}$, (b) DOTA-PEG_{2k}-b-poly(AAm-*co*-AN)$_{8k}$, (c) Gd^{3+}/DOTA-b-PEG_{2k}-poly(-AAm-*co*-AN), (d) Gd^{3+}/DOX loaded DOTA-PEG_{2k}-b-poly(AAm-*co*-AN)$_{8k}$ (scale bar: 200 nm). C: IC_{50} of DOX-loaded micelles on BEL-7402 cells: (a) DOX-loaded PEG_{2K}-b-poly(AAm-*co*-AN)$_{8k}$ micelles, (b) DOX-loaded DOTA-PEG_{2K}-b-poly(AAm-*co*-AN)$_{8k}$ micelles, (c) Gd^{3+}/DOX-loaded DOTA-PEG_{2K}-b-poly(AAm-*co*-AN)$_{8k}$ micelles, (d) free DOX. Reproduced from ref. 13 with permission from The Royal Society of Chemistry.

damage. It is noteworthy that due to the presence of the EGFR-targeting ligand at the outside of the nanodroplets, this formulation is actively targeting tumor tissue.

7.2.3 Nanovesicles

Nanovesicles are another class of hollow, spherical assemblies, which are structurally related to nanomicelles. Liu *et al.* developed the PEG-b-PLGA diblock copolymer (Poly(ethylene glycol)-block-poly(L-lactic-*co*-glycolic acid)) based nano-sized vesicles, which spontaneously form in THF/water solution at

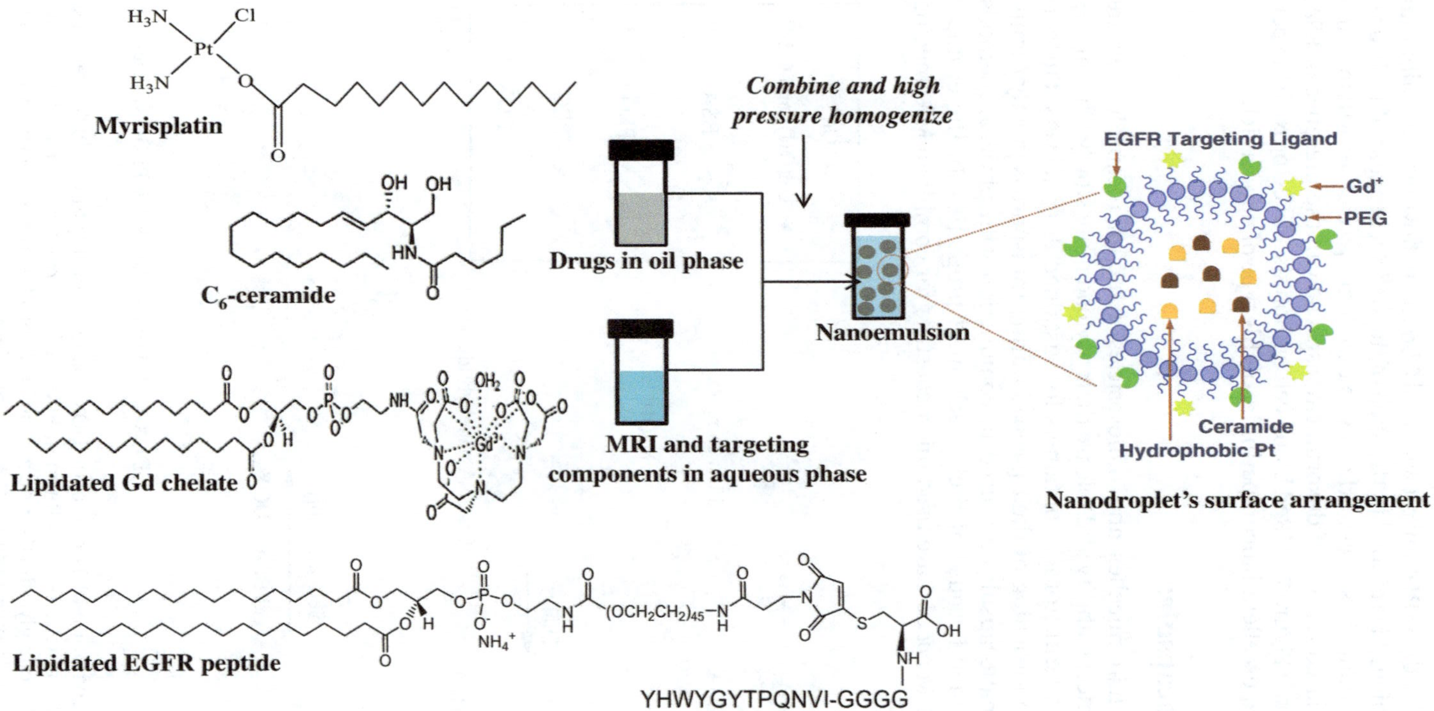

Figure 7.4 Schematic showing encapsulation of myrisplatin and C6-ceramide into the lipid core of the nanoemulsion. Its surface is modified with gadolinium and EGFR-binding peptide for MRI functionality and targeting, respectively. Reproduced from Pharmaceutical Research, Development of EGFR-Targeted Nanoemulsion for Imaging and Novel Platinum Therapy of Ovarian Cancer, 31 (9), 2014, 2490–2502, S. Ganta, A. Singh, N. R. Patel, J. Cacaccio, Y. H. Rawal, B. J. Davis, M. M. Amiji, T. P. Coleman, © Springer Science+Business Media New York 2014, with permission of Springer.[19]

room temperature.[21] The interior of the resulting theranostic nanovesicle has been loaded with the MRI contrast agent Gd^{3+} bound to bovine serum albumin (BSA). Doxorubicin[11] is trapped in the shell of the vesicle, because of its hydrophobicity. This nanoconstruct worked as an excellent T_1-weighted MR imaging agent while delayed release of doxorubicin, especially in the presence of BSA, was observed (see Figure 7.5). Red blood cell hemolysis and cytotoxicity experiments revealed excellent biocompatibility of this type of nanovesicle.

7.2.4 Nanocapsules

Compared to nanomicelles and nanovesicles, nano/microcapsules feature higher mechanical stability and release their payload more slowly. This makes them ideal long-term storage containers for sustained drug delivery. Furthermore, unwanted leaching of their payload during transport is greatly diminished. Often, nanocapsules are used in conjunction with physical methods (*e.g.* hyperthermia) because of their mechanical properties. In the example presented here Ma *et al.* have used microcapsules for combined MR imaging

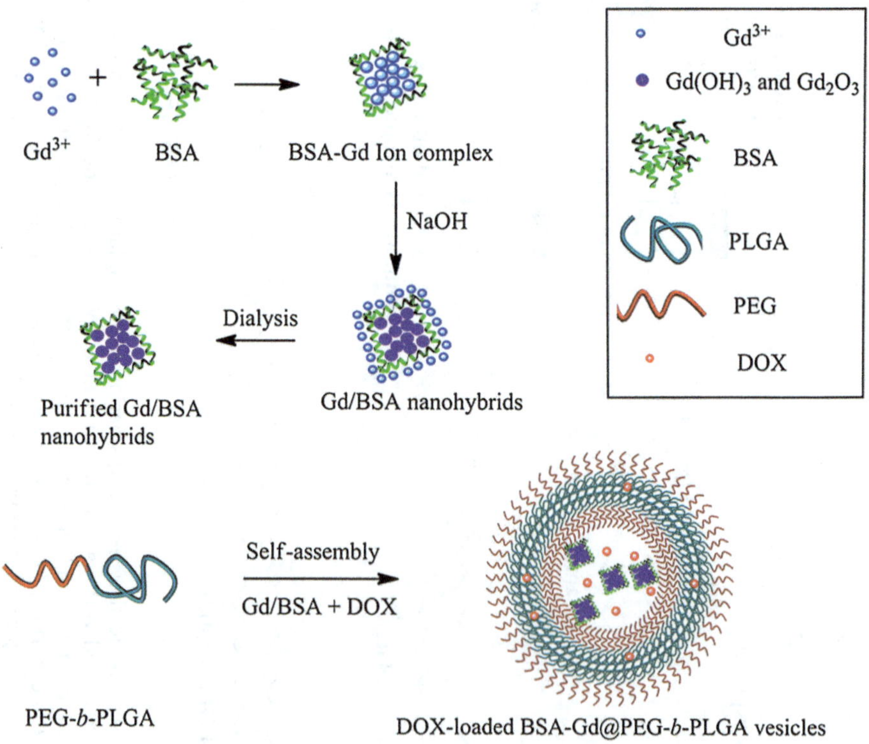

Figure 7.5 Preparation of Theranostic DOX-Loaded BSAGd@PEG-*b*-PLGA vesicles. BSA: bovine serum albumin, PLGA: L-lactic-*co*-glycolic acid, PEG: polyethylene glycol, DOX: doxorubicin. Reproduced with permission from ref. 21. Copyright (2014) American Chemical Society.

and thermal ablation by applying high-intensity focused ultrasound (HIFU).[22] Fe_3O_4 nanoparticles (particle sizes below 10 nm) are being used as an MRI contrast agent (T_2-weighted imaging) and isopentyl acetate (PeA) as HIFU synergistic liquid.[23] It is noteworthy that the authors have established a template-free strategy for synthesizing a hollow/mesoporous organosilica nanocapsule (OSNC), which is summarized in Figure 7.6. Synthetic key steps are the surfactant-free emulsification process of an oil phase isopentyl acetate/(trimethoxysilyl)propyl methacrylate (TMP) mixture in a positively charged Fe_3O_4 nanoparticle aqueous dispersion, as well as the subsequent radical polymerization of TPM. The resulting organosilica nanocapsules are capable of having both a liquid isopentyl acetate core and Fe_3O_4 in their shell. It is noteworthy that in contrast to gadolinium probes, iron oxides do not require direct contact with their surrounding water to exhibit T_2 shortening.[16] This novel material exhibits virtually uniform particle size (~82 nm), a high BET surface area (~534 $m^2\ g^{-1}$), as well as excellent colloidal stability in aqueous solution.

7.2.5 Magnetoliposomes

Liposomes were first described in 1965 as the first closed phospholipid bilayer systems.[24] They are now widely known as promising drug delivery vehicles for small molecular drugs and genes.[24] These multilamellar, or unilamellar, spherical lipid vesicles can be prepared by using two-chain lipid

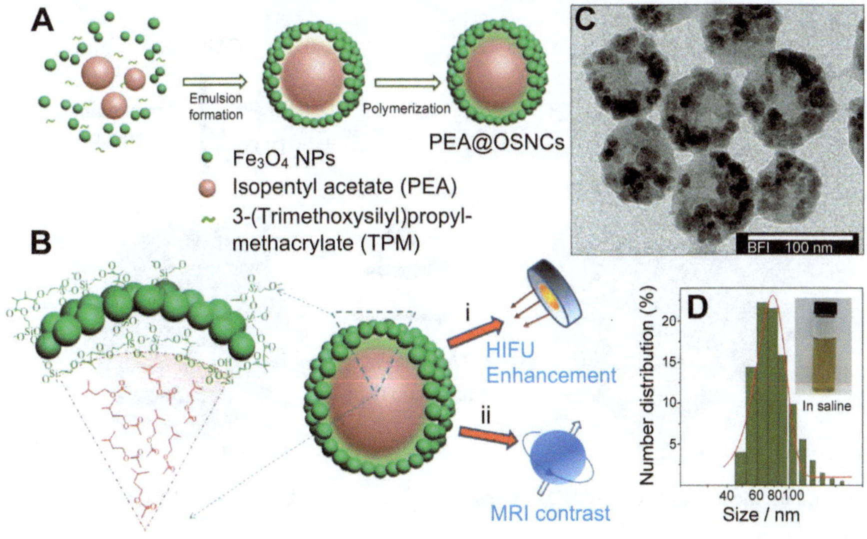

Figure 7.6 (A) Organosilica emulsion polymerization process. (B) Chemical composition and theranostic applications of PeA@ OSNC, including (i) HIFU (high-intensity focused ultrasound) and (ii) MRI. (C) TEM images of PeA@OSNC. (D) Size distribution of PeA@OSNC in saline. Reproduced with permission from ref. 22. Copyright (2016) American Chemical Society.

molecules with a hydrophilic head and hydrophobic chains. Whereas single-chain surfactants would form micelles, two-chain lipids will form bilayers or liposomes, depending on their shape.[25] Many characteristics, such as targeted delivery, triggered drug release, sustained release, long circulation time, biomarker-targeting and imaging capabilities can be incorporated into liposomes by introducing new chemical entities or nanoparticles, as for instance shown in Figure 7.7.

7.2.5.1 Ferri-Liposomes

Mikhaylov *et al.* developed lipidated magnetic nanocarriers named "ferri-liposomes" to target tumors and their microenvironment.[26] They used L-a-phosphatidylcholine and 1,2-distearoyl-*sn*-glycero-3- phosphoethanolamine-*N*-[methoxy(polyethylene glycol)-2000] (95% : 5%) to prepare liposomes.

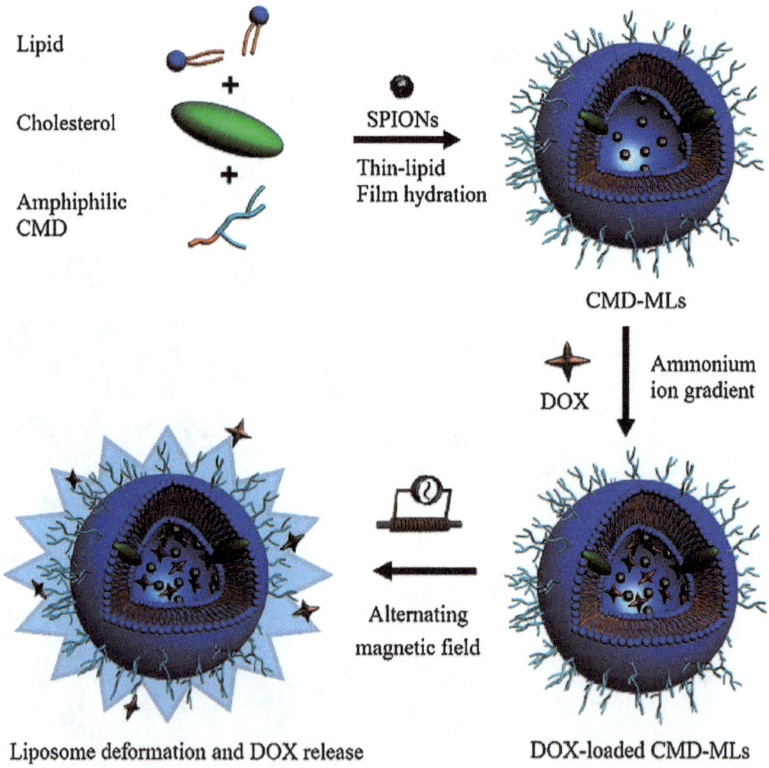

Figure 7.7 Small paramagnetic iron oxide nanoparticle (SPION) encapsulation in liposomes, followed by loading with doxorubicin (DOX) and subsequent release in alternating magnetic field. Reprinted from Carbohydrate Polymers, 118, H. Guo, W. Chen, X. Sun, Y.-N. Liu, J. Li, J. Wang, Theranostic magnetoliposomes coated by carboxymethyl dextran with controlled release by low-frequency alternating magnetic field, 209–217, Copyright (2015) with permission from Elsevier.[27]

Ferrimagnetic iron oxide (magnetite, Fe_3O_4; FMIO) nanoparticles were encapsulated in polyethylene glycol (PEG)-coated liposomes (PEGylated, Stealth Liposomes) to obtain ferri-liposomes. The average diameter of the resulting supramolecular assembly is about 95 nm. PEG coating diminishes opsonization of liposomes and their subsequent clearance by the mononuclear- phagocytic system and thereby increases the circulation half-life of the liposomes. Targeted delivery was demonstrated by setting up an external magnetic field at the tumor site. Preliminary studies showed that ferri-liposomes containing DOX can eliminate 90% of the tumor after two weeks upon single-dose administration. With standard DOX administration the maximum value was 60%. Furthermore, the authors have evaluated the effectiveness of JPM-565 containing ferri-liposomes against MMTV-PyMT transgenic female mice (breast cancer model). Promising results were obtained for the cysteine cathepsins inhibitor, JPM-565 ($C_{20}H_{28}N_2O_6$), when delivered by ferri-liposomes compared to systemic delivery, due to improved bioavailability.[26]

7.2.5.2 SPION-Containing Magnetoliposomes

Guo *et al.* developed theranostic magnetoliposomes for the sustained release of doxorubicin.[27] The thin-lipid film hydration method was followed to prepare SPION-encapsulated carboxymethyl dextran (CMD)-coated liposomes. The lipid shell consists of phosphatidyl-choline: cholesterol: amphiphilic CMD (molar ratio of 55 : 40 : 0.5). Encapsulated SPION and DOX release under a low-frequency alternating magnetic field was observed. It is noteworthy that there was no hyperthermia effect observed under this condition. This enables the delivery and sustained release of drugs to tumors located near or in highly heat sensitive organs, such as the brain. As discussed in Chapter One of this book, the most likely cause for DOX release is the generation of ultrasound by the SPIONs. Furthermore, SPIONs make this magnetoliposome MRI-visible due to iron-concentration dependent T_2-relaxation time shortening. This ultimately makes it a promising theranostic agent for cancer treatment.[27]

7.2.6 Polymer-Based Theranostic Agents

Novel PEGylated Mn^{2+}-chelated polydopamine (PMPDA) nanoparticles were developed by Miao *et al.* for MRI-guided photothermal therapy.[28] First, polydopamine (PDA) nanoparticles were prepared by using dopamine hydrochloride and Mn^{2+}, which was incorporated by vigorous stirring of PDA with $MnSO_4{\cdot}H_2O$. Functional catechol groups of polydopamine chelate Mn^{2+} ions.[29] Further, Mn^{2+}-PDA was reacted with thiol-polyethylene glycol to obtain negatively charged, highly stable PMPDA colloidal nanoparticles. The paramagnetic Mn^{2+} ions permit MR imaging while biocompatible polydopamine is utilized as the photosensitizing agent for photothermal therapy. The photothermal effect of prepared particles was evaluated by irradiating treated HeLa cells with an NIR laser (808 nm) beam for 10 minutes, followed by the MTT assay and fluorescence microscopy (Figure 7.8). A 10 minute irradiation after the administration of 80 μg ml^{-1} concentration destroyed 84% of HeLa cells. 4T1 (metastasizing breast cancer) tumor bearing mice were then used to assess and validate *in vivo* T_1-weighed MR imaging. Again, the results obtained were excellent. Therefore, the novel theranostic agent is a promising candidate for further biomedical applications.[28]

Paramagnetic, pH and temperature-sensitive polymeric particles (PPPs) were synthesized by Liu *et al.* for anticancer drug delivery and brain tumor imaging. Soap-free emulsion polymerization was carried out in the presence of methacrylic acid, *N*-isopropylacrylamide, hydrophobic styrene, and aqueous Gd^{3+}-monomer solutions. Drug loading was performed by directly mixing an aqueous solution of DOX·HCl with diluted polymer emulsion. In PPPs, methacrylic acid and *N*-isopropylacrylamide are the pH-sensitive monomer and temperature-sensitive monomers respectively. The hydrodynamic diameters of prepared particles ranged from 215–277 nm depending on the Gd(III)-monomer contents.[30] The dual stimuli response of the PPPs was assessed by dynamic light scattering (DLS). As the pH increased from 2 to 8, the hydrodynamic diameter of the PPPs changed by about 50 nm due to variations in hydrogen bonding and the increasing ionization state of carboxyl groups. The temperature effect was examined by varying the temperature from 25 °C to 43 °C. The particles shrink when the temperature increases

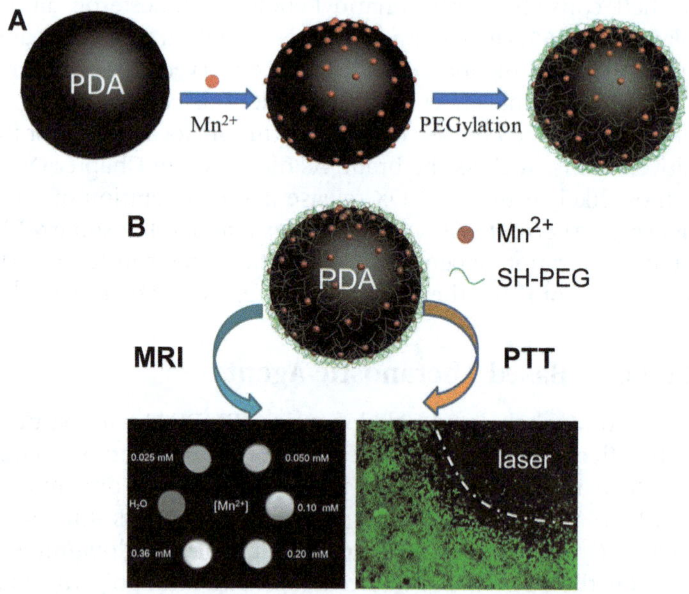

Figure 7.8 A: Schematic illustration of the fabrication of PEGylated Mn^{2+}-polydopamine (PMPDA) nanoparticles (SH-PEG: thiol-modified polyethylene glycol). B: PMPDAs are suitable T_1-weighted MRI agents and potent photothermal therapy (PTT) agents when excited with an NIR laser (808 nm). Reproduced with permission from ref. 28. Copyright (2015) American Chemical Society.

due to polymer wettability changes. The particles collapse above 32 °C (lower critical solution temperature of PNIPAM) due to hydrogen bond disruption.[31] Glioma-xenograft-bearing rats were utilized to study *in vivo* T_1-weighted MR imaging of the brain tumor and to perform biodistribution studies. These experiments suggested that these receptor-free, dually responsive polymer particles are potential agents for combined glioma diagnosis and therapy.[30]

7.2.7 Dendrimers

Dendrimers are a class of monodisperse structurally defined macromolecules.[32] All dendrimers possess a central core, a low-density inner region, which is formed by repetitive branching units, and a high-density exterior, which is defined by the choice of surface groups. Numerous dendrimer scaffolds have been utilized for combined targeting and imaging of solid tumors. General design considerations of dendrimer-based theranostic agents are discussed here, together with a couple of specific examples (see Figure 7.9). The general discussion includes major challenges and potential obstacles to using dendrimer-based theranostic platforms.[32]

Besides their chemical building principles (*e.g.* PAMAM- (polyamidoamine[33]), Triazine-,[34] PPI- (polypropyleneimine[34]), 5-ALA- (5-aminolevulinic acid) based dendrimers[32]), the generation of a dendrimer plays a decisive role with respect

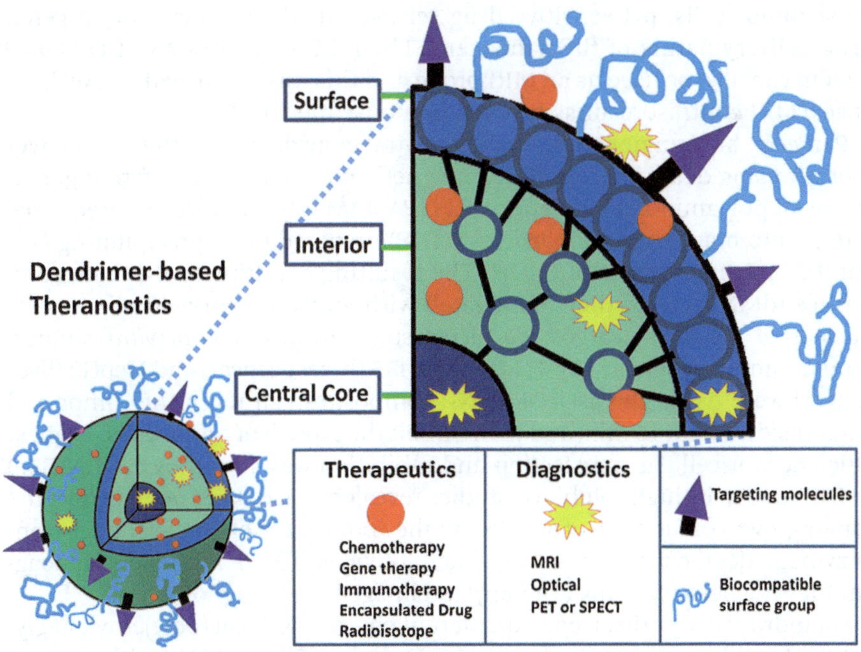

Figure 7.9 Dendrimer-based theranostics: general design considerations and possibilities. Reproduced with permission from ref. 32. Copyright (2013) American Chemical Society.

to tailored applications. "Generation" is defined as the number of repetitive additions of defined branching units to the dendrimer core. The same type of dendrimer often has tunable biophysical properties, depending on its size and the number of available surface groups. In Figure 7.10, the PAMAM-DTPA(Gd^{3+}) system is depicted, which was developed as a MRI contrast agent.[32]

As shown in Figure 7.10, dendrimers conjugated with paramagnetic ions or, alternatively, magnetic nanoparticles, are excellent candidates for many biomedical applications, such as MRI and fluorescence imaging, magnetofection, and drug delivery.[35] Chang *et al.* have pioneered a pH-responsive nanosystem containing doxorubicin (DOX) loaded polyamidoamine dendrimers (PAMAM, generation 2.5) in 2011. First they developed the PAMAM–DOX conjugate, which then was tethered onto oleylamine stabilized superparamagnetic iron oxide nanoparticles (IONPs) to obtain a water soluble nanocarrier capable of chemotherapeutic treatment and tumor imaging.[36]

Drug release studies showed that 80% of DOX released at pH 5.2 whereas only 8% released at pH 7.4 due to the use of a hydrazide/hydrazone linker featuring pH-dependent stability. Thus, this formulation is ideal for cancer treatment as it can minimize the side effects by preventing drug leaking to the healthy tissues. The nanosystem enters the tumor site *via* EPR effect and is taken up by tumor cells *via* endocytosis.[37] The acidic pH inside the endosome (pH 4.5–6.5) triggers a sufficient amount of DOX release within a short period of time to be competitive with drug efflux pumps that are active in most tumor cells. pH sensitive drug release greatly improves the targeted drug delivery nature of the nanoagent. The mPEG-G2.5PAMAM–DOX-IONP system can also be used as an MRI probe and this was confirmed by excellent *in vivo* MRI contrast results observed in mouse models.[36]

Pan *et al.* have engineered a novel nonviral gene delivery system for cancer therapy using dendrimer-modified magnetic nanoparticles. Different generations of polyamidoamine dendrimers (PAMAM G1–G5) were tethered separately, onto magnetic nanoparticles (MNP) prepared by co-precipitating Fe^{2+} and Fe^{3+} in the presence of NaOH. The resulting dendrimer-MNP nanocomposites (dMNPs) were then conjugated with antisense survivin oligodeoxynucleotide (asODNs) cargos *via* electrostatic means. For *in vitro* studies, human tumor cell lines MCF-7, MDA-MB-435 (breast cancer) and HepG2 (liver cancer) were used. The asODN-dMNP composites can reach the tumor and enter inside within 15 min, and then inhibit the growth of tumor cells. Highest efficiency for cellular transfection and inhibition was shown by asODN-G5.0 dMNPs due to its high solubility. Studies revealed that dMNPs do not affect the tumor growth or function of asODN but they protect the oligonucleotides from enzymatic degradation. This dendrimer-conjugated MNP is a potent nanoagent for gene delivery, cancer therapy and image-based diagnosis.[38]

Chandra *et al.* synthesized two different chain lengths of (poly)ethyleneglycol-PAMAM dendrimers (L6-PEG-PAMAM and S6-PEG-PAMAM with six end-grafted ethyleneglycol ether-tentacles of type $CH_2CH_2C(O)O(CH_2CH_2O)_9CH_3$ and $CH_2CH_2C(O)O(CH_2CH_2O)_2C_2H_5$, respectively), which can be used to stabilize mesoporous, superparamagnetic Fe_3O_4 (MNP) nanoparticles. The dendrimer-MNP complex was studied for suitability as drug carrier. It was

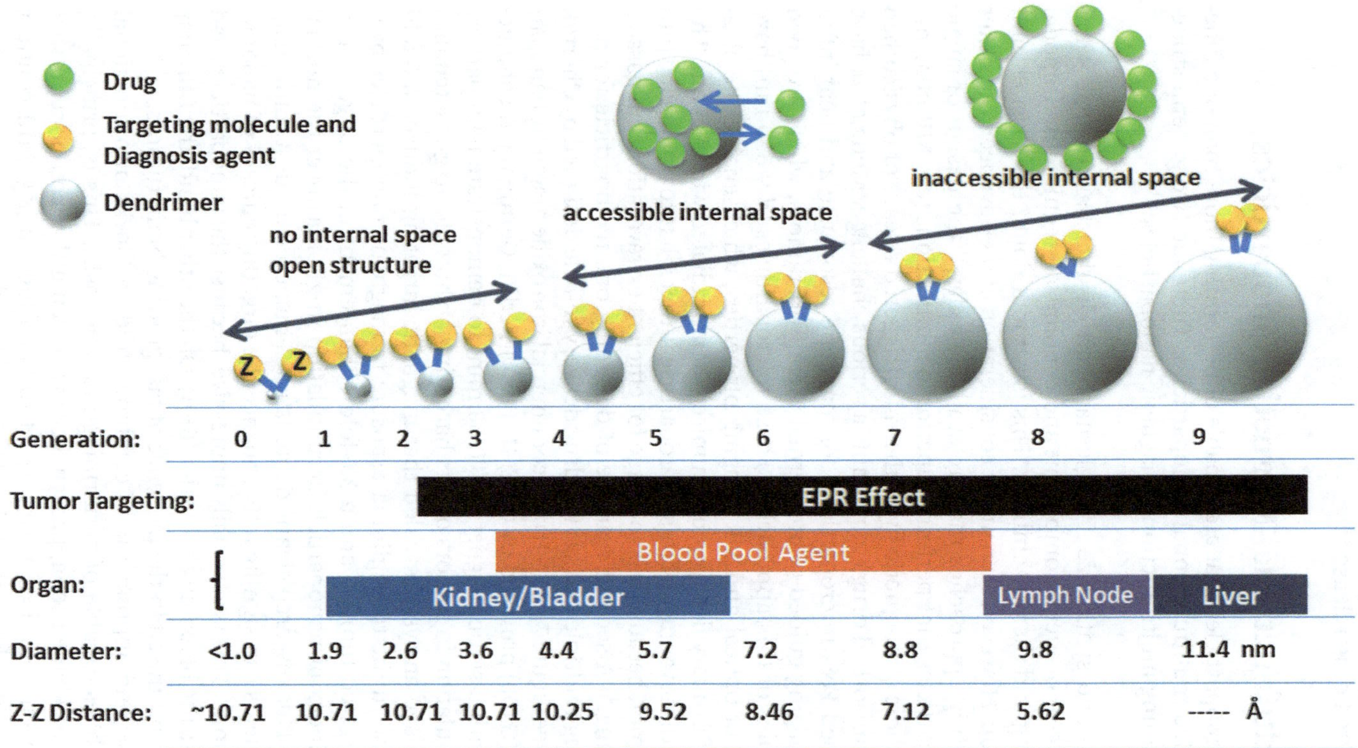

Figure 7.10 PAMAM-DTPA (Gd^{3+}) system (G0–G9) developed as series of MRI contrast agents. The dendrimer generation defined the mechanism of possible tumor targeting, as well as the organs that will be preferentially targeted. PAMAM: poly(amidoamine); DTPA: diethylene triamine pentaacetic acid. Reproduced with permission from ref. 32. Copyright (2013) American Chemical Society.

found that the loading efficiency of DOX to the "nanoshuttle" is >95%. The resulting nanoshuttle is capable of pH-, temperature-, and protease (cathepsin B)-triggered drug release.[39]

7.3 Multifunctional Magnetic Nanoparticles

Magnetic nanoparticles that are capable of performing more than one function are called multifunctional nanoparticles. There are two general strategies in designing multifunctional magnetic nanoparticles:[40]

(1) Design and synthesize a single nanoparticle that has a single semiconductor or metal component and contains all required multimodal imaging and therapy abilities. This is without doubt the most elegant solution. However, even under consideration of the advanced computer modelling approaches that are available today,[41] the design of single nanoparticles that perform superior in all functions (*e.g.* MR imaging, plasmonic or magnetic hyperthermia, triggered release of physisorbed or chemisorbed drugs...) remains a serious challenge. Meng *et al.* have designed FeS_2 microparticles for magnetic resonance and near-infrared imaging guided photothermal therapy of tumors, which are good examples of this approach.[42] These microparticles (d = 350 nm) have excellent optical and MR imaging properties, which enable the ablation of primary liver tumors in mice when irradiated with 915 nm. The major disadvantage of this microparticle is that it has to be injected into the tumor, because it is too large for efficient extravasation and passive targeting.[42] A second example of designing one nanoparticle for performing multiple tasks is the drug delivery vehicle designed by Chertok *et al.* (see Figure 7.11).[43] It uses commercial iron oxide magnetic nanoparticles (G100) provided by CHEMICELL® (Berlin, Germany) in conjunction with a static magnet (0.4 T, 30 min) to demonstrate nanoparticle accumulation in tumors of rats harboring orthotopic 9L-gliosarcomas. MR image analysis revealed that magnetic targeting induced a 5-fold increase in the total glioma exposure to magnetic nanoparticles over nontargeted tumors, and a 3.6-fold enhancement in the target selectivity index of nanoparticle accumulation in gliomas over the normal brain. This research clearly demonstrates that even modest static magnetic fields can significantly enhance the targeting efficacy of nanoformulations. This is especially impressive, because the nanoparticles had to cross the blood–brain barrier (BBB). On the other hand, the BBB may be compromised in the presence of fast growing brain tumors.

(2) The second approach is to combine building blocks into functional assemblies capable of performing multiple tasks. This is, principally, the somewhat easier approach, because it can utilize components that have been optimized with respect to one function (*e.g.* MRI contrast or plasmonic hyperthermia).[44,45] The major challenge to this strategy arises from the requirement of fabricating nanoformulations reproducibly in the amount required for applications in the clinic. A second

challenge is the result of statistical variations between individual multifunctional particles in one batch. Principally, the attachment of smaller particles or ligands to the surface of a larger particle follows a Poisson distribution. This effect limits the number of components that can be utilized to design a multifunctional theranostic nanoplatform.

Wang *et al.* have offered a most interesting approach to avoiding these pitfalls by designing multifunctional 1D magnetic and fluorescent nanoparticle chains for combined MRI and fluorescence imaging and combined

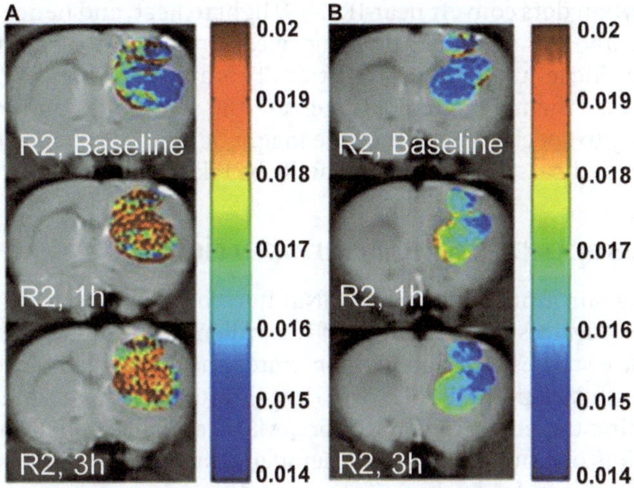

Figure 7.11 MR images of time dependent accumulation of nanoparticles in a brain tumor in the presence (a) and absence (b) of an external magnetic field (0.4 T, 30 min). The MR images utilize the same relative intensity scale. Reprinted from Biomaterials, 29 (4), B. Chertok, B. A. Moffat, A. E. David, F. Yu, C. Bergemann, B. D. Ross, V. C. Yang, Iron oxide nanoparticles as a drug delivery vehicle for MRI monitored magnetic targeting of brain tumors, 487–496, Copyright (2008) with permission from Elsevier.[43]

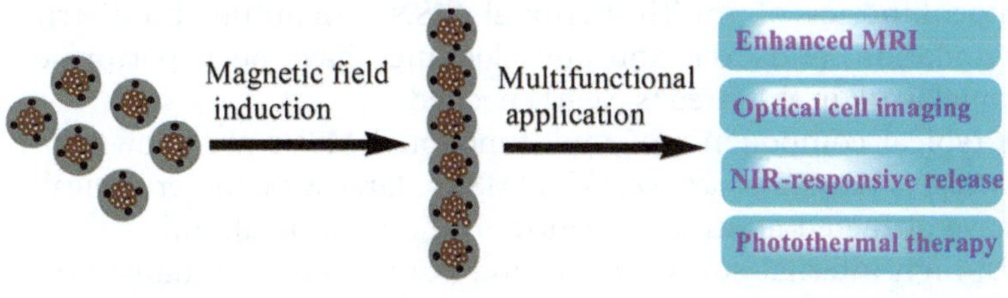

Figure 7.12 Assembling multifunctional 1D magnetic and fluorescent nanoparticle chains for combined MRI, fluorescence imaging and combined photothermal therapy and chemotherapy from mesoporous carbon particles containing Fe_3O_4 nanoparticles in their centers and fluorescent carbon dots in their shells. Reproduced with permission from ref. 46. Copyright (2014) American Chemical Society.

photothermal therapy and chemotherapy (see Figure 7.12).[46] The building blocks of the 1D assemblies are clustered Fe_3O_4 nanocrystals in the core of mesoporous fluorescent carbon dots that feature hydroxyl/carboxyl groups on their surfaces. Due to their mesoporous structure, drugs can be adsorbed into the carbon-based framework. These hybrid nanoparticles assemble under external magnetic fields and form stable 1D chains ~90 nm in diameter and ~3 μm in length. The formation of 1D chains enhanced the ability of the material to act as an MRI contrast agent. The 1D chains are capable of penetrating cell clusters and enrich in the intracellular space. While the mesoporous carbon shell provides high drug loading capacity, the embedded fluorescent carbon dots convert near-IR (NIR) light to heat, and hence kill tumor cells efficiently and enhance the drug release rate to further improve the therapeutic efficacy under NIR irradiation. This approach is most promising, because it uses the magnetic field to create the assemblies and to enhance drug delivery to specific sites. Once the magnetic field is switched off, the 1D structures will disassemble and eventually be taken up by endocytosis.

7.4 Mesoporous Silica Nanoparticles

Mesoporous silica nanoparticles (MSNs) have been intensely discussed as drug delivery devices since the turn of the millennium. Today, there is a history of *in vitro* studies that have demonstrated that MSNs can be designed to exhibit stimulus responsive release of drugs, suitable cellular uptake, as well as cell-specific targeting.[47] Furthermore, MSNs are able to deliver relatively large payloads of a broad variety of therapeutic agents, ranging from hydrophobic drug molecules to therapeutic peptides and imaging agents.[48] However, in spite of all reported successes, MSNs have not performed exceedingly well in meaningful animal models to date.[47,48] As expressed in a recent review by Argyo *et al.*, "Researchers are still faced with many challenges, especially *in vivo* applicable stimuli-responsive release mechanisms, targeting specificity, and biosafety issues, which need to be fully understood to achieve efficient and safe drug delivery. A complete understanding of the mechanisms for intracellular uptake, trafficking, and the fate of multifunctional MSNs in the body would be highly desirable. Up to now, only a few *in vivo* studies of the pharmacokinetics of multifunctional MSNs including biodistribution, biodegradation, and excretion and clearance have been performed, and additional work in this area is clearly needed."[49]

The typical components of multifunctional MSNs are shown in Figure 7.13. *Mesoporous silica nanoparticles* (MSNs, largest diameter typically <500 nm) have attracted attention because of their colloidal and structural stabilities, high internal surface areas and pore volumes, tunable pore sizes, and the possibility to chemically functionalize the inner pore system and/or the external particle surface. These highly attractive features make MSNs a promising and versatile platform for biomedical applications. Their *cargo* can consist of any drug and/or fluorescent dye molecule that can be absorbed in the pores or adsorbed on the MSN's surface. *Gatekeepers* have the crucial function of retaining cytotoxic drugs within the mesoporous silica network during transport to the tumor cells. Selecting the right gatekeeper is crucial. Around 5% leakage is acceptable during passive and active tumor targeting from MSNs while circulating in the bloodstream. Gatekeepers must be activatable upon reaching the delivery site. Triggering mechanisms can be either external (*e.g.* enzymatic cleavage) or internal stimuli (*e.g.* pH-sensitive bonds). The actual coating mechanisms can be Molecular Pore Gating, Particle Pore Gating, and Surface Coating.

Examples for Molecular Pore Gating are the photochemical cleavage of thymidine dimers,[50] the dethreading of cyclodextrines,[51] cucurbit[6]uril[52] or nitroveratryl carbamate.[53] Surface Coating of MSNs was performed by polymer shells[54] and supported lipid bilayers.[55] All of these applications require UV/Vis light, which make them difficult to apply for the treatment of deep seated tumors.[56] Particle Pore Gating can be achieved by combining a superparamagnetic nanoparticle (*e.g.* SPION[57]) in an A/C-magnetic or Rf-field. This can also be achieved with a plasmonic gold nanoparticle[58] featuring a bond, molecule or macromolecule that can be thermally cleaved.[49] *Tracking Markers* enable the 2D- and 3D-imaging of MSNs by means of (confocal) microscopy, optical tomography, or Magnetic Resonance Imaging. *Targeting Ligands* enable active targeting of cancer cells and other targets (Figure 7.13), whereas *Endosomal Escape Triggers* ensure rapid escape from endosomes after uptake. Otherwise, the presence of cathepsins and other proteases, as well as the low pH in endosomes will cause the (partial) digestion of MSNs and their surface decoration and payload. *Spacers* are required for enhanced circulation time (polyethylene glycol, PEG) or enhanced mobility of targeting ligands at the MSN surfaces.

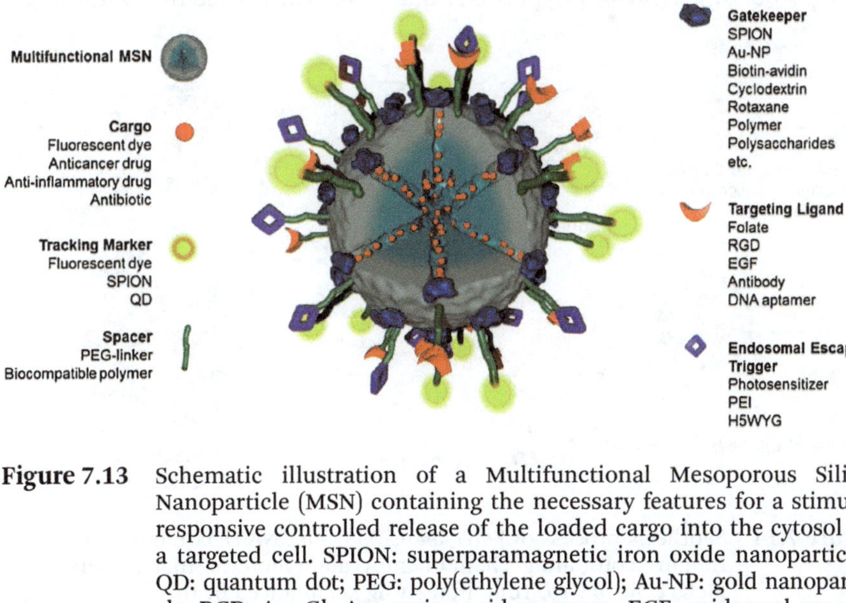

Figure 7.13 Schematic illustration of a Multifunctional Mesoporous Silica Nanoparticle (MSN) containing the necessary features for a stimuli-responsive controlled release of the loaded cargo into the cytosol of a targeted cell. SPION: superparamagnetic iron oxide nanoparticle; QD: quantum dot; PEG: poly(ethylene glycol); Au-NP: gold nanoparticle; RGD: Arg-Gly-Asp amino acid sequence; EGF: epidermal growth factor; PEI: poly(ethylene imine); H5WYG: endosomolytic peptide. Reproduced with permission from ref. 49. Copyright (2014) American Chemical Society.

7.4.1 Composite Structures Utilizing Mesoporous Silica Nanoparticles

Li *et al.* designed a novel multifunctional nanoparticle for the purpose of near-infrared (NIR)-triggered drug release for various biomedical applications. The core of the highly innovative structure consists of a Fe_3O_4-Au heterostructure with polyhedral core and mesoporous SiO_2 shell.

Oligonucleotides (referred to as dsDNA) were used as pore blockers of the mesoporous silica shell that permit the controlled release, resulting in an NIR-responsive DNAgated Fe_3O_4@Au@MSN@DOX nanocarrier (Figure 7.14). Taking advantage of the magnetic properties of the Fe_3O_4-containing cores, magnetically aided targeting in conjunction with NIR-triggered doxorubicin-release was successfully established using nude mice (BALB/cAnN) bearing HeLa cell xenografts. Fe_3O_4@Au@MSN nanocarriers without pre-loaded doxorubicin served as negative control. These experiments clearly indicate that mesoporous silica nanolayers can be successfully integrated in composite structures for treating cancer and other diseases. The resulting theranostic nanoplatforms are very versatile and capable of magnetically aided targeting, MRI diagnosis, and combination therapy through the manipulation of a magnet and a NIR laser. The results verified the significant therapeutic effects on tumor models. Therefore, MSN-based systems are excellent candidates for translation into the clinic.

Wang *et al.* have engineered a *Magnetic Graphene-Based Mesoporous Silica* material capable of dual tumor targeting and photothermal chemotherapy.[60] The novel material comprises peptide-modified magnetic graphene-based mesoporous silica (MGMSPI) particles that were synthesized in several steps.

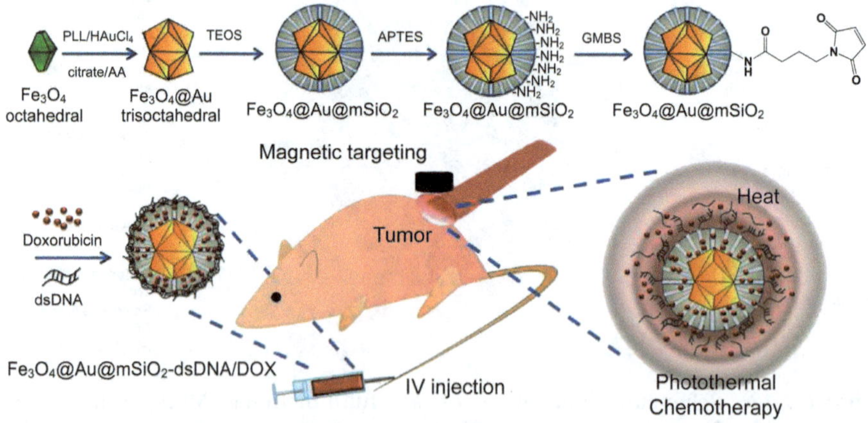

Figure 7.14 Synthesis process of Fe_3O_4@Au@MSN-dsDNA/DOX nanoparticles for therapy combining chemotherapy and photothermal treatment of HeLa cells *in vivo*. Tumor targeting is enhanced by the presence of an external magnet. Reproduced with permission from ref. 59. Copyright (2014) American Chemical Society.

First, magnetic graphene was obtained by one-step solvothermal reduction of graphene oxide nanosheets and Fe^{3+} in the presence of acetate reducing agents. Then these particles were reacted with APTES/TEOS-ethanol (aminopropyltriethoxysilane/tetraethyl orthosilicate) solution followed by the application of the hydrothermal method to obtain amido-mesoporous silica coated nanoparticles. Transmission electron microscopy images clearly showed the even distribution of iron nanoparticles on graphene nanosheets. These particles were then treated with bifunctional PEG (NHS-PEG3400-maleimide) in PBS to obtain PEGylated particles. They were then modified with the glioma-targeting interleukin-13-based peptide to facilitate targeted delivery. The classic anticancer agent doxorubicin (DOX) was loaded into the MGMSPI nanoparticles to obtain the final mesoporous, magnetic, nanotheranostic agent MGMSPI. Drug release studies showed that low pH releases DOX by weakening the electrostatic interaction between DOX and mesoporous silica. Furthermore, heat (NIR laser irradiation—808 nm) in conjunction with low pH diminishes the hydrophobic interaction between DOX and graphene to subsequently release the drug. However, it is noteworthy that the cumulative release was less than 25% after 24 h. Therefore, this formulation is ideal for sustained release of DOX because of its high loading capacity and controlled released ability. Since this nanoplatform contains Fe_3O_4 nanoparticles, it is a T_2-weighted MRI contrast agent. At the same time, targeted and magnetically enhanced drug delivery under external magnetic field is possible. Strong NIR absorbance and the possibility of using the photothermal effect for local cancer treatment is provided by the graphene nanosheets within the structure.[60] The resulting theranostic agent combines the following characteristics:

(1) Water dispersibility and virtual nontoxicity resulting from the biocompatible mesoporous silica and PEG coating, (2) Strong near-infrared absorbance of the reduced graphene oxide (photothermal therapy is possible), (3) High loading capacity and sustained release of the anticancer drug doxorubicin due to adsorption within mesoporous silica and the π–π stacking interaction with graphene, (4) Large transverse relaxivity, which is caused by dispersed Fe_3O_4 nanoparticles, which enables MRI tracking of the nanoparticles, (5) High tumor-specificity achieved by receptor-mediated targeting in combination with magnetic targeting, and (6) pH- and photothermally triggered release of doxorubicin. This makes it one of the most advanced theranostic nanoplatforms to date.

7.5 Conclusions

As discussed in this chapter, Magnetic Theranostic Agents have gone through an intense evolution process during the last decade. Consequently, they have reached an unprecedented structural and functional diversity. Nanomicelles, Nanoemulsions, Nanovesicles, Microcapsules, Magnetoliposomes, Polymer-based Theranostic Agents, Dendrimers, Multifunctional Magnetic Nanoparticles, and Mesoporous Silica Nanoparticles have been discussed in this particular sequence, because it demonstrates the increasing complexity of theranostic structures and functions. It is the conviction of the authors that this evolution will proceed on the academic level, whereas a strong selection process will occur during translation of theranostic materials and procedures in the clinic.

References

1. J. Xie, S. Lee and X. Chen, *Adv. Drug Delivery Rev.*, 2010, **62**, 1064–1079.
2. H. Gong, Z. L. Dong, Y. M. Liu, S. N. Yin, L. Cheng, W. Y. Xi, J. Xiang, K. Liu, Y. G. Li and Z. Liu, *Adv. Funct. Mater.*, 2014, **24**, 6492–6502.
3. B. E. Smith, P. B. Roder, X. Zhou and P. J. Pauzauskie, *Nanoscale*, 2015, **7**, 7115–7126.
4. H. Y. Yoon, S. Jeon, D. G. You, J. H. Park, I. C. Kwon, H. Koo and K. Kim, *Bioconjugate Chem.*, 2017, **28**, 124–134.
5. H. Huang and J. F. Lovell, *Adv. Funct. Mater.*, 2017, **27**, DOI: 10.1002/adfm.201603524.
6. Y. Feng, N. Panwar, D. J. H. Tng, S. C. Tjin, K. Wang and K.-T. Yong, *Coord. Chem. Rev.*, 2016, **319**, 86–109.
7. D. Chen, C. A. Dougherty, K. Zhu and H. Hong, *J. Controlled Release*, 2015, **210**, 230–245.
8. S. Kunjachan, J. Ehling, G. Storm, F. Kiessling and T. Lammers, *Chem. Rev. (Washington, DC, U. S.)*, 2015, **115**, 10907–10937.
9. P. Xing and Y. Zhao, *Adv. Mater. (Weinheim, Ger.)*, 2016, **28**, 7304–7339.
10. N. D. Thorat, R. A. Bohara, M. R. Noor, D. Dhamecha, T. Soulimane and S. A. M. Tofail, *ACS Biomater. Sci. Eng.*, 2016, DOI: 10.1021/acsbiomaterials.6b00420.
11. M. K. Goftar, N. A. Rayeni and S. Rasouli, *Int. J. Adv. Biol. Biomed. Res.*, 2014, **2**, 2431–2436.
12. H. Wang, T. B. Shrestha, M. T. Basel, R. K. Dani, G.-M. Seo, S. Balivada, M. M. Pyle, H. Prock, O. B. Koper, P. S. Thapa, D. Moore, P. Li, V. Chikan, D. L. Troyer and S. H. Bossmann, *Beilstein J. Nanotechnol.*, 2012, **3**, 444–455, 412 pp.
13. G. Tong, Z. Fang, G. Huang, Y. Jing, S. Dai, Q. Jiang, C. Zhang, S.-T. Feng and Z.-P. Li, *RSC Adv.*, 2016, **6**, 50534–50542.
14. F. Yaqub, *Lancet Oncol.*, 2013, **14**, e296.
15. G. J. Stasiuk and N. J. Long, *Chem. Commun. (Cambridge, U. K.)*, 2013, **49**, 2732–2746.
16. C. Burtea, S. Laurent, L. Vander Elst and R. N. Muller, *Handb. Exp. Pharmacol.*, 2008, **185**, 135–165.
17. M. Jaiswal, R. Dudhe and P. K. Sharma, *3 Biotech*, 2015, **5**, 123–127.
18. A. Gupta, H. B. Eral, T. A. Hatton and P. S. Doyle, *Soft Matter*, 2016, **12**, 2826–2841.
19. S. Ganta, A. Singh, N. R. Patel, J. Cacaccio, Y. H. Rawal, B. J. Davis, M. M. Amiji and T. P. Coleman, *Pharm. Res.*, 2014, **31**, 2490–2502.
20. S. Ganta, A. Singh, P. Kulkarni, A. W. Keeler, A. Piroyan, R. R. Sawant, N. R. Patel, B. Davis, C. Ferris, S. O'Neal, W. Zamboni, M. M. Amiji and T. P. Coleman, *Pharm. Res.*, 2015, **32**, 2753–2763.

21. Q. Liu, H. Zhu, J. Qin, H. Dong and J. Du, *Biomacromolecules*, 2014, **15**, 1586–1592.
22. M. Ma, F. Yan, M. Yao, Z. Wei, D. Zhou, H. Yao, H. Zheng, H. Chen and J. Shi, *ACS Appl. Mater. Interfaces*, 2016, **8**, 29986–29996.
23. K. Saravanakumar, R. Baskaran and T. R. Kubendran, *Asian J. Chem.*, 2011, **23**, 2643–2647.
24. T. M. Allen and P. R. Cullis, *Adv. Drug Delivery Rev.*, 2013, **65**, 36–48.
25. S. Tripathy, N. N. Palei and M. K. Das, *Int. Res. J. Pharm.*, 2014, **5**, 232–238, 237 pp.
26. G. Mikhaylov, U. Mikac, A. A. Magaeva, V. I. Itin, E. P. Naiden, I. Psakhye, L. Babes, T. Reinheckel, C. Peters, R. Zeiser, M. Bogyo, V. Turk, S. G. Psakhye, B. Turk and O. Vasiljeva, *Nat. Nanotechnol.*, 2011, **6**, 594–602.
27. H. Guo, W. Chen, X. Sun, Y.-N. Liu, J. Li and J. Wang, *Carbohydr. Polym.*, 2015, **118**, 209–217.
28. Z.-H. Miao, H. Wang, H. Yang, Z.-L. Li, L. Zhen and C.-Y. Xu, *ACS Appl. Mater. Interfaces*, 2015, **7**, 16946–16952.
29. L. Burlamacchi, A. Lai, M. Monduzzi and G. Saba, *J. Magn. Reson.*, 1983, **55**, 39–50.
30. R. Q. Liu, S. Liang, C. Jiang, X. Wang, Y. Gong, P. H. Li, Z. Xu, H. B. Xu and P. K. Chu, *RSC Adv.*, 2015, **5**, 87512–87520.
31. F. M. Winnik, M. F. Ottaviani, S. H. Bossmann, W. Pan, M. Garcia-Garibay and N. J. Turro, *Macromolecules*, 1993, **26**, 4577–4585.
32. S.-T. Lo, A. Kumar, J.-T. Hsieh and X. Sun, *Mol. Pharmaceutics*, 2013, **10**, 793–812.
33. S. L. Mekuria, T. A. Debele and H.-C. Tsai, *RSC Adv.*, 2016, **6**, 63761–63772.
34. A. J. Perise-Barrios, D. Sepulveda-Crespo, D. Shcharbin, B. Rasines, R. Gomez, B. Klajnert-Maculewicz, M. Bryszewska, F. J. De La Mata and M. A. Munoz-Fernandez, *RSC Nanosci. Nanotechnol.*, 2014, **34**, 246–279.
35. S. Chandra, S. Nigam and D. Bahadur, *J. Biomed. Nanotechnol.*, 2014, **10**, 32–49.
36. Y. L. Chang, X. L. Meng, Y. L. Zhao, K. Li, B. Zhao, M. Zhu, Y. P. Li, X. S. Chen and J. Y. Wang, *J. Colloid Interface Sci.*, 2011, **363**, 403–409.
37. N. Oh and J.-H. Park, *Int. J. Nanomed.*, 2014, **9**, 51–63, 13 pp.
38. B. F. Pan, D. X. Cui, Y. Sheng, C. G. Ozkan, F. Gao, R. He, Q. Li, P. Xu and T. Huang, *Cancer Res.*, 2007, **67**, 8156–8163.
39. S. Chandra, G. Noronha, S. Dietrich, H. Lang and D. Bahadur, *J. Magn. Magn. Mater.*, 2015, **380**, 7–12.
40. W. Aadinath, T. Ghosh and C. Anandharamakrishnan, *J. Magn. Magn. Mater.*, 2016, **401**, 1159–1172.
41. T. N. Lan, T. H. Hai, L. H. Phuc, L. K. Vinh, T. T. Kieu and N. Q. Hien, *Nanotechnol. Res. J.*, 2012, **5**, 297–317.
42. Z. Meng, F. Wei, W. Ma, N. Yu, P. Wei, Z. Wang, Y. Tang, Z. Chen, H. Wang and M. Zhu, *Adv. Funct. Mater.*, 2016, **26**, 8231–8242.
43. B. Chertok, B. A. Moffat, A. E. David, F. Yu, C. Bergemann, B. D. Ross and V. C. Yang, *Biomaterials*, 2007, **29**, 487–496.
44. K. Hayashi, Y. Sato, W. Sakamoto and T. Yogo, *ACS Biomater. Sci. Eng.*, 2017, **3**, 95–105.

45. A. Cervadoro, M. Cho, J. Key, C. Cooper, C. Stigliano, S. Aryal, A. Brazdeikis, J. F. Leary and P. Decuzzi, *ACS Appl. Mater. Interfaces*, 2014, **6**, 12939–12946.
46. H. Wang, A. Mararenko, G. Cao, Z. Gai, K. Hong, P. Banerjee and S. Zhou, *ACS Appl. Mater. Interfaces*, 2014, **6**, 15309–15317.
47. R. Roggers, S. Kanvinde, S. Boonsith and D. Oupicky, *AAPS PharmSciTech*, 2014, **15**, 1163–1171.
48. Z. Tao, *RSC Adv.*, 2014, **4**, 18961–18980.
49. C. Argyo, V. Weiss, C. Braeuchle and T. Bein, *Chem. Mater.*, 2014, **26**, 435–451.
50. D. He, X. He, K. Wang, J. Cao and Y. Zhao, *Langmuir*, 2012, **28**, 4003–4008.
51. D. P. Ferris, Y.-L. Zhao, N. M. Khashab, H. A. Khatib, J. F. Stoddart and J. I. Zink, *J. Am. Chem. Soc.*, 2009, **131**, 1686–1688.
52. J. Croissant and J. I. Zink, *J. Am. Chem. Soc.*, 2012, **134**, 7628–7631.
53. N. Z. Knezevic, B. G. Trewyn and V. S. Y. Lin, *Chem.-Eur. J.*, 2011, **17**, 3338–3342, S3338/3331–S3338/3336.
54. R. Liu, Y. Zhang and P. Feng, *J. Am. Chem. Soc.*, 2009, **131**, 15128–15129.
55. M. P. Dobay, A. Schmidt, E. Mendoza, T. Bein and J. O. Raedler, *Nano Lett.*, 2013, **13**, 1047–1052.
56. A. L. Vahrmeijer, M. Hutteman, J. R. van der Vorst, C. J. H. van de Velde and J. V. Frangioni, *Nat. Rev. Clin. Oncol.*, 2013, **10**, 507–518.
57. S. Giri, B. G. Trewyn, M. P. Stellmaker and V. S. Y. Lin, *Angew. Chem., Int. Ed.*, 2005, **44**, 5038–5044.
58. E. Aznar, M. D. Marcos, R. Martinez-Manez, F. Sancenon, J. Soto, P. Amoros and C. Guillem, *J. Am. Chem. Soc.*, 2009, **131**, 6833–6843.
59. W.-P. Li, P.-Y. Liao, C.-H. Su and C.-S. Yeh, *J. Am. Chem. Soc.*, 2014, **136**, 10062–10075.
60. Y. Wang, R. Huang, G. Liang, Z. Zhang, P. Zhang, S. Yu and J. Kong, *Small*, 2014, **10**, 109–116.

CHAPTER 8

Image-Guided Cancer Thermal Therapies

PEGAH FARIDI[a], STEFAN H. BOSSMANN[b] AND PUNIT PRAKASH*[a]

[a]Department of Electrical and Computer Engineering, Kansas State University, Manhattan, KS 66506, USA; [b]Department of Chemistry, Kansas State University, Manhattan, KS 66506, USA
*E-mail: prakashp@ksu.edu

8.1 Introduction

Extensive investigations of the biological effects of therapeutic heating have led to clinical translation of technologies for treatment of cancer. Several Phase II/III clinical trials have demonstrated the benefits of mild hyperthermia, moderate heating in the range of $39 < T < 45$ °C for 20–60 min, as an adjuvant to ionizing radiation and chemotherapy of select cancers.[1–3] Thermal ablation, heating to temperatures in excess of 50 °C, is increasingly being used for treatment of surgically unresectable tumors in the liver, kidney, lung, and other organs.[4] This chapter provides an overview of the biological rationale for thermal therapy and a review of the physics of tissue–energy interactions, and surveys technologies for delivering controlled thermal doses under image guidance.

Figure 8.1 illustrates the components of a generic image-guided thermal therapy system. Minimally/non-invasive applicators deliver therapeutic energy from an external source to the targeted tissue. Real-time imaging

Smart Materials No. 26
Magnetic Nanomaterials: Applications in Catalysis and Life Sciences
Edited by Stefan H. Bossmann and Hongwang Wang

Published by the Royal Society of Chemistry, www.rsc.org

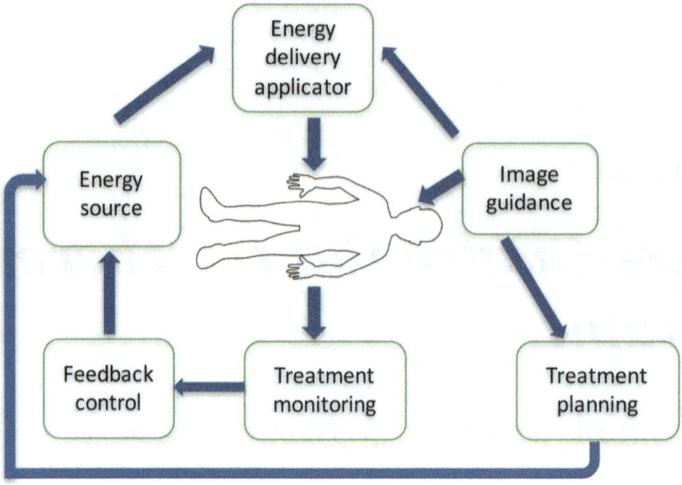

Figure 8.1 Overview of thermal therapy systems.

facilitates accurate positioning of the applicators relative to the targeted tumors, to maximize the likelihood of delivering safe and effective treatments. Imaging data may also serve as input to simulation-based predictive models that assist physicians in selecting energy delivery parameters tailored for individual patient anatomies. Techniques for measuring spatial temperature profiles in the targeted tissue provide a means for real-time monitoring of treatment progress, which is particularly important because unknown blood perfusion profiles serve as a heat sink, and may yield highly variable thermal dose profiles for a fixed set of energy delivery parameters. These can be coupled with feedback-control techniques to afford conformal delivery of prescribed thermal doses on a patient-specific basis. This chapter will detail the technical foundations of each of the components summarized in Figure 8.1.

8.2 Biological Rationale for Thermal Therapy

8.2.1 Review of Physiological Effects of Heating and Thermal Dosimetry

Figure 8.2 summarizes the physiological effects of heating over the temperature range 37–100 °C. Mild heating at temperatures in the range of 37–43 °C yields increases in local blood flow. The response of vasculature to heating differs in healthy and tumor tissues, yielding biophysical effects that can be targeted for therapeutic applications. Normal tissues have more structured vasculature than tumors, resulting in greater convective heat loss during heating. Furthermore, vessels in healthy tissue dilate in

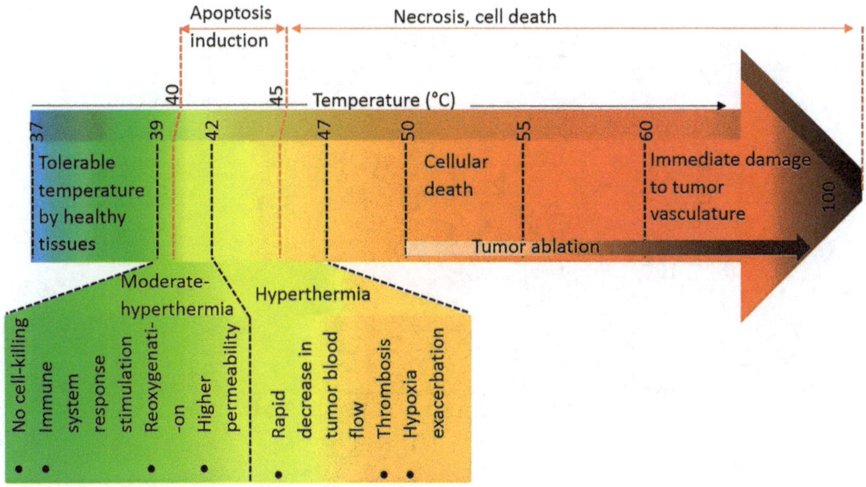

Figure 8.2 Physiological changes induced by heating.

response to heating, thereby increasing blood flow rate which in turn leads to greater heat dissipation, temperature alleviation, and less damage to normal tissues when compared to tumors.[5] Consequently, exposure to a fixed amount of therapeutic energy may yield increased heating in tumors relative to healthy tissue. Thermosensitivity of tumors is also influenced by hypoxic cancer cells and the acidic tumor microenvironment. Low pH reduces blood flow rate which in turn leads to a lower amount of oxygen delivery to the tumor cells. At higher temperatures, in excess of 45 °C, blood flow rate decreases causing thrombosis, heat trapping and tissue necrosis.[6]

Thermotolerance, an impediment to thermal cytotoxicity, is regulated by the release of Heat Shock Protein (HSP). HSPs are chaperone proteins which are upregulated during heat stress, and transport misfolded proteins to the cell membrane.[7] However, this phenomenon occurs only up to approximately 43–43.5 °C.[8]

Thermal damage induced by heating is a function of the time–temperature history during heating. Figure 8.3 illustrates an Arrhenius plot for Chinese hamster ovary (CHO) cells exposed to isothermal heating at various temperatures. In Figure 8.3, D_0 defines the time at a particular temperature for cells to reach the 63% survival rate. Cellular deactivation due to heating can be calculated using eqn (8.1):

$$B = Ae^{(-E/RT)} \tag{8.1}$$

where B is the rate constant ($B = 1/D_0$), E is the activation energy [kcal mol^{-1}], R is the universal gas constant, A is the frequency factor, and T is the temperature in K. A clear breakpoint in rate of deactivation is observed at ~43 °C.

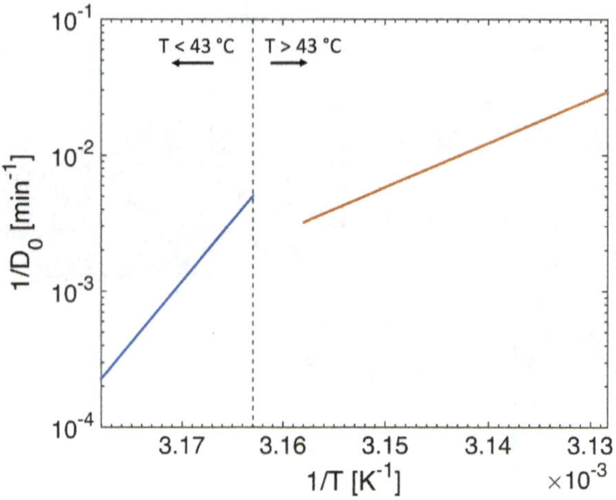

Figure 8.3 Arrhenius plot for CHO cells (Arrhenius coefficients for generating this plot are taken from ref. 9).

Sapareto and Dewey[58] developed a thermal isoeffective dose model for comparing non-isothermal exposures to an equivalent isothermal exposure at a reference temperature. 43 °C is typically taken as the reference temperature due to the breakpoint in deactivation at this temperature. The Sapareto–Dewey thermal dose model is given by eqn (8.2):

$$\text{CEM43} = \int_0^t R^{\left(T_{\text{break}} - T(\tau)\right)} \mathrm{d}\tau \tag{8.2}$$

where CEM43 is the cumulative equivalent minutes at 43 °C, R represents the rate of damage below and above break point, and T stands for the measured temperature during heat exposure. For most cells, at temperatures below 43 °C, $R = 0.25$; at temperatures greater than 43 °C, $R = 0.5$. This thermal dose model has found widespread applications in research and clinical studies for comparative assessment of the effects of therapeutic hyperthermia exposures.[10]

8.2.2 Moderate Heating/Hyperthermia

Moderate heating in the range of 40–45 °C for 20–90 min has the following effects:

8.2.2.1 Mild Cytoxicity

Heating in the range of 40–45 °C for 20–90 min has a cytotoxic effect on tumor cells. While early clinical applications of hyperthermia aimed at achieving cell kill *via* heating, this proved to be ineffective due to the technical challenges of maintaining tumor temperatures within the desired change in the absence of real-time volumetric thermometry.

8.2.2.2 Increased Efficacy of Chemotherapy and/or Ionizing Radiation

In addition to cytotoxic effects, heating at temperatures in the range of 40–45 °C synergizes with chemotherapy and ionizing radiation therapy. Large tumors have reduced blood flow compared to healthy tissues due to the disordered vasculature structure. Consequently, the effect of chemotherapy in large tumors is hampered, because the poor blood flow limits the amount of drug delivered to the tumor. During hyperthermia, physiological changes such as higher blood flow rates and higher vascular permeability due to increased vessel pore size, result in enhanced delivery of chemotherapeutic drugs.[11] Furthermore, increased blood flow also alleviates hypoxia in large tumors, which increases the efficacy of ionizing radiation therapy. Radiation therapy takes advantage of reoxygenation of the hypoxic tumor environment because molecular oxygen (O_2) possesses a high potential for radio-sensitization.[12–16] Mild temperature hyperthermia also influences a part of the cell cycle which is resistant to radiation (*i.e.*, the S-phase).[17]

8.2.2.3 Stimulation of the Immune Response

Tumor cells possess antigens that are recognizable by the immune system.[18] Hyperthermia navigates a high range of these unique antigens to the tumor surface, thus enhancing the performance of the immune system response to tumors.[19] Moreover, hyperthermia initiates a negative feedback response with the induction of HSPs on the tumor site during heat exposure, which magnifies the immune response.[20] HSPs activate dendritic cells (DCs) which are antigen-restricted and do not recognize normal cells, thereby inhibiting collateral damage to normal cells.[18] Additionally, hyperthermia modifies blood flow in tumor sites which in turn facilitates a higher rate of immune cell transmission to the tumor. Specific treatment conditions that maximize immune stimulation remain the subject of active research.[21]

8.3 Tumor Ablation

Thermal tumor ablation refers to the destruction of tumors by exposing tumor tissue to $T > 50$ °C for 10–15 min. Cell death at temperatures in excess of ~60 °C occurs within a few seconds. Several energy modalities have been applied for thermal ablation, including: radiofrequency current, microwaves, lasers, and ultrasound. Cryoablation refers to tissue freezing which has been clinically applied for tumor ablation. Non-thermal techniques for thermal ablation, such as irreversible electroporation have also been developed.

During an ablation procedure, a minimally invasive applicator (typically, less than 2.5 mm in diameter) is inserted into the target tumor under image guidance. Energy delivered by the device is absorbed by surrounding tissue leading to heating. In proximity to the ablation device, the majority of heating is due to direct power absorption from the energy source. Due to steep thermal gradients, heat spreads radially outward from the ablation applicator, leading to enhancement of the ablation zone by passive thermal conduction.

As illustrated in Figure 8.4, the ablation zone can be divided into three sub-sections around the applicator, each displaying a variety of physiological changes based on the extent of tissue heating. In the central zone, the tissue in closest contact with the ablation device, necrosis is the dominant feature. Tissue in the central zone is at temperatures in excess of 50 °C, and peak temperatures could be greater than 100 °C depending on the energy modality used. In the transition zone, tissue temperature is in the range of 42–47 °C; in this region, passive heat transfer due to thermal conduction is the major source of heating. Apoptosis and some forms of sub-lethal thermal damage are observed in the transition zone. Beyond the transition zone, the tissue temperature in the mild hyperthermia zone ranges from 39–42 °C.[4,22]

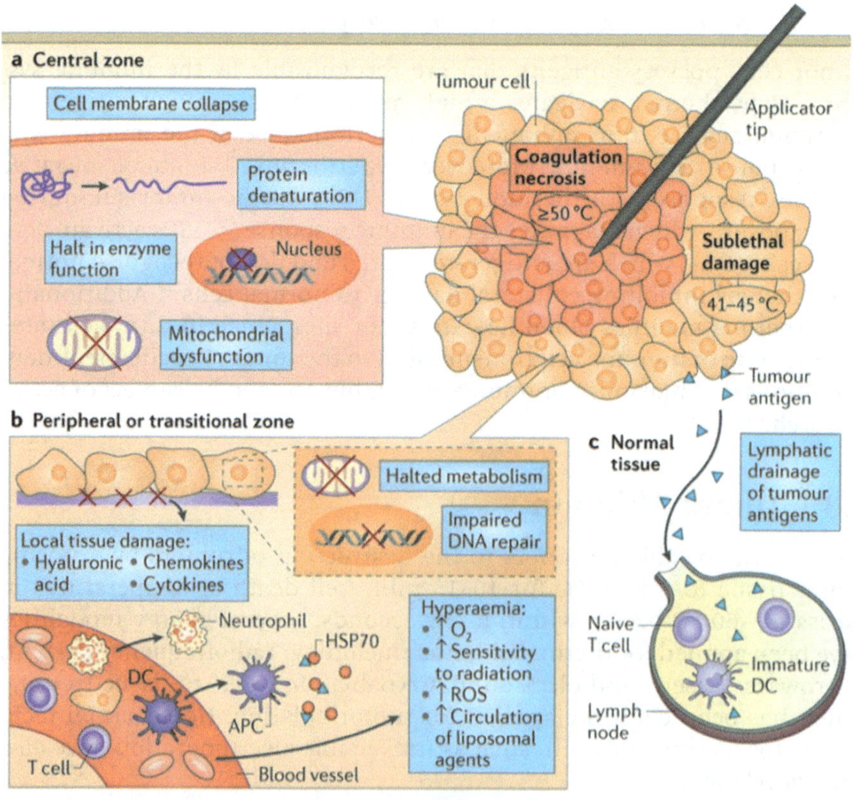

Figure 8.4 Sub-sections of ablation zone, (A) central zone, (B) transitional zone, (C) mild hyperthermia zone. (Reprinted with permission from Macmillan Publishers Ltd: Nature Reviews Cancer,[23] copyright 2014)

8.4 Thermally-Triggered Release of Therapeutic Agents

There are significant barriers obstructing drug delivery to cancer cells which limits the specificity of chemotherapy.[24] To alleviate this problem, therapeutic agents may be transported and delivered through carriers such as polymers, liposomes and particles. Using drug delivery systems employing carriers in conjunction with heat exposure has led to an improvement in increasing the quantity of drug delivered to tumors.

Polymers, macromolecular carriers, are biocompatible materials used for delivering chemotherapeutic agents to the target. Responsive polymers are a specific type of polymer that respond to external stimuli (physical or chemical). One class of polymers, referred to as thermosensitive polymers, undergo a phase transition at a critical temperature. Using heat as a trigger for responsive polymers permits the use of hyperthermia for localized drug release.[25] Temperature sensitive responsive polymers are subjected to phase transition based on distinctively selected temperature thresholds according to the drug delivery system. The hydrophilic structure of responsive polymers is activated below its lower critical solution temperature (LCST) where it remains in aqueous solution; however, above LCST it collapses and releases therapeutic agents. A typical critical temperature is in the range of 39–40 °C as this value is above normothermic temperature, and at the same time, lower than hyperthermic temperatures where vascular stasis occurs, thereby limiting drug release within the heated target.[26]

Liposomes for drug delivery are vesicles containing a bilayer lipid surrounding aqueous solution which establishes the context of injecting and transporting hydrophilic and hydrophobic therapeutic drugs at the same time. Temperature sensitive liposomes (TSLs) exploit the contribution of hyperthermia in conjunction with liposomal drug delivery to transport, aggregate and preserve chemotherapeutic agents. TSLs are triggered above a temperature threshold, which is associated with temperature change during hyperthermia procedure, to decrease the possibility of drug extravasation into healthy cells. Figure 8.5 illustrates the temperature dependent structure of TSLs.[27]

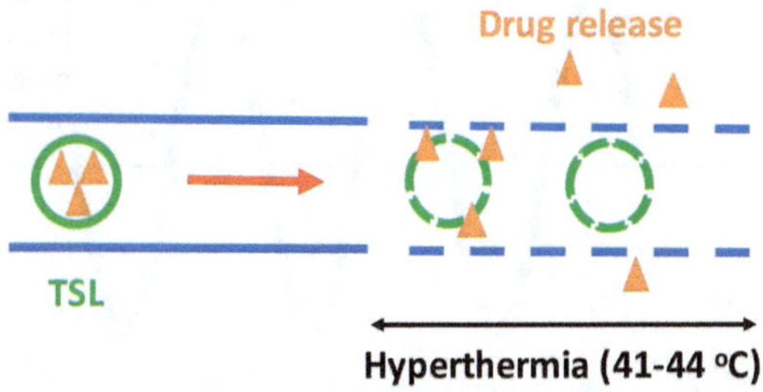

Figure 8.5 Liposomal drug release during hyperthermia (HT) in tumors. At temperatures in the range of 41–44 °C, the drug is released from temperature sensitive liposomes (TSL) within the vasculature. Extravasation is enhanced by increased vessel permeability due to heating.

8.5 Methods of Tissue Heating

8.5.1 Microwave Heating

Microwaves, electromagnetic waves with frequency ranging from 30–300 MHz have been widely used for thermal therapy (both hyperthermia and thermal ablation). Polar molecules in tissue, such as water and protein, attempt to align with the polarity of the electric field, leading to heating by dielectric hysteresis (see Figure 8.6).[28]

Microwave propagation and absorption within tissue is governed by the complex dielectric properties of tissue $\varepsilon(\omega) = \varepsilon'(\omega) - \varepsilon''(j\omega)$, where ω is angular frequency. The real part, $\varepsilon'(\omega)$, is a measure of a material's ability to store the electric field, while the imaginary part, $\varepsilon''(j\omega)$, is a measure of loss. The effective electrical conductivity is given by $\sigma_{\text{eff}} = \sigma_{\text{s}} + \omega\varepsilon''$. Frequency-dependent dielectric properties for a range of tissue types have been measured by several groups and reviewed in, ref. 29,30. From these fundamental properties, the absorption coefficient α can be computed using:

$$\alpha = \omega\sqrt{\mu\varepsilon}\left\{\frac{1}{2}\left[\sqrt{1+\left(\frac{\sigma}{\omega\varepsilon}\right)^2}-1\right]\right\}^{1/2} \tag{8.3}$$

The penetration depth, $d[m]$, is defined as the reciprocal of the attenuation coefficient and is a measure of the distance within a material which a wave propagates before it is attenuated to $1/e$ (*i.e.* 36.8%) of its original intensity.

$$d = \frac{1}{\alpha} = \frac{1}{\omega\sqrt{\mu\varepsilon}\left\{\frac{1}{2}\left[\sqrt{1+\left(\frac{\sigma}{\omega\varepsilon}\right)^2}-1\right]\right\}^{1/2}} \tag{8.4}$$

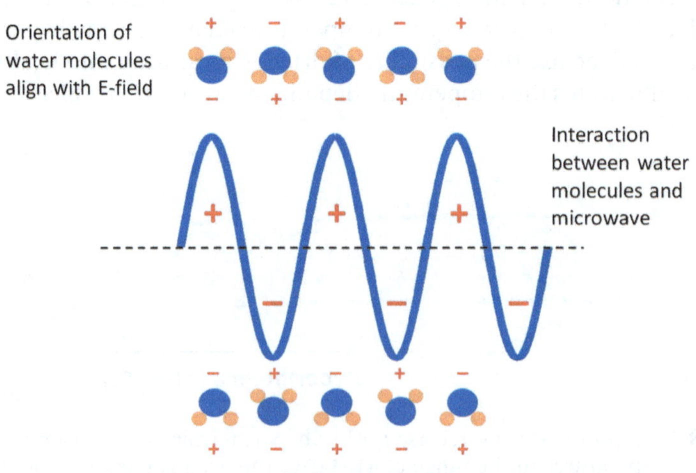

Figure 8.6 Schematic illustration of the interaction between water molecules and microwaves.[28]

Electromagnetic penetration depth, d, decreases with increasing frequency. Thus, lower frequencies are better suited for non-invasive energy delivery to deep-seated tissues. However, smaller wavelengths at higher frequencies facilitate the use of smaller antennas and ability to focus power deposition patterns to smaller regions. The choice of operating frequency for specific applications trades off these two considerations.

8.5.1.1 *Microwave Hyperthermia Systems*

Microwave thermal therapy systems consist of a high power signal source, power delivery cables, and microwave antennas that radiate energy to the target tissue. While early systems employed magnetrons for signal sources, contemporary systems largely employ solid state amplifiers. For hyperthermia, non-invasive applicators operating at 100–400 MHz have been developed for regional heating of pelvic,[31] limb,[32] and head-and-neck targets. At these frequencies, wavelengths are relatively long, leading to large focal spots, with limited precision of heating.[33] A challenge with non-invasive delivery is the unintended heating ("hot spots") due to dielectric differences at tissue interfaces. Invasive applicators, also referred to as interstitial applicators, operating at 915 MHz have been developed and clinically applied for non-invasive heating of tumors in the prostate and brain.

Several microwave applicator designs have been presented in the literature and are in clinical use for hyperthermia and tumor ablation.[33] Systems and devices for microwave ablation have evolved from those originally developed for hyperthermia, as recently reviewed by Ryan *et al.*[34]

8.5.1.2 *Microwave Ablation Systems*

Figures 8.7 and 8.8 illustrate available clinical microwave ablation systems and applicators. As illustrated in these figures, microwave devices are thin cylindrical antennas, attached to a microwave generator to produce an

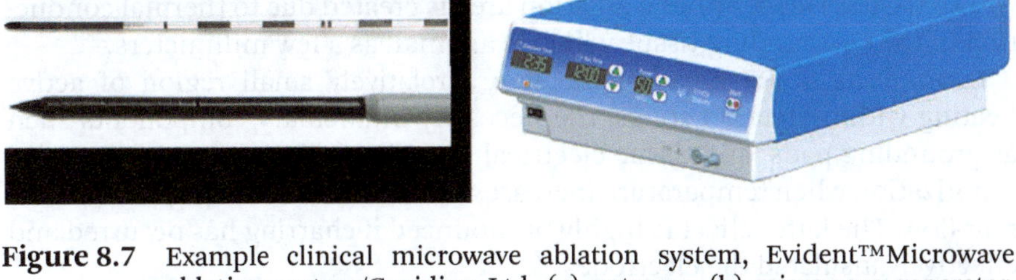

Figure 8.7 Example clinical microwave ablation system, Evident™Microwave ablation system/Covidien Ltd, (a) antenna, (b) microwave generator. Reprinted from International Journal of Surgery, 6, G. Carrafiello, D. Laganà, M. Mangini, F. Fontana, G. Dionigi, L. Boni, F. Rovera, S. Cuffari, C. Fugazzola, Microwave tumors ablation: Principles, clinical applications and review of preliminary experiences, S65–S69, Copyright (2008) with permission from Elsevier.[43]

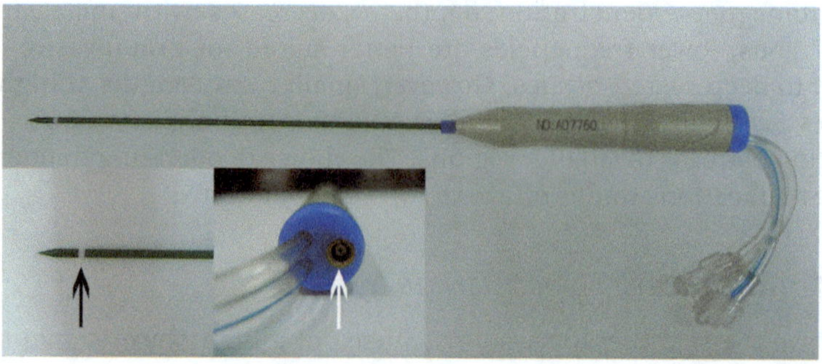

Figure 8.8 Example of a 2.45 GHz cooled-shaft microwave ablation device from Nanjing Yigao Microwave Electric Institute, Nanjing, China. The black arrow displays the radiating section of the applicator. The microwave generator is connected to applicator *via* the connector indicated with the white arrow. (Image reproduced with permission from Zhou W. *et al.*, (2013) Comparison of ablation zones among different tissue using 2450-MHz cooled-shaft microwave antenna: results in *ex vivo* porcine models. PLOS One 8(8), e71873. DOI: 10.1371/journal.pone.0071873).[44]

adequate amount of power at a certain frequency; 915 and 2450 MHz are most widely used in clinical systems, though higher frequency systems are under investigation.[35,36] Early devices consisted of coaxial-based antennas. Recent developments include modified microwave devices to overcome the challenges with unintended heating along the applicator feedline. For instance, sleeve/choke antennas are designed to control the heating along the antenna shaft.[37–40] Active cooling along the antenna shaft limits unintended heating to adjacent tissue due to losses within coaxial cables.[41,42]

8.5.2 Radiofrequency Current

RF energy for tissue heating is delivered *via* electrodes inserted to the tissue, whether percutaneously or during open surgery. RF ablation typically employs currents operating at 460–500 kHz.[45,46] The mechanism of heating during RF ablation is due to the agitation, oscillation and collision of ions within tissue. The peripheral ablation area is created due to thermal conduction in the surrounding tissue which is as small as a few millimeters.[28]

Limitations of RF ablation include: a relatively small region of active heating within ablation zones, the necessity of ancillary components such as grounding pads, and tissue electrical impedance limitation due to water vaporization when temperature increases up to 100 °C impeding further current flow. The latter effect is highly pronounced if charring has occurred and effectively insulated the electrodes.[47]

RF applicators can be used for treating superficial lesions such as skin tumors, lesions in the bladder or gastrointestinal system which are accessible *via* body cavities[48] or deep-seated tumors.[49] Examples of RF applicators in clinical use are displayed in Figure 8.9.

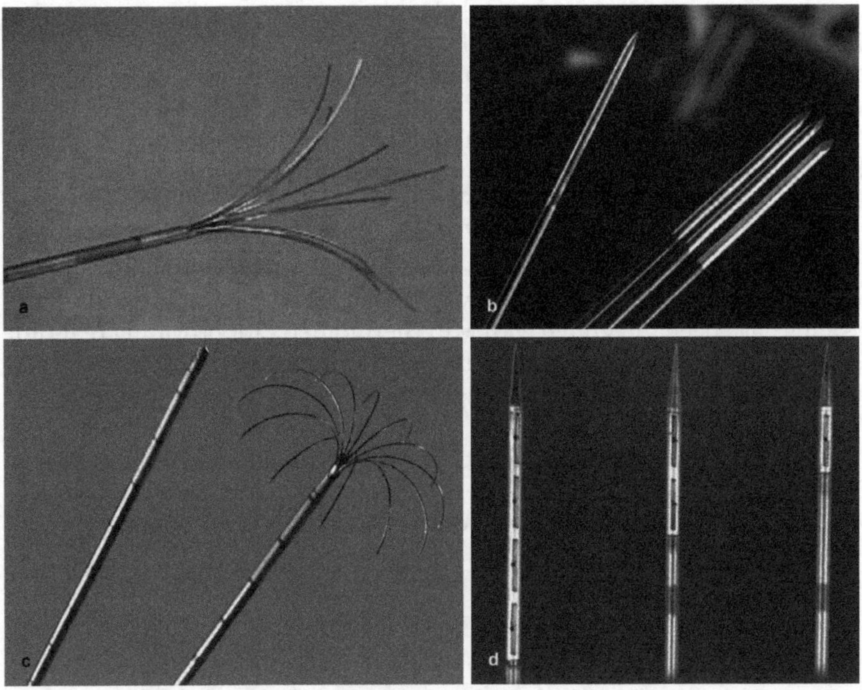

Figure 8.9 Various commercially available RF devices, (a) RITA 14G Starburst XL needle, (b) cooled-tip Radionics needle, (c) LeVeen pronged needle and (d) the Berchtold needle with holes at the end of the needle. Reproduced from European Radiology, Radio-frequency tissue ablation of the liver: *in vivo* and *ex vivo* experiments with four different systems, 13 (10), 2003, 2346–2352, A. L. Denys, © Springer-Verlag 2003 with permission of Springer.[50]

8.5.3 Lasers

Laser-tissue interaction is based on the physical mechanism of photon penetration, absorption, and scattering in tissue. Absorption occurs *via* electron excitation and transmission to a higher state. A chromophore is the light-absorbing component of the tissue that absorbs the incident laser beam resulting in heating. Penetration depth reduces rapidly with increased water content.[51] The penetration depth of laser within tissue is given by $1/\mu_a$. The total energy attenuation coefficient μ_t, is defined in eqn (8.5):

$$\mu_t = \mu_s + \mu_a \tag{8.5}$$

The higher absorptive coefficient of the tissue compared to scattering coefficient results in keeping the energy in line with the laser beam diameter. Greater scattering coefficient diffuses the energy outside the beam light diameter.[52] Taking into account all the above laser principles and optical characteristics, μ_s, μ_a and g, several laser tissue interactions relevant to medical applications occur, including: photochemical reactions, heating and tissue ablation, and photodisruption.[53]

In addition to interstitial tumor ablation, laser thermal therapy has applications in dermatology and ophthalmology.[54] Laser beams are either applied through optical fibers non-invasively or inserted invasively into the target. They produce thermal ablation of tissue through light delivered *via* the inserted tip. One of the applications of interstitial laser ablation is focal heating and image guided drug delivery (IGDD) as shown in Figure 8.10. Laser systems have been integrated with MRI thermometry for providing real-time assessment of target heating, and to avoid lethal damage to healthy tissue while affording delivery of the drug to the target and temperature profile during the experiment.

Figure 8.11 shows an example laser device, in clinical use for thermal ablation, incorporating a cooling catheter to restrict peak tissue temperatures within surrounding tissue and reduce the risk of applicator damage.

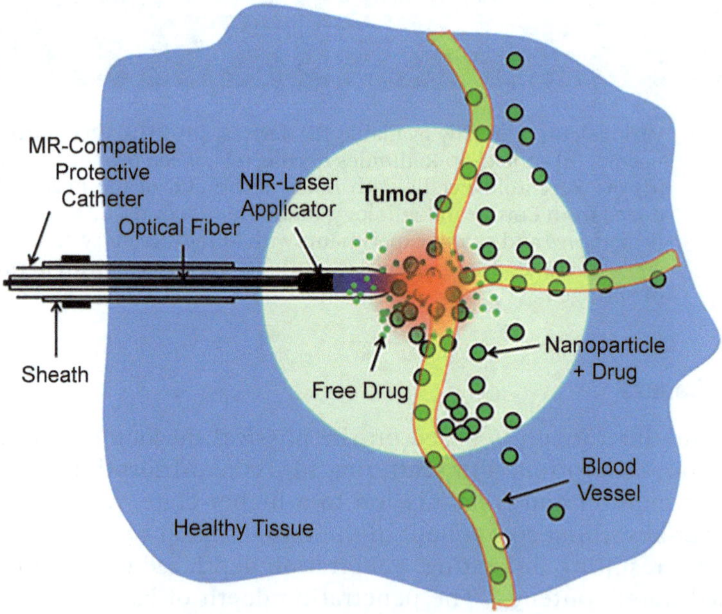

Figure 8.10 MR compatible laser applicator used for thermal ablation and drug delivery. Nanoparticles carrying drugs are disassembled during treatment releasing drugs within the tumor. Reproduced from Pharmaceutical Research, MRI-Guided Monitoring of Thermal Dose and Targeted Drug Delivery for Cancer Therapy, 30 (11), 2013, 2709–2717, R. Fernando, © Springer Science+Business Media New York 2013, with permission of Springer.[55]

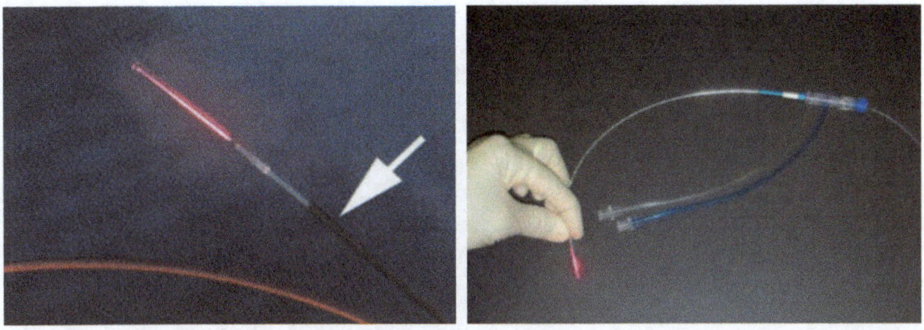

Figure 8.11 Laser induced thermal therapy (LITT) device, (left) An illuminated fiber tip and shaft covered with magnetite particles for visibility during MR Imaging, (right) Visualase laser device with cooling shaft. (Image reproduced from ref. 56)

The size of ablation zone achievable with lasers is limited (~1–2 cm) because of the rapid absorption and scattering features of small tissue ablation zones. These ablation zone sizes are relatively small compared to the other energy modalities described previously.[54]

8.5.4 Ultrasound

Ultrasound (US) is a mechanical pressure wave, with frequency in the range of 20 kHz–10 MHz, causing molecular oscillation around their mean position in the direction of wave propagation. Ultrasound propagates through the tissue which attenuates exponentially as energy moves away from the source. Acoustic intensity is defined as the time average of energy flow through a unit area. In general, US waves are transmitted from a transducer, made from piezoelectric crystals which vibrate in response to the applied input voltage thereby generating high frequency sound pressure waves, focused in a desirable spot with a diameter less than the applicator diameter at a distance 5–20 times larger than the applicator diameter (Figure 8.12).[57]

US can be used for either hyperthermia with long duration heat exposures and low temperature rise to sensitize tissue to radiotherapy/chemotherapy as discussed in Section 8.1, or ablation with short duration exposures yielding larger temperature rises. When temperature increases to 100 °C, tissue water evaporates and changes its state to gas. US energy interaction with gas bubbles inside the tissue significantly differs from liquid water molecules, due to different density and sound speed, and, therefore, temperature distribution is modified based on higher rate of reflection and scattered beams in the gas bubbles. High pressure ultrasound produces microbubbles, in a process called acoustic cavitation. Inertial cavitation is a phenomenon caused by microbubble explosion due to high amplitude pressure and consecutive expansion and contraction of microbubbles, which results in mechanical tissue destruction.[57]

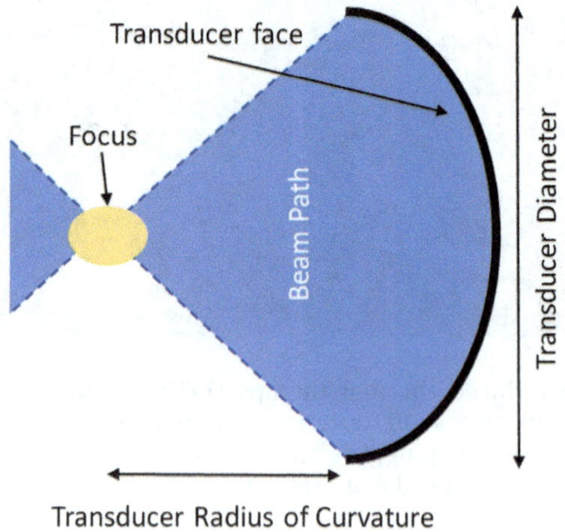

Figure 8.12 Illustration of beam path and focal spot from a curved ultrasound transducer.[57]

High intensity focused ultrasound (HIFU) ablates the tumor based on two strategies: (a) thermal effect and (b) mechanical effect. Thermal effects lead to tissue heating and cell death by coagulative necrosis, while the mechanical effects are primarily due to acoustic cavitation.[58] High frequency mechanical acoustic waves delivered to the localized target cause tissue dilution leading to periodic elevation and alleviation of tissue pressure. As a consequence of alternative changes in tissue pressure, gas bubbles are created inside tissue which can interact with US waves to eradicate tumor cells *via* necrosis. A clinical HIFU system is shown in Figure 8.13.

Single element HIFU systems may not be capable of heating large regions of interest and require mechanical steering of the beam to treat the entire target region. This displacement may cause heterogeneous temperature distribution and may lead to relatively long duration treatments.[55] To overcome these challenges, phased array transducers are employed allowing electronic steering of the ultrasound beam, by applying a time delay or change in phase angle to the signal exciting each transducer, such that an optimized amount of energy is projected to the desired spot. Phased array transducers are used to enable heating of the target with varying beam shapes and sizes, adjustable focal point size and position, to avoid discontinuities between the focal points of all the transducers. Phased array transducers may be excited to yield single or multiple focal points to heat up a larger area at the same time to reduce the treatment time.[57,60–63] As such, phased array transducers provide strong flexibility for adjusting focal spot size, shape, and location, to minimize unintended heating of non-targeted tissues.[64]

Although thermal HIFU has been investigated for more than a decade, mechanical HIFU is still under development[10] and the corresponding

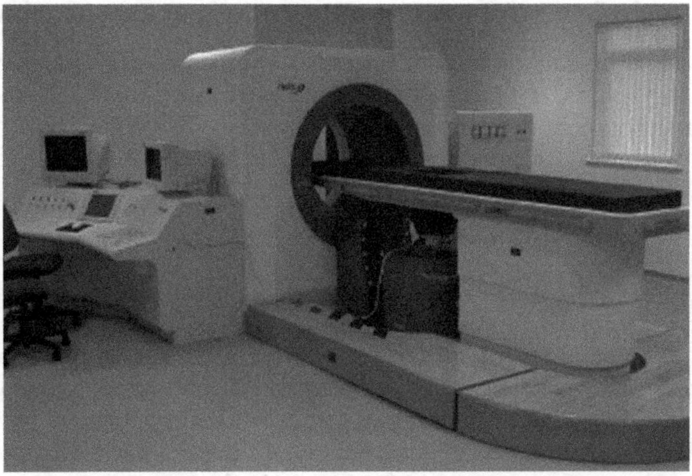

Figure 8.13 JC HIFU system used for patient monitoring and therapy (HAIFU Technology Company, Chongqing, PR China) (Reproduced from Ultrasonics, 42, 1–9, J. E. Kennedy, F. Wu, G. R. ter Haar, F. V. Gleeson, R. R. Phillips, M. R. Middleton, D. Cranston, High-intensity focused ultrasound for the treatment of liver tumours, 931–935, Copyright (2004) with permission from Elsevier).[59]

physiological changes occurring during HIFU ablation are still not well understood.[65,66] Integrating HIFU with temperature monitoring techniques, such as magnetic resonance imaging thermometry enables delivery of feedback control techniques for achieving prescribed thermal doses.[67]

8.6 Image Guidance and Monitoring for Thermal Therapies

8.6.1 Significance of Image Guidance

Thermal therapy regimes aim to target tumor sites with negligible damage to healthy tissue. Heterogeneity in tissue biophysical properties may lead to highly variable delivered thermal doses during thermal therapies. One of the most significant changes is the blood perfusion which varies based on tissue architecture. In order to verify the amount of thermal dose delivered to the treated target and to evaluate the performance of the procedure, real-time monitoring and image guidance is of high importance. Image-based thermometry provides a thermal map to monitor the progress of thermal therapy procedures and offers the therapeutic system an opportunity to control the amount of heat delivered to the tumor site, while sparing surrounding tissue.

8.6.2 Techniques for Monitoring Treatment Progression

Non-invasive thermal dose monitoring techniques can be integrated with heating modalities, to offer real-time monitoring and treatment guidance. These imaging techniques do not require the use of the invasive insertion of thermocouples, thermistors and fiber optic temperature probes.

8.6.2.1 MR Thermometry

Magnetic resonance imaging (MRI) is one of the most prevalent, reliable and comprehensive thermal monitoring methods for visualizing the thermal profile during thermal therapy procedures. MR images consist of two sets of data: *Magnitude Imaging* conveys anatomic information such as the tumor site and the position and orientation of the thermal applicator, while *Phase Imaging* provides a relative measurement of temperature change with the proton resonance frequency shift (PRFS) method.[81,82]

Eqn (8.6) describes the conversion of phase change to temperature change:

$$\Delta T = \frac{\varphi(T) - \varphi(T_0)}{\alpha\gamma B_0 TE} \tag{8.6}$$

where each phase image during heating experiment, $\varphi(T)$, is subtracted from the initial phase image which contains a specified temperature information, $\varphi(T_0)$ so that the phase change is proportional to temperature change. In eqn (8.6), α is the PRF temperature-dependent change coefficient, γ is the gyromagnetic ratio, B_0 is the magnetic field and TE indicates the echo time defined in the imaging sequences. Figure 8.14 illustrates an anatomic MR image of a live mouse and tissue mimicking phantom with

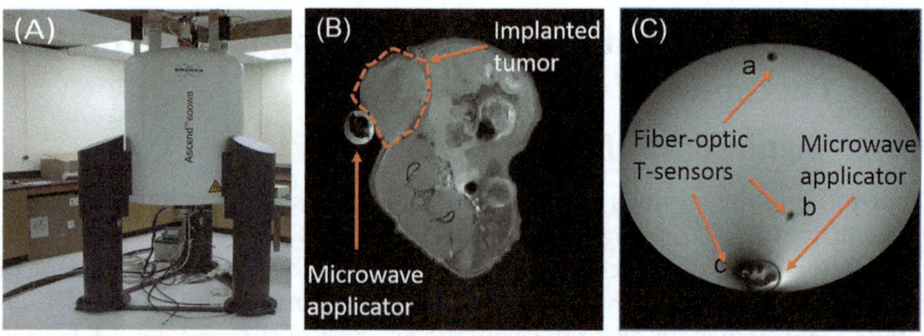

Figure 8.14 (A) Photograph of a vertical MRI scanner, (B) *in vivo* anatomical image of mouse, (C) tissue-mimicking phantom, a and b are fiber optic temperature probes inserted inside the phantom to validate the temperature measurement with PRFS method, and c shows the heating applicator.

a 14 T vertical small-animal scanner (Bruker Ascend III, 30 mm diameter micro-imaging probe).

MRI guided thermal therapy provides a non-invasive temperature monitoring method which can be integrated with different heating modalities to improve the accuracy of treatment. As shown in Figure 8.14, MR can locate the tumor site, heating applicator position and orientation along with monitoring a spatio-temporal temperature map during heating. Figure 8.15 shows an MR temperature map during breast cancer ablation using a Magnetic Resonance guided Focused Ultrasound System (MRgFUS). MR thermometry has also been investigated for guiding liver RF ablation as illustrated in Figure 8.16.

A major drawback of the PRF phase mapping method is the sensitivity to motion inside the MR scanner. For *in vivo* procedures, movement of the subject inside the scanner while scanning causes motion artifacts such as blurring and ghosting in anatomical images and it's also an impediment for temperature measurements.[70] According to eqn (8.6), the subtraction of phase images is proportional to temperature change, however, movement of the object alters the position of the specific area in each image with respect to the baseline image. Therefore, to circumvent this problem, different methods such as referenceless and multi-baseline methods have been investigated.[71–73]

8.6.2.2 *Ultrasound Thermometry*

The goal of US imaging guidance, the same as other imaging techniques, is to localize the tumor site and its boundaries inside the organs. US imaging may not be repeatable due to its operator dependency in finding the

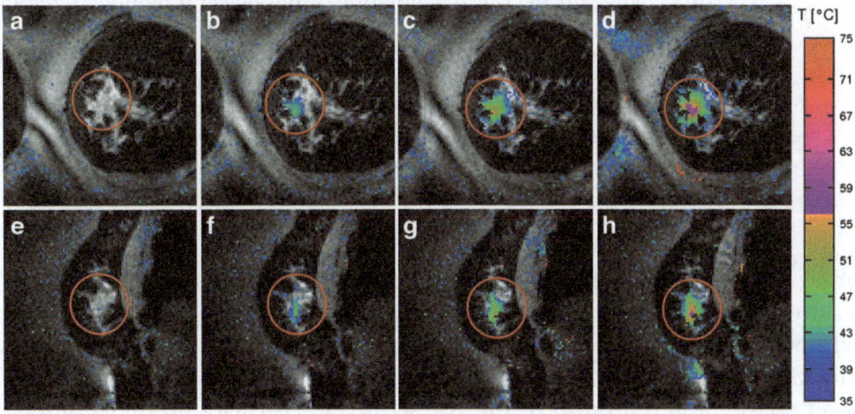

Figure 8.15 Temperature superimposition on MR magnitude images of breast cancer using HIFU with temporal resolution of 2.5 s, (a)–(d) display the coronal and (e)–(h) display the sagittal images[68]).

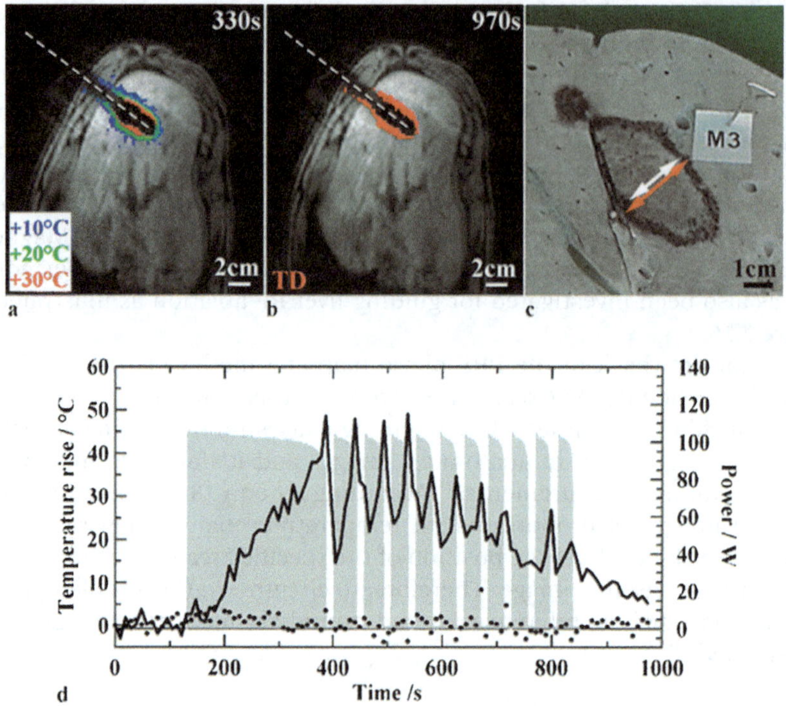

Figure 8.16 *In vivo* RF ablation on the liver of a pig, (a) MR temperature overlay on the anatomic image after 330 s, (b) Thermal dose map after 970 s at the end of treatment showing thermal damage reached tissue surrounding the RF applicator, (c) Macroscopic image of ablation zone after liver fixation in which diameter of the ablated zone (white and red arrows) is measured perpendicular to RF applicator, (d) Temperature change during thermal ablation, input power is demonstrated in gray, black dashed and dotted lines indicate the temporal evolution of temperature at 10 mm and 40 mm away from the applicator which is illustrated by a dashed white line. Reproduced from European Radiology, Real time monitoring of radiofrequency ablation based on MR thermometry and thermal dose in the pig liver *in vivo*, 18 (2), 408–416, O. Seror, © European Society of Radiology 2007 with permission of Springer.[69]

target tumor, inconsistency and artifacts caused by bone and air interaction with US beams that decrease the quality of imaging. US is better defined for imaging superficial soft tissue. In comparison to MR images and MRI-HIFU guidance, organs and abnormal tissue visualization is of lower spatial and temporal resolution with US.[57] Ultrasound thermometry techniques are under development for monitoring and guiding HIFU ablation procedures.[74,75]

Despite all the weaknesses of US compared to MR treatment guidance discussed above, US imaging has its own merits and strengths. Because of its real-time monitoring ability, the whole tumor treatment procedure can be visualized along with a variety of changes that occur in tumors, marginal tissues and also blood flow. US thermometry employs different image acquisition geometries including passive, pulse-echo, and hybrid which are briefly described below.[76]

In *passive* geometry, thermal strain and cavitation are influenced by temperature change dependent characteristics including speed of sound and beam attenuation, leading to the ability of ablation temperature monitoring. A diagnostic transducer receives the acoustic beams emitted from the HIFU treated ablation zone.[76] In some cases, the beam generating transducer, or active transducer, and the diagnostic one, or passive transducer, are the same.[77]

As for *pulse-echo* imaging, the same transducer is utilized for generating and receiving signals. During the pulse–tissue interaction, temperature dependent parameters may be altered and produce some acoustic echoes which can be used for temperature estimation. In both of the above geometries, single or array elements can be used for monitoring. Array elements can monitor a wider range of tissues.[76]

The *hybrid* method combines ultrasonic imaging with an alternative imaging modality to improve the performance of thermometry. Photoacoustics is of great interest, using US images with high spatial resolution and low contrast combined with optical imaging with reciprocal features, where the inner chromophores result in high contrast images. In summary, an optical laser produces a high energy low bandwidth pulse, leading to temperature distribution and elevation and consequently generating acoustic waves. These acoustic waves are received by the diagnostic transducer, generating photoacoustic images.[76]

8.6.2.3 CT Thermometry

X-ray computed tomography (CT) measures Hounsfield Unit, which is a temperature dependent parameter, and can thus be used to provide a thermal map during thermal therapy procedures. A major drawback of CT compared to MRI-based thermometry is the patient's exposure to radiation for acquiring CT scans. However, CT scans are available at lower costs compared to MRI scanners. The Hounsfield Unit is a function of the density of the treated tissue which in turn changes with temperature modification.[78] During thermal therapy, the temperature of the target region increases leading to tissue expansion and density reduction. Figure 8.17 illustrates the dependency of CT scans on the temperature.

8.7 Feedback Control Techniques

Due to inter-individual variations in tissue physical properties, and unknown fluctuations in blood perfusion profiles, thermal therapy procedures employing the same energy delivery parameters may yield highly

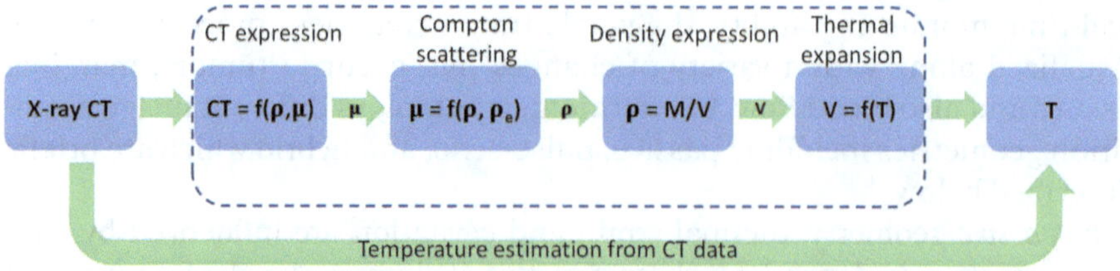

Figure 8.17 The dependency of CT value on temperature.[78]

variable heating patterns, particularly during long duration procedures such as mild hyperthermia. This necessitates the use of feedback control techniques to adjust energy delivery parameters so as to yield the desired thermal dose profiles.

Real-time monitoring is a prerequisite for non-invasive feedback control. By way of real-time monitoring, thermal dose delivery to the correct region is displayed and, consequently, feedback control algorithms can be used to adjust energy delivery parameters to meet the treatment goal. Closed loop feedback control systems can be designed based on the temperature and heat energy distribution and thermal zone estimation.[79–83] Controllers such as PI,[83,84] PID, multipoint adaptive control[85] and linear quadratic regulator[86] can be optimized with the aid of simulations. Estimating temperature change based on the pattern of derived temperature profiles throughout the whole procedure may also be helpful for optimizing the model predictive controller, especially in the case of continuous thermal dose effects after turning off the thermal source.[87]

A small animal Magnetic Resonance guided Focused Ultrasound System (MRgFUS) is displayed in Figure 8.18. In this system, a 16 element annular array is coupled to the animal resting position while performing feedback control maintaining temperature in the desired range.

Conjunction of MRI-HIFU with feedback control in hyperthermia experiments possesses a variety of advantages:

- Providing anatomic information to manage the system settings
- Real-time temperature monitoring to limit damage to healthy tissue
- Applying feedback control on the input power in correspondence to temperature change to keep the temperature in the mild hyperthermia range (39–42 °C)

In Figure 8.19 the temporal evolution of temperature is shown. It quickly rises and is then maintained as a constant inside the target utilizing a binary feedback control system that is integrated with the MRI-HIFU hyperthermia system. The thermal map illustrates how the temperature is well focused within the target.

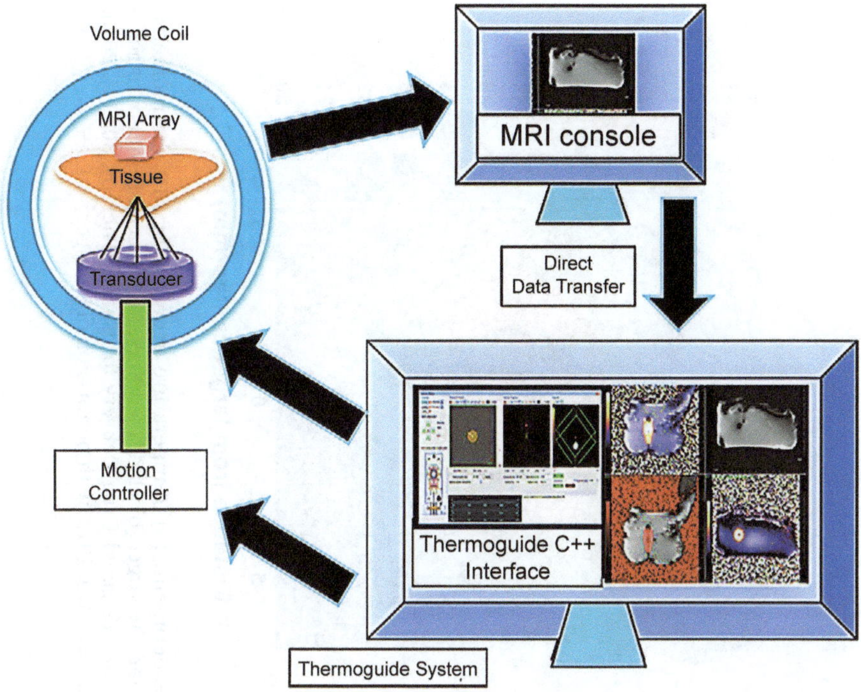

Figure 8.18 An example of feedback controlled MRgFUS procedure (Image reproduced from Fite B. Z. *et al.*, (2012). Magnetic resonance thermometry at 7T for real-time monitoring and correction of ultrasound induced mild hyperthermia. PLOS One 7(4), e35509. DOI: 10.1371/journal.pone.0035509 [88])

8.8 Post Treatment Verification

Post processing has always been an important system in computer aided diagnosis (CAD) systems. Post processing techniques improve the quality of the image obtained for a clearer visualization and higher analysis performance. The most common post processing methods include contrast enhancement,[89–94] image denoising[95–98] and filtering. Contrast increase can also occur during the experiment, therefore obtaining an enhanced image as the outcome of the scanning.

Temperature sensitive contrast agents can be injected into the sample, *ex vivo* or *in vivo*, to enhance the thermal therapy image guidance. Gadolinium has gained interest among researchers as an MRI contrast agent.[99] Gadolinium is a composition of paramagnetic thermosensitive liposomes which go through a phase transition based on a defined temperature. *In vivo* applications of contrast agents are also investigated in which utilizing non-toxic agents holds a great significance.[100,101]

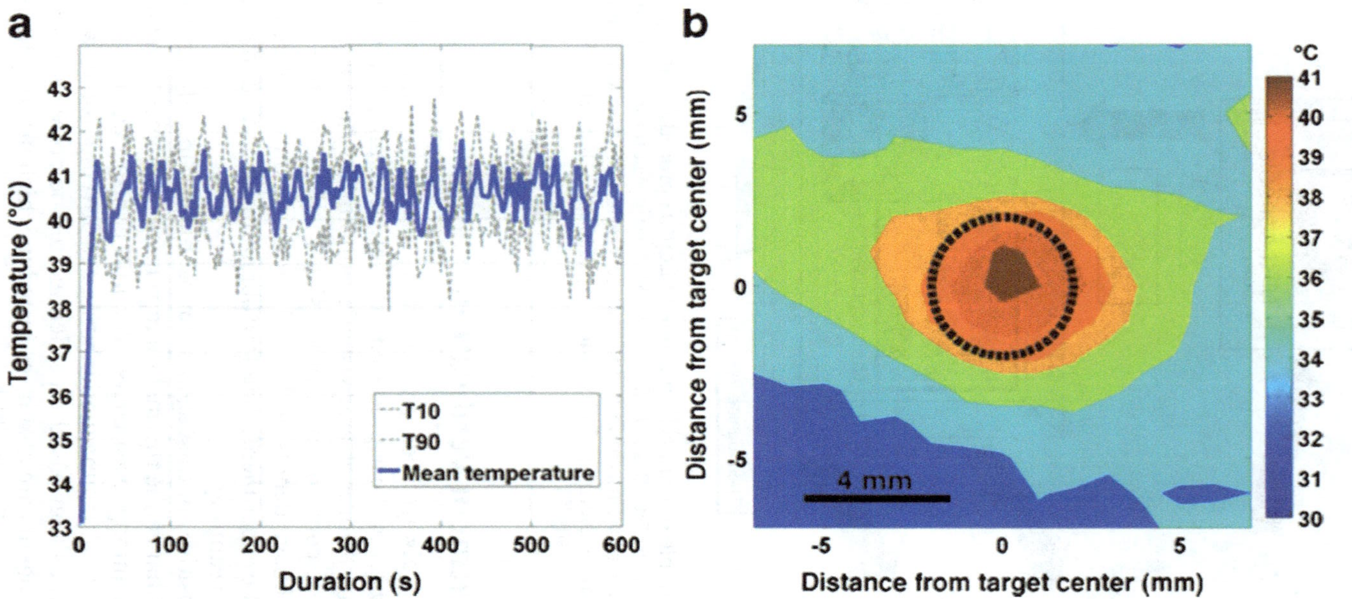

Figure 8.19 Temperature profile during MRI-HIFU hyperthermia experiment, (a) Temperature acquired by MR thermometry, (b) Time averaged temperature distribution inside and around the target (black dashed line). Reproduced from Pharmaceutical Research, MRI-Guided Monitoring of Thermal Dose and Targeted Drug Delivery for Cancer Therapy, 30 (11), 2013, 2709–2717, R. Fernando, © Springer Science+Business Media New York 2013, with permission of Springer.[55]

References

1. R. D. Issels, *et al.*, *Lancet Oncol.*, 2010, **11**(6), 561–570.
2. E. L. Jones, *et al.*, *J. Clin. Oncol.*, 2005, **23**(13), 3079–3085.
3. P. K. Sneed, *et al.*, *Int. J. Radiat. Oncol., Biol., Phys.*, 1998, **40**(2), 287–295.
4. K. F. Chu and D. E. Dupuy, *Nat. Rev. Cancer*, 2014, **14**(3), 199–208.
5. V. Muralidharan, C. Malcontenti-Wilson and C. Christophi, *J. Surg. Res.*, 2002, **103**(2), 165–174.
6. C. W. Song, *Cancer Res.*, 1984, **44**(suppl. 10), 4721s–4730s.
7. K. Liberek, A. Lewandowska and S. Ziętkiewicz, *EMBO J.*, 2008, **27**(2), 328–335.
8. R. H. Burdon, A. Slater, M. McMahon and A. C. Cato, *Br. J. Cancer*, 1982, **45**(6), 953.
9. S. A. Sapareto, L. E. Hopwood, W. C. Dewey, M. R. Raju and J. W. Gray, *Cancer Res.*, 1978, **38**(2), 393–400.
10. M. W. Dewhirst, B. L. Viglianti, M. Lora-Michiels, M. Hanson and P. J. Hoopes, *Int. J. Hyperthermia*, 2003, **19**(3), 267–294.
11. R. D. Issels, *Eur. J. Cancer*, 2008, **44**(17), 2546–2554.
12. S. B. Field and J. W. Hand, *An Introduction to the Practical Aspects of Clinical Hyperthermia*, Taylor & Francis Group, 1990.
13. G. P. Raaphorst, Fundamental aspects of hyperthermic biology, *An Introduction to the Practical Aspects of Clinical Hyperthermia*, Taylor & Francis Group, 1990, pp. 10–54.
14. J. van der Zee, *Annu. Oncol.*, 2002, **13**(8), 1173–1184.
15. C. W. Song, A. Shakil, R. J. Griffin and K. Okajima, *Semin. Oncol.*, 1997, **24**, 626–632.
16. S. Rockwell, I. T. Dobrucki, E. Y. Kim, S. T. Marrison and V. T. Vu, *Curr. Mol. Med.*, 2009, **9**(4), 442–458.
17. A. Westra and W. C. Dewey, *Int. J. Radiat. Biol. Relat. Stud. Phys., Chem. Med.*, 1971, **19**(5), 467–477.
18. J. J. Skitzki, E. A. Repasky and S. S. Evans, *Curr. Opin. Investig. Drugs*, 2009, **10**(6), 550.
19. H. Shi, *et al.*, *J. Immunol.*, 2006, **176**(4), 2134–2141.
20. S. M. Todryk, A. A. Melcher, A. G. Dalgleish and R. G. Vile, *Immunology*, 2000, **99**(3), 334–337.
21. J. A. Dickson and S. K. Calderwood, *Ann. N. Y. Acad. Sci.*, 1980, **335**(1), 180–205.
22. S. K. Calderwood, *Tumor Ablation*, Springer, 2013, pp. 29–37.
23. K. F. Chu and D. E. Dupuy, *Nat. Rev. Cancer*, 2014, **14**(3), 199–208.
24. R. K. Jain, *Sci. Am.*, 1994, **271**(1), 58–65.
25. E. S. Gil and S. M. Hudson, *Prog. Polym. Sci.*, 2004, **29**(12), 1173–1222.
26. A. Chilkoti, M. R. Dreher, D. E. Meyer and D. Raucher, *Adv. Drug Delivery Rev.*, 2002, **54**(5), 613–630.
27. G. Kong, *et al.*, *Cancer Res.*, 2000, **60**(24), 6950–6957.
28. C. J. Simon, D. E. Dupuy and W. W. Mayo-Smith, *Radiographics*, 2005, **25**(suppl. 1), S69–S83.

29. C. Gabriel, S. Gabriel and E. Corthout, *Phys. Med. Biol.*, 1996, **41**(11), 2231–2249.
30. S. Gabriel, R. W. Lau and C. Gabriel, *Phys. Med. Biol.*, 1996, **41**(11), 2271–2293.
31. D. Fatehi, J. van der Zee, M. de Bruijne, M. Franckena and G. C. van Rhoon, *Int. J. Hyperthermia.*, 2007, **23**(8), 623–643.
32. K.-S. Cheng, V. Stakhursky, O. I. Craciunescu, P. Stauffer, M. Dewhirst and S. K. Das, *Phys. Med. Biol.*, 2008, **53**(6), 1619–1635.
33. P. R. Stauffer, *Int. J. Hyperthermia*, 2005, **21**(8), 731–744.
34. T. P. Ryan, P. F. Turner and B. Hamilton, *Int. J. Hyperthermia*, 2010, **26**(5), 415–433.
35. S. Curto, M. Taj-Eldin, D. Fairchild and P. Prakash, *Med. Phys.*, 2015, **42**(11), 6152–6161.
36. J. F. Sawicki, J. D. Shea, N. Behdad and S. C. Hagness, *Int. J. Hyperthermia*, 2016, 1–8.
37. M. Cavagnaro, C. Amabile, P. Bernardi, S. Pisa and N. Tosoratti, *IEEE Trans. Biomed. Eng.*, 2011, **58**(4), 949–959.
38. C. L. Brace, P. F. Laeseke, L. A. Sampson, T. M. Frey, D. W. van der Weide and F. T. Lee Jr, *Radiology*, 2007, **242**(2), 435–440.
39. D. Yang, *et al.*, *IEEE Trans. Biomed. Eng.*, 2006, **53**(3), 533–537.
40. A. M. Ierardi, *et al.*, *Updates Surg.*, 2015, **67**(1), 39–45.
41. B. T. McWilliams, E. E. Schnell, S. Curto, T. M. Fahrbach and P. Prakash, *IEEE Trans. Biomed. Eng.*, 2015, **62**(9), 2144–2150.
42. M. Kuang, *et al.*, *Radiology*, 2007, **242**(3), 914–924.
43. G. Carrafiello, *et al.*, *Int. J. Surg.*, 2008, **6**, S65–S69.
44. W. Zhou, *et al.*, *PLoS One*, 2013, **8**(8), e71873.
45. L. W. Organ, *Stereotact. Funct. Neurosurg.*, 1976, **39**(2), 69–76.
46. J. W. Hand and G. ter Haar, *Br. J. Radiol.*, 1981, **54**(642), 443–466.
47. A. S. Wright, L. A. Sampson, T. F. Warner, D. M. Mahvi and F. T. Lee, *Radiology*, 2005, **236**(1), 132–139.
48. W. Clark, *J. Adv. Ther.*, 1911, **29**, 169–183.
49. Y. Ni, S. Mulier, Y. Miao, L. Michel and G. Marchal, *Abdom. Imaging*, 2005, **30**(4), 381–400.
50. A. L. Denys, *et al.*, *Eur. Radiol.*, 2003, **13**(10), 2346–2352.
51. L. Carroll and T. R. Humphreys, *Clin. Dermatol.*, 2006, **24**(1), 2–7.
52. A. J. Welch, J. H. Torres and W.-F. Cheong, *Tex. Heart Inst. J.*, 1989, **16**(3), 141.
53. R. Steiner, Laser-tissue interactions, *Laser and IPL Technology in Dermatology and Aesthetic Medicine*, Springer, 2011, pp. 23–36.
54. M. Ahmed, C. L. Brace, F. T. Lee Jr and S. N. Goldberg, *Radiology*, 2011, **258**(2), 351–369.
55. R. Fernando, J. Downs, D. Maples and A. Ranjan, *Pharm. Res.*, 2013, **30**(11), 2709–2717.
56. S. E. Norred and J. A. Johnson, *BioMed Res. Int.*, 2014, **2014**, 761312.
57. C. M. Tempany, N. J. McDannold, K. Hynynen and F. A. Jolesz, *Radiology*, 2011, **259**(1), 39–56.

58. S. A. Sapareto and W. C. Dewey, *Int. J. Radiat. Oncol., Biol., Phys.*, 1984, **10**(6), 787–800.
59. J. E. Kennedy, *et al.*, *Ultrasonics*, 2004, **42**(1), 931–935.
60. X. Fan and K. Hynynen, *Phys. Med. Biol.*, 1996, **41**(4), 591.
61. H. Wan, P. VanBaren, E. S. Ebbini and C. A. Cain, *IEEE Trans. Ultrason. Ferroelectr. Freq. Control*, 1996, **43**(6), 1085–1098.
62. T. Fjield and K. Hynynen, *IEEE Trans. Ultrason. Ferroelectr. Freq. Control*, 1997, **44**(5), 1157–1167.
63. E. B. Hutchinson and K. Hynynen, *Med. Phys.*, 1998, **25**(12), 2392–2399.
64. M. O. Köhler, *et al.*, *Med. Phys.*, 2009, **36**(8), 3521–3535.
65. M. Hoogenboom, *et al.*, *J. Ther. Ultrasound*, 2015, **3**(1), 1.
66. M. Hoogenboom, D. Eikelenboom, M. H. den Brok, A. Heerschap, J. J. Fütterer and G. J. Adema, *Ultrasound Med. Biol.*, 2015, **41**(6), 1500–1517.
67. N. Hijnen, S. Langereis and H. Grüll, *Adv. Drug Delivery Rev.*, 2014, **72**, 65–81.
68. L. G. Merckel, *et al.*, *Eur. Radiol.*, 2016, **26**(11), 4037–4046.
69. O. Seror, *et al.*, *Eur. Radiol.*, 2008, **18**(2), 408–416.
70. V. Rieke and K. Butts Pauly, *J. Magn. Reson. Imaging*, 2008, **27**(2), 376–390.
71. V. Rieke, K. K. Vigen, G. Sommer, B. L. Daniel, J. M. Pauly and K. Butts, *Magn. Reson. Med.*, 2004, **51**(6), 1223–1231.
72. A. Seginer, R. Schmidt, A. Leftin, E. Solomon and L. Frydman, *Magn. Reson. Med.*, 2014, **72**(6), 1687–1695.
73. K. K. Vigen, B. L. Daniel, J. M. Pauly and K. Butts, *Magn. Reson. Med.*, 2003, **50**(5), 1003–1010.
74. R. Seip, P. VanBaren, C. A. Cain and E. S. Ebbini, *IEEE Trans. Ultrason. Ferroelectr. Freq. Control*, 1996, **43**(6), 1063–1073.
75. L. Curiel, R. Chopra and K. Hynynen, *Ultrasound Med. Biol.*, 2009, **35**(1), 65–78.
76. M. A. Lewis, R. M. Staruch and R. Chopra, *Int. J. Hyperthermia*, 2015, **31**(2), 163–181.
77. B. A. Rabkin, V. Zderic and S. Vaezy, *Ultrasound Med. Biol.*, 2005, **31**(7), 947–956.
78. F. Fani, E. Schena, P. Saccomandi and S. Silvestri, *Int. J. Hyperthermia*, 2014, **30**(4), 219–227.
79. V. Salgaonkar, *et al.*, *J. Ther. Ultrasound*, 2015, **3**(1), 1.
80. J. K. Enholm, M. O. Kohler, B. Quesson, C. Mougenot, C. T. Moonen and S. D. Sokka, *IEEE Trans. Biomed. Eng.*, 2010, **57**(1), 103–113.
81. D. Arora, *et al.*, *Int. J. Hyperthermia*, 2006, **22**(1), 29–42.
82. B. Z. Fite, *et al.*, *PLoS One*, 2012, **7**(4), e35509.
83. D. Haemmerich and J. G. Webster, *Biomed. Eng. Online*, 2005, **4**(1), 1.
84. R. Staruch, R. Chopra and K. Hynynen, *Radiology*, 2012, **263**(1), 117–127.
85. P. VanBaren and E. S. Ebbini, *IEEE Trans. Biomed. Eng.*, 1995, **42**(8), 818–827.
86. E. Hutchinson, M. Dahleh and K. Hynynen, *Int. J. Hyperthermia*, 1998, **14**(1), 39–56.

87. D. Arora, M. Skliar and R. B. Roemer, *IEEE Trans. Biomed. Eng.*, 2002, **49**(7), 629–639.
88. B. Z. Fite, *et al.*, *PLoS One*, 2012, 7(4), e35509.
89. A. Jackson, *Br. J. Radiol.*, 2004, **77**, S154–S166.
90. S.-D. Chen, *Digital Signal Process.*, 2012, **22**(4), 640–647.
91. M. Kitano, H. Sakamoto and M. Kudo, *Gastrointest. Endosc. Clin. N. Am.*, 2012, **22**(2), 349–358.
92. L. D. Chen, *et al.*, *Br. J. Radiol.*, 2008, **81**, 881–889.
93. S.-C. Huang and C.-H. Yeh, *Eng. Appl. Artif. Intell.*, 2013, **26**(5), 1487–1492.
94. G. Jiang, *et al.*, *J. Mod. Opt.*, 2015, **62**(7), 536–547.
95. V. N. Varghees, M. S. Manikandan and R. Gini, *Advances in Engineering, Science and Management (ICAESM), 2012 International Conference on*, 2012, pp. 506–511.
96. J. V. Manjón, P. Coupé, L. Concha, A. Buades, D. L. Collins and M. Robles, *PLoS One*, 2013, **8**(9), e73021.
97. H. M. Golshan, R. P. Hasanzadeh and S. C. Yousefzadeh, *Magn. Reson. Imaging*, 2013, **31**(7), 1206–1217.
98. J. Klosowski and J. Frahm, *Magn. Reson. Med.*, 2017, **77**, 1340–1352.
99. S. L. Fossheim, K. A. Il'yasov, J. Hennig and A. Bjørnerud, *Acad. Radiol.*, 2000, 7(12), 1107–1115.
100. N. McDannold, S. L. Fossheim, H. Rasmussen, H. Martin, N. Vykhodtseva and K. Hynynen, *Radiology*, 2004, **230**(3), 743–752.
101. L. Frich, A. Bjørnerud, S. Fossheim, T. Tillung and I. Gladhaug, *Magn. Reson. Med.*, 2004, **52**(6), 1302–1309.

CHAPTER 9

Magnetic Nanoformulations for Enhanced Drug Delivery and Retention

ARUNI P. MALALASEKERA[a], STEFAN H. BOSSMANN[a] AND GAOHONG ZHU*[b]

[a]Department of Chemistry and Johnson Cancer Center, Kansas State University, Manhattan, KS 66506, USA; [b]First Affiliated Hospital of Kunming Medical University, Kunming, Yunnan Province, China
*E-mail: sbossman@ksu.edu, 1026909611@qq.com

9.1 Introduction

Poor delivery of drugs to the targeted site and failure of the drug to accumulate in the targeted tissue are two of the central obstacles to successfully treating cancers and infectious diseases. Poor selectivity of drug targeting causes cytotoxic "bystander effects",[1] which cause malfunctioning of healthy tissues. During the last decade, numerous researchers have focused on utilizing therapeutically effective and noninvasive magnetic nanoformulations to promote drug efficacy by enhancing both delivery and retention.[2–8] As already discussed in Chapters 1, 2, 7, and 8, the "switch-on" magnetism of superparamagnetic nanoparticles[9] in magnetic fields and the straightforward functionalization of their surfaces at the molecular level make them good candidates for enhanced drug delivery.[10] At the same time, superparamagnetic nanoparticles permit quantitative T_2- and T_2*-weighted MRI mapping.[9,11–13] In this

Smart Materials No. 26
Magnetic Nanomaterials: Applications in Catalysis and Life Sciences
Edited by Stefan H. Bossmann and Hongwang Wang

Published by the Royal Society of Chemistry, www.rsc.org

chapter, we will go beyond that in discussing the effects of the nanoparticles' shapes and protein corona formations on their uptake into cells and tissue, as well as the combination of therapeutic radionucleides with (super)paramagnetic and ferromagnetic delivery platforms. The utilization of radionucleides for therapeutic purposes is closely related to their use for imaging purposes. Taking advantage of the core expertise of Gaohong Zhu, MD in Nuclear Medicine, we will also discuss how to use Positron Emission Spectroscopy (PET)[14,15] and Single-Photon Emission Computed Tomography (SPECT or SPET)[14,15] of radionucleide-bearing nanoformulations to access tumor growth, the formation of metastases, and the efficacy of drug delivery. We anticipate that especially PET has great potential in enhancing the research on nanoformulations by providing accurate quantitative data on *in vivo* nanoparticle distribution.

In general, compatibility of organic, inorganic, and hybrid nanoparticles is achieved by fine-tuning their sizes, shapes, biophysical properties (especially surface charges and hydrophilicity *vs.* hydrophobicity), and the attachment of ligands with specific functions, such as providing stealth properties and targeting of specific biological structures.[16] Nanoformulations for targeted drug delivery have several advantages such as lowering dosage, effective pharmaceutical activity, efficient drug accumulation, the minimizing of side effects and improved drug stability.[17] However, although nanoparticle-based formulations are efficacious in biological systems, their recognition and clearance by the reticuloendothelial system that is a part of the immune system and consists of phagocytic cells located in reticular connective tissue, is a major limiting factor competing with the efficacy of drug delivery.[18]

9.2 Barriers Towards Drug Delivery

According to Elsabahy and Wooley,[16] the barriers that exist with respect to drug delivery by nanoparticle-based drug formulations can be classified into external barriers (skin and mucosa), barriers occurring en-route, and, finally, cellular barriers, such as limited cellular uptake, endosomal/lysosomal degradation of the nanoformulation, and inefficient transport to cellular targets (*e.g.* nucleus or mitochondria).

9.2.1 External Barriers

Skin consists of several layers, stratum corneum (SC), epidermis, dermis and subcutaneous tissues.[19] The hydrophobic SC is the major permeability barrier for delivery of nanoformulations through the skin.[19] Disrupting the lipid bilayers of the SC is utilized commonly to enhance permeation through the skin. Disruption of the SC's lipid bilayers (*e.g.* by microneedles,[20] by ultrasound[21] or chemically enhanced)[22] is frequently applied in transdermal delivery. Besides being a physical barrier, the skin has a rich biochemistry, which leads to enzymatic degradation/modification of the nanoformulations. Furthermore, skin has an active immune system, for instance Langerhans cells in the epidermis and dendritic cells in the dermis. The major (skin-free) mucus coat regions are the gastrointestinal tract, eyes, lung airways, nasal, rectal, and vaginal cavities. Mucus is a viscoelastic hydrogel of varying chemical compositions, which is secreted by mucosal glands. Its chemical main components are crosslinked mucin fibers, which feature alternating hydrophobic domains and glycosylated hydrophilic domains, resulting in a "sponge-like" structure. The presence of *N*-acetylneuramic acids (sialic acids) and sulfated monosaccharides in mucin causes its overall negative surface charge.[23] Unlike the skin, mucus gels are very nonsterile environments. Nanoparticle transport through mucus coat regions is typically hindered by its cross-linked nanoscopic networks, nanoparticle-adhesion (especially of positively charged nanoparticles), and biochemical modifications/degradation.[16,23]

In both skin and mucus coat regions, corona formation (intense, time-variable peptide binding to the surface of nanoparticles) will occur, changing their biophysical characteristics.[24] Corona formation and the consequences for drug delivery will be discussed in Section 9.2.2.2.

9.2.2 En-Route Barriers

A commonly utilized strategy to circumvent external barriers to drug delivery is the injection of nanoformulations into the diseased tissue or, alternatively, intraarterially,[25] intravenously,[26] or intraperitoneally.[27] It must be noted that the injection of nanoformulations is an invasive procedure, which has to be precisely regulated.[16] Once the nanoparticles are en-route to their destination and circulating in the bloodstream, they have to be sufficiently stable against aggregation, chemical modification (enzymatic and *via* corona formation), renal and hepatic clearance, as well as opsonization and consequent clearance by the reticuloendothelial system.[16] At the sites of tumors, and—to a lesser extent—also infected tissue, nanoparticles undergo extravasation (also known as passive targeting), taking advantage of the dysfunctional structure of the microvasculature.[28] It is noteworthy that usually less than 10 percent of the nanoformulations reach their intended target sites.[29] The remaining nanocarriers are distributed into various other tissues and organs, often causing collateral damage.[29]

9.2.2.1 Renal and Hepatic Clearance

The size, shape and especially the density of nanoparticles critically affects their renal clearance and, therefore, circulation time and passive targeting efficacy.[30–32] In general, smaller, rod-shaped and especially denser nanoparticles favor margination (adhesion at the walls of blood vessels) in laminar blood flow. Lower-density nanoparticles are more easily distributed throughout the body, but have shorter retention times in tissue. Typically, the cut-off for renal clearance varies from 5 nm (metallic NPs, metal oxide NP or quantum dots) to 30 nm (light organic nanoparticles).[30–32] However, several studies[28,33] have shown that polymeric carrier nanoparticles that clearly exceeded the established renal cut-off were excreted renally. This is a clear indication for either partial enzymatic degradation or disassembling of the polymer nanoparticles. A similar phenomenon has been observed in immune-competent mice where silica-coated Fe/F_3O_4-based nanoparticles of approx. 25 nm in diameter were readily excreted renally, thus indicating effective biocorrosion processes.[34]

Hepatic clearance of nanoparticles into bile, followed by fecal excretion, is the second major pathway of nanoparticle clearance.[16] The third general pathway of nanoparticle clearance is the biocorrosion/cellular digestion of nanoparticles that occurs within cells, which leads to low molecular weight products (such as Fe(II) salts from iron-containing nanoparticles or organic acids from polymer nanoparticles), which are then easily excreted.[35]

9.2.2.2 *Corona Formation*

Immediately after intravenous injection or any other internalization pathway, nanoparticles get in contact with complex protein-rich environments. This is especially pronounced for nanoformulations that are intended to circulate in the bloodstream. The plasma proteome comprises more than 2000 proteins,[36] which bind to the high-energy surfaces of nanoscopic objects. As indicated in Figure 9.1, a highly complex protein corona is established as soon as 30 s after contact. The corona composition changes predominantly quantitatively rather than qualitatively over time. Historically, these changes were attributed to the "Vroman effect".[37] This paradigm assumed that more abundant proteins (especially albumin in blood) quickly bind to the nanoparticles and then are slowly replaced by proteins featuring higher binding constants.[37] However, newer studies have shown that the observed binding patterns are even more complex.[24] As a consequence of corona formation, the effective size, surface charge, and hydophilicity *vs.* hydrophobicity of nanoparticles can be completely altered, thus making many nanoengineering approaches obsolete. These corona effects are not all negative, because some adsorbed proteins are known to enable desired "behavior" of nanoformulations.[38] For instance, some apolipoproteins can facilitate nanoparticle transport across the blood–brain barrier.[39] Numerous immunoglobulins and other proteins (see Section 9.2.2.3) trigger opsonization, whereas dysopsonins, like albumin, inhibit phagocytic uptake.[40]

As shown in Figure 9.2, generally unpredictable corona formation is able to mask the initial surface charges of nanoparticles, which has severe consequences for their biological fate. The major approach towards avoiding corona formation and potentially undesired effects is PEGylation, the attachment of *p*oly*e*thylene *g*lycol-based chains of various lengths.[41] The underlying paradigm is to minimize interaction of the nanoparticle-attached PEG with the proteome of its environment, as well as with the PEG units of other nanoparticles, leading to enhanced dispersion.[42] Although this approach was able to generate nanoparticles that are quite stable in biological media,

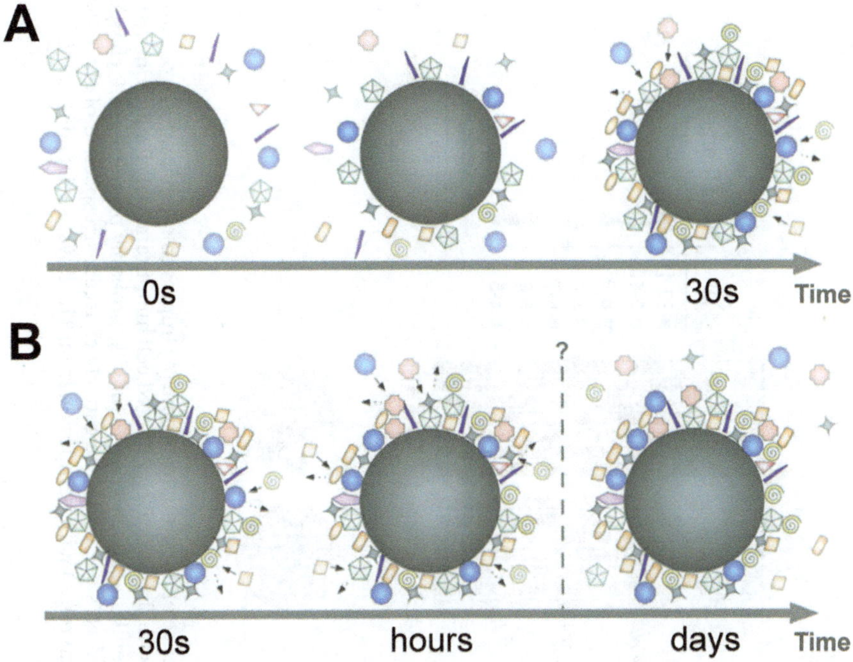

Figure 9.1 Complexity and evolution of the biomolecule corona: (A) The early phase: a highly complex corona is established already in 30 s, which may be composed of multiple core–shell structures. (B) The late phase: corona composition *ex situ* remains stable and changes predominantly quantitatively rather than qualitatively over time with Vroman effect dependent and independent binding kinetics. Note that the objects are not drawn to scale. Reproduced from ref. 24 with permission from The Royal Society of Chemistry.

it was not able to completely prevent corona formation.[24] Reduced corona formation is observed,[43–45] leading to enhanced circulation times in the bloodstream and enhanced biodistribution,[43–45] potentially due to decreased opsonization.

9.2.2.3 *Opsonization and Phagocytotic Uptake*

Depending on the size and chemical composition of the injected nanoformulations, opsonization is a major en-route barrier to drug delivery. Albumin, immunoglobulins, complement proteins, apolipoprotein and fibrinogen have been identified as efficient opsonins, which target the nanoparticles and tag them for rapid uptake by phagocytic cells, such as blood monocytes/macrophages in lymph nodes, spleen, liver (Kupffer cells) and other tissues.[46] Furthermore, clearance by the reticuloendothelial system (also known as the mononuclear phagocyte system (MPS)) can start severe immunological reactions against nanoformulations in blood, lymphatic fluid and tissue.[16]

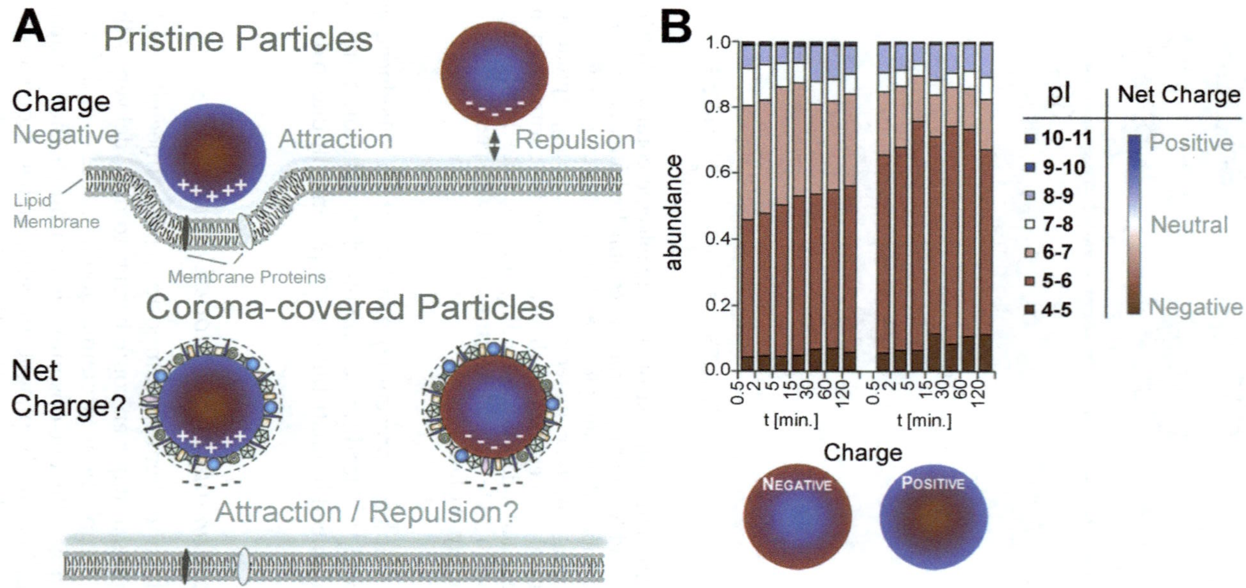

Figure 9.2 Impact of NP charge on cellular uptake in the absence or presence of the protein corona: (A) Improved cellular uptake of positively charged NPs may be mediated by enhanced interaction with the negatively charged cell membrane only for pristine NPs (upper panel). In contrast, plasma corona covered NPs are overall negatively charged *in situ*, probably preventing NP-charge driven cell membrane interaction. (B) Plasma corona covered NPs are overall negatively charged, irrespective of the NPs' negative or positive surface functionalization. Reproduced from ref. 24 with permission from The Royal Society of Chemistry.

9.2.3 Cellular Barriers

In general, the cellular uptake of nanomaterials is often inefficient.[16] The first required step is the interaction of the nanoparticles with the outer surface of cellular membranes. Internalization proceeds *via* the formation of membrane-bound vesicles surrounding the nanoparticle. Depending on the interaction of the highly heterogeneous cellular membrane with the nanoparticles, which precedes vesicle formation, the latter differ in their chemical compositions and internal environments. All of these processes are referred to as endocytosis. However, endocytotic pathways depend on size, morphology, and surface chemistry of the nanostructures, as well as the cell types. Depending on the internalization pathway, endosomes, phagosomes, or macropinosomes are formed.[47,48] To date, there are five principal mechanisms for nanoparticle uptake by mammalian cells: phagocytosis, macropinocytosis, clathrin-mediated, caveolin-mediated, and clathrin/caveolin-independent endocytosis.[49] The sixth pathway, known as "energy-independent" or "receptor-independent" uptake, is still under discussion.[50] As shown for the class of cell-penetrating peptides, this may become a very advantageous uptake pathway by cells, because it avoids pH-dependent and/or enzymatic (proteolytic) degradation after endocytosis. In fact, all endocytosis-mediated uptake processes require escape from the initially formed vesicles to avoid destruction. Once the nanoparticles have escaped, they reach the cytosol, which again has numerous active cytosolic enzymes. The cytosol is characterized by high viscosity, which slows down nanoparticle diffusion to potential cellular targets. Furthermore, passive translocation to organelles, such as nuclei or mitochondria, is very difficult, because they are membrane-protected. The best option for targeting these organelles is the utilization of nuclear[51,52] or mitochondrial localization sequences,[52] which provide access to active transport mechanisms in the cells' interior. Finally, the term "exocytosis" describes all cellular processes that permit the excretion of nanoparticles from the cells' interior. It leads to a decrease of nanoparticle concentration in targeted and other tissue.[53]

9.2.4 Drug Delivery Strategies

During the last two decades, the tumor-homing abilities of nanoparticle-based formulations have been intensely studied.[54] In angiogenic tissues, nanoparticles reach the interstitium *via* passive transport through endothelial gaps, typically requiring a size of less than 80–100 nm. This effect is commonly referred to as Enhanced Permeability and Retention (EPR).[55] Recently, serious doubts have been cast as to whether the extent of the EPR effect, which was originally established in rodent models, is sufficient in humans.[56] Three possible strategies are discernible in the light of these recent findings: (1) Cell-based delivery strategies will become more important, because they rely on active transport of anticancer drugs in stem cells or defensive cells.[34,57–59] This approach is described in Chapter 10 of this book. (2) Active targeting strategies will have to compensate for possible shortcomings of the EPR effect. Their efficacy will depend on their ability to overcome physiological barriers, which are typically found in tumors. (3) The accumulation of magnetic nanoformulations can be enhanced by magnets, which are able to retain the nanodelivery drug in the targeted tissue. These approaches will be discussed later in this chapter.

Active targeting strategies rely on the presence of a chemically attached "targeting moiety", which corresponds to a unique cellular target.[54,60] The efficacy of a wide variety of (bio)chemical targeting moieties have been investigated, such as monoclonal antibodies and antibody fragments,[61–63] small molecule ligands,[64,65] peptidomimetrics,[65,66] aptamers,[67,68] and polysaccharides (Figure 9.3).[69,70]

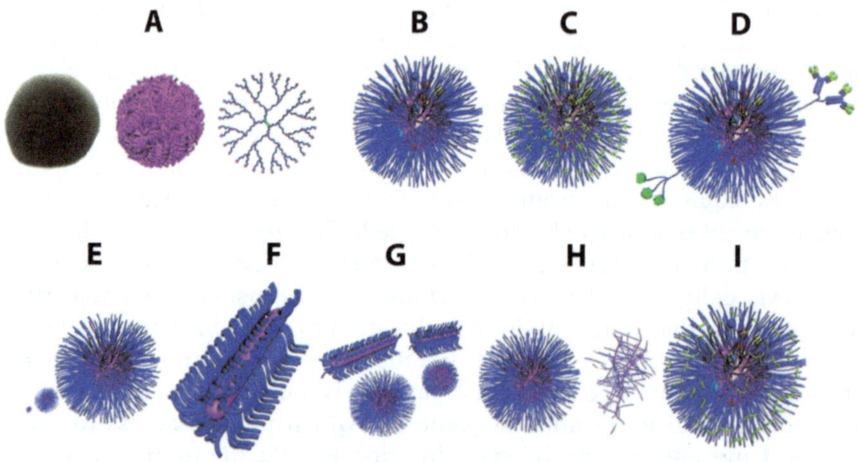

Figure 9.3 (A) Plain nanoparticles (metal nanoparticle/polymer nanoparticle/dendrimer): high aggregation, low drug protection, rapid opsonization, rapid clearance. (B) Stealth nanoparticles: steric stability, high drug protection, slow opsonization, biocompatibility. (C) Charge: High cellular uptake, high toxicity, low specificity. (D) Targeting: High cellular uptake, high specificity, clearance and immunogenicity depend on the targeting ligands. (E) Nanoparticle <10 nm, rapid clearance 10–200 nm: enhanced permeation and retention effects optimal; >200 nm: enhanced opsonization and potentially increased toxicity. (F) Morphology: expanded shapes cause low cellular uptake, low clearance, but permit high drug loading. (G) Aspect ratio "core *vs.* shell" and "length *vs.* diameter": cellular uptake is significantly influenced by shape, as well as clearance. Both parameters are crucial for drug loading and release, as well as toxicity. (H) Assembly *vs.* unimolecular: previously assembled drug carriers can disassemble, leading to fast drug delivery and rapid clearance. (I) Stability: High: obstacle to drug release. Low: Drug leaching and premature release likely. Crosslinking enhances stability of assembled drug carriers. Reproduced from ref. 16 with permission from The Royal Society of Chemistry.

9.3 Noninvasive Imaging Techniques in Nanomedicine

Quantitative monitoring of the entire drug delivery process comprised of pharmacokinetics, biodistribution, target site accumulation, local distribution across physiological barriers at the target site, localization in presumably healthy tissue and collateral effects, kinetics of drug release, and, most importantly, therapeutic efficacy, is mandatory to optimizing treatment outcomes.[71] Noninvasive imaging techniques, such as Positron Emission Tomography (PET), Single-Photon Emission Computed Tomography (SPECT), Computed Tomography (CT), Magnetic Resonance Imaging (MRI), Optical Imaging (OI), and Ultrasound (US) have been optimized to monitor the "fate" of nanoparticle-based drug formulations *in vivo*, as can be seen in Figure 9.4.[28] It is noteworthy that CT, MRI and US can be utilized either with or without contrast agents. They require background level scans before the administration of contrast agents. For the "hot-spot" techniques PET and SPECT no background scans are required. For OI it usually depends on the investigated wavelength region, whether a background scan is necessary. Hybrid imaging techniques, such as PET-CT, SPECT-CT and PET-MRI, are used to provide anatomical information for the correct allocation of the hot-spot information. This chapter will focus on the use of PET, SPECT and MRI for tracking nanoformulations.

9.3.1 Positron Emission Tomography (PET)

Positron-emitting radionucleides are visualized and quantified in PET. The emitted positrons annihilate neighboring electrons, thus generating 511 keV photons, which can be quantified by sensitive detectors. PET is characterized by high sensitivity, unlimited penetration depth and a broad range of suitable and available radionucleides. Examples of routinely used positron emitting isotopes are ^{11}C, ^{13}N, ^{15}O, ^{18}F, ^{44}Sc, ^{62}Cu, ^{64}Cu, ^{68}Ga, ^{72}As, ^{74}As, ^{76}Br, ^{82}Rb, ^{86}Y, ^{89}Zr, and ^{124}I.[72–78] Today, PET is routinely used in the clinic for disease diagnostic, staging, and therapy monitoring.[72,79,80]

Besides clinical applications, PET is also a very valuable research tool, because it permits the recording of pharmacokinetics, biodistribution, and target site accumulation of nanoformulations. For this purpose, PET probes are tethered to or encapsulated by nanomedicines. Widely used chelators for cationic PET isotopes are 1,4,7,10-tetraazacyclododecane-1,4,7,10-tetraacetic acid (DOTA), 2-[Bis[2-[bis(carboxymethyl)amino]-ethyl]amino]acetic acid (DTPA), or hydrazinonicotinic acid (HYNIC). ^{11}C and ^{18}F are often bound to drug molecules, which are used as payloads.

The ever-increasing understanding of the interactions between nanomedicines and the hosts led to a significant increase in their circulation times. Consequently, PET isotopes with enhanced lifetimes are required to study pharmacokinetics and EPR processes. Besides ^{64}Cu ($t_{1/2}$ = 13 h), ^{72}As ($t_{1/2}$ = 26 h), ^{74}As ($t_{1/2}$ = 17.7 d), ^{124}I ($t_{1/2}$ = 4.1 d) are emerging isotopes in animal

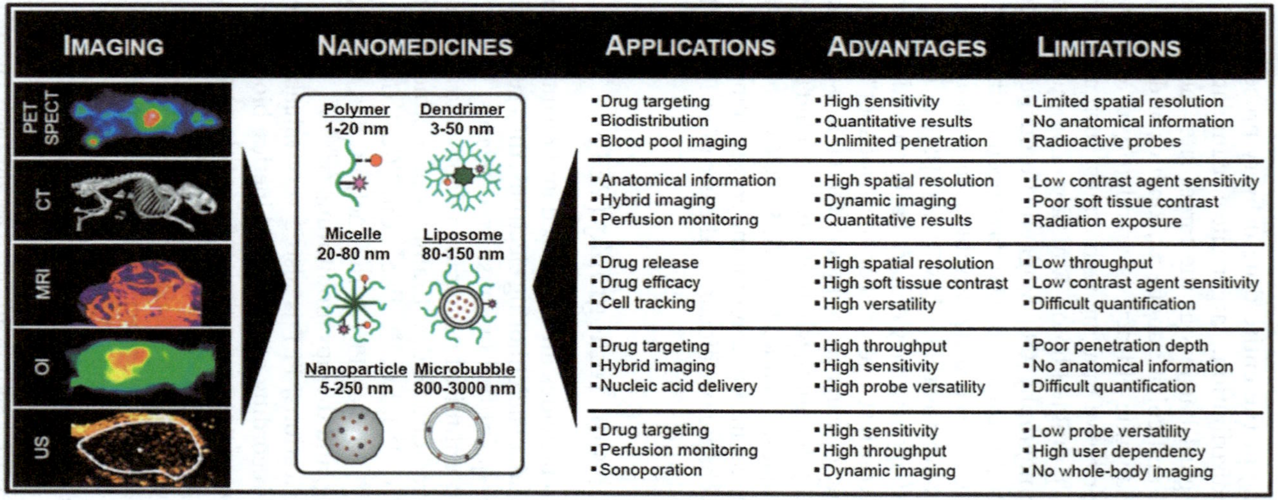

Figure 9.4 Imaging techniques applied in nanomedicine: PET: Positron Emission Tomography, SPECT: Single-Photon Emission Computed Tomography, MRI: Magnetic Resonance Tomography, OI: optical imaging, US: ultrasound, together with their typical applications, advantages, and limitations. Reproduced with permission from ref. 28. Copyright (2013) American Chemical Society.

studies.[81] ^{89}Zr ($t_{1/2}$ = 78 h) was successfully used in tumor-targeting,[82,83] the tracking of tumor-associated macrophages[84] and of tumor-targeting monoclonal antibodies.[85–87]

Besides the necessity for using a radioactive probe, albeit in the (sub)-picomolar concentration range, the major disadvantages of PET, as for all "hot-spot" techniques, are the absence of anatomical information and a relatively low spatial resolution.

In Figure 9.5, an example for the advantages of using PET is shown. Sentinel lymph nodes are—generally—the first sites that are invaded by metastasizing cancer. They then act as a platform for further metastatic events. Therefore, identifying sentinel lymph nodes is mandatory for surgical treatment of cancer, as well as correct cancer staging. Immune-competent mice were injected with highly metastatic 4T1 breast carcinoma tumors into the ankle region of the left hind limb. The sentinel lymph nodes were imaged with ^{64}Cu, Gd(III), and near-infrared dye-labeled mesoporous silica nanoparticles (MSNs).[88] As shown in Figure 9.5, PET imaging was performed at four different times after administration of the MSNs into the foot sole (1 h, 6 h, 24 h, and 48 h) to visualize sentinel lymph node targeting *vs.* overall biodistribution. A very strong signal was observed in the sentinel lymph node, whereas a presumably normal contralateral lymph node virtually did not accumulate the MSN. PET signals in the sentinel lymph node were 35- and 7-fold higher than in the contralateral lymph node at 24 h and 48 h, respectively.[88]

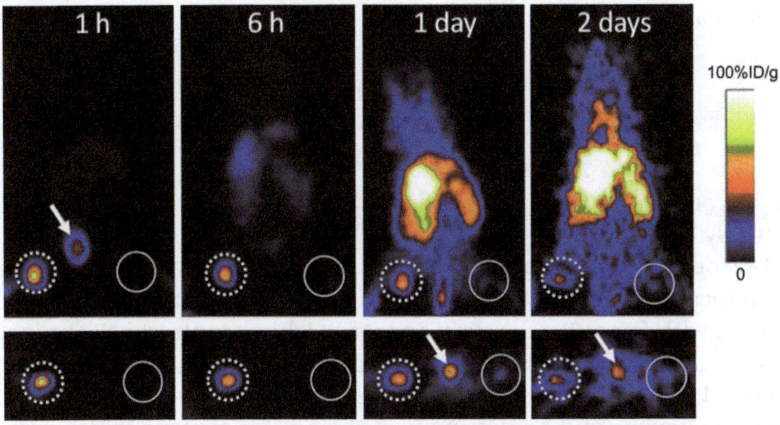

Figure 9.5 PET imaging of sentinel lymph nodes in a 4T1 tumor metastatic model: (A) PET imaging of T-SLN (square dot) and N-SLN (solid line) after injection of particles for 1 h, 6 h, 1 day and 2 days. Arrow denotes bladder. Top: Cross section; Bottom: Transverse section. See Figure 9.6 for comparison. Reprinted from Biomaterials, 33(17), X. Huang, F. Zhang, S. Lee, M. Swierczewska, D. O. Kiesewetter, L. Lang, G. Zhang, L. Zhu, H. Gao, H. S. Choi, G. Niu, X. Chen, Long-term multimodal imaging of tumor draining sentinel lymph nodes using mesoporous silica-based nanoprobes, 4370–4378, Copyright (2012) with permission from Elsevier.[88]

9.3.2 Single Photon Emission Computed Tomography (SPECT)

PET and SPECT are, principally, related techniques. Both permit the detection of "hot-spots".[28] The major difference is that whereas PET utilizes coincident γ-photons of exactly 511 keV, SPECT detects noncoincident γ-photons emitted by ^{99m}Tc, ^{111}In, ^{123}I, and ^{201}Tl. The sensitivity of SPECT is approx. 10-fold lower than that of PET. Both techniques share the same advantages and disadvantages.[28]

9.3.3 Magnetic Resonance Imaging (MRI)

Classic proton-based MRI converts proton relaxation measurements in biological systems into 3D tomographic images.[89] Under the influence of a strong external magnetic field B_0, a small fraction of the proton nuclei align with B_0. These aligned nuclei are under precession with a net magnetic moment m and a Larmor precession frequency of $\omega_0 = \gamma B_0$ ($\gamma = 2.67 \times 10^8$ rad s^{-1} T^{-1} for ^{1}H). When a resonant radio frequency (RF) transverse pulse is applied to B_0, it causes resonant excitation of the preceding magnetic moments into the perpendicular plane. Once the RF pulse is completed, the magnetic moment gradually relaxes to the equilibrium by realigning with B_0. There are two principal relaxation pathways: longitudinal relaxation (*via* energy exchange with the lattice) and transverse relaxation (loss of phase coherence of the precessing nuclei spins in the xy plane due to spin–spin interaction). These principal processes can be expressed as follows:

$$m_z = m(1 - \mathrm{e}^{-1/T_1}) \text{ (longitudinal)} \tag{9.1}$$

$$m_{xy} = m \sin(\omega_0 t + \phi)\mathrm{e}^{-t/T_2} \text{ (transverse)} \tag{9.2}$$

where T_1 and T_2 are the longitudinal and transverse relaxation times, respectively.

MRI records these relaxation processes in 3D volume elements and then mathematically reconstructs them to obtain anatomical information. Classically, areas with faster T_1 relaxation appear as whiter contrast in T_1-weighted MRI, whereas areas with faster T_2 relaxation are shown as darker contrast in T_2-weighted MRI. The physiological basis of MRI is that biological tissues differ in the density and homogeneity of their water content, which provide different contrasts during MR imaging.

Magnetic nanoparticles enhance contrast differences by increasing the efficacy of either T_1- or T_2-relaxation. Furthermore, they are capable of providing molecular imaging capabilities.[90,91] (Super)paramagnetic nanoparticles are magnetized with the magnetic moment μ when subjected to the magnetic field B_0. The induced magnetic field generates a dipolar field, which causes a significant local perturbation. The local magnetic field inhomogeneity accelerates the dephasing rate of the Larmor precession of the surrounding water molecules. According to the outer sphere spin–spin relaxation approximation, the relaxivity $R_2(1/T_2)$ of the magnetic nanoparticle can be described as:[92]

$$R_2 = \left(\frac{32\pi}{405}\right)\gamma_1^2\mu^2\frac{N_A}{1000}\left(\frac{[M]}{r_D}\right)\left[6.5j_2(\omega_s) + 1.5j_1(\omega_1) + 2j_1(0)\right] \tag{9.3}$$

where M is the molality of the magnetic nanoparticle, r the nanoparticle radius, $j_n(\omega,\tau)$ the spectral density function, μ the transverse component of the magnetic moment of the nanoparticle, γ_1 the gyromagnetic ratio of the proton, N_A Avogadro's number, and ω_s and ω_1 the Larmor angular precession frequencies of the nanoparticle and water proton magnetic moments, respectively.

From eqn (9.3), it can be discerned that the spin–spin relaxation time (T_2) is inversely proportional to the square of the magnetic moment μ of the nanoparticle. Furthermore, T_2 is proportional to the diameter of the nanoparticle. In Chapter 2 of this book, synthetic strategies are discussed to fine-tune the chemical compositions, sizes, shapes, and the resulting magnetic moments of magnetic nanoparticles. MR contrast agents, as opposed to radionucleides, depend on proximity to freely diffusing water molecules. This enables the spatio-temporal observation of drug release events. Furthermore, MRI is highly suited for multimodal imaging (PET-MRI or SPECT-MRI),[28] in which MRI provides the required anatomical information to accurately monitor the biodistribution and target site accumulation of a nanocarrier.[54]

Finally, when magnetic nanoparticles are close to each other, they form agglomerates, in which their magnetic moments are coupled to generate stronger magnetic fields. This magnetic coupling effect, which influences the relaxivity R_2, can be expressed as:[93]

$$R_2 = \frac{1}{T_2} = \left(\frac{64\pi}{135}\right)\left(\mu N_g \frac{L(x)}{4\pi}\right)^2 \frac{N_A C_a}{R_a D} \tag{9.4}$$

where μ is the magnetic moment of the nanoparticle, N_g is the number of nanoparticles in an agglomerate, D is the diffusion coefficient of water, N_A is Avogadro's number and $L(x)$ is the Langevin function. R_2 is proportional to μ and N_g.

In Figure 9.6, the same experiment as in Figure 9.5 is shown: immune-competent mice were injected with highly metastatic 4T1 breast carcinoma tumors into the ankle region of the left hind limb. Then, the sentinel lymph nodes were imaged.[88] However, the imaging results shown in Figure 9.6 were obtained by using a 7.05 T small animal MR scanner, taking advantage of the presence of Gd(III)-chelated H_4DTTA in the MSN probes that is known to provide high T1 signals in MRI, especially when a 3D gradient-echo-scan (FLASH) sequence is used.[89,94] The enlarged MR images of the tumor-bearing sentinel lymph nodes (T-SLNs) indicate that the MSN probes did not

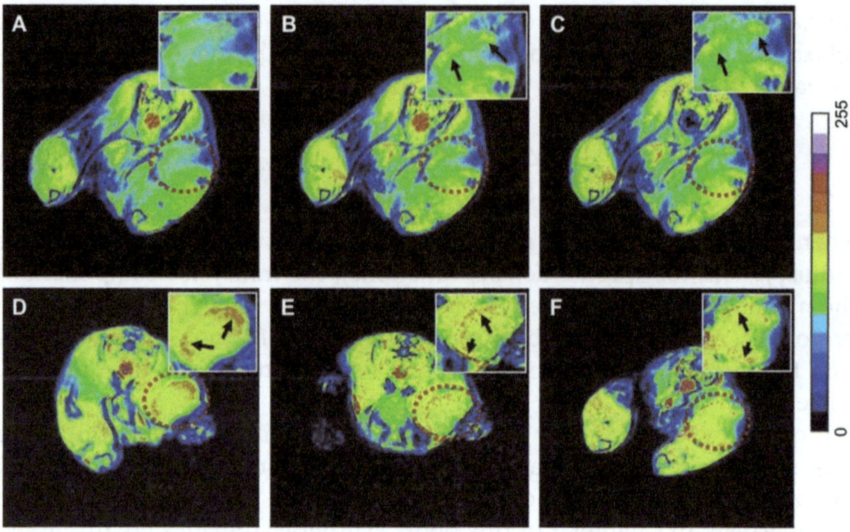

Figure 9.6 MRI of sentinel lymph nodes in a 4T1 tumor metastatic model. MRI of lymph nodes were shown (A) before and after injection of MSN-nanoprobes for (B) 1 h, (C) 6 h, (D) 1 day, (E) 5 days and (F) 15 days. Arrows show the accumulation area of MSN. See Figure 9.5 for comparison. H_4DTTA: (H_4DTTA = diethylenetriaminetetraacetic acid = *N*,*N*′-[iminobis(ethane-2,1-diyl)]bis[*N*-(carboxymethyl)glycine]). Reprinted from Biomaterials, **33**(17), X. Huang, F. Zhang, S. Lee, M. Swierczewska, D. O. Kiesewetter, L. Lang, G. Zhang, L. Zhu, H. Gao, H. S. Choi, G. Niu, X. Chen, Long-term multimodal imaging of tumor draining sentinel lymph nodes using mesoporous silica-based nanoprobes, 4370–4378, Copyright (2012) with permission from Elsevier.[88]

significantly penetrate the lymph nodes, but were mainly distributed at their margins. This distribution pattern is consistent with *ex vivo* microscopy. This information is complementary to the data shown in Figure 9.5. It demonstrates that the MSNs did not disintegrate prior to imaging. The MRI data clearly indicate the presence of a physiological barrier within the lymph node, which was not discernible in PET imaging.

9.4 Magnetically Enhanced Drug Delivery—Magnetic Targeting of Tumors

The magnetic targeting of solid tumors is an emerging strategy that holds great promise.[95–100] As shown in Figure 9.7,[99] a permanent magnet is used to attract magnetic nanoparticles to a well-defined region. Targeting is enhanced by the following mechanisms: (A) The local concentration of the magnetic nanoparticle is enhanced in the region where extravasation occurs due to leaky microvasculature around tumors.

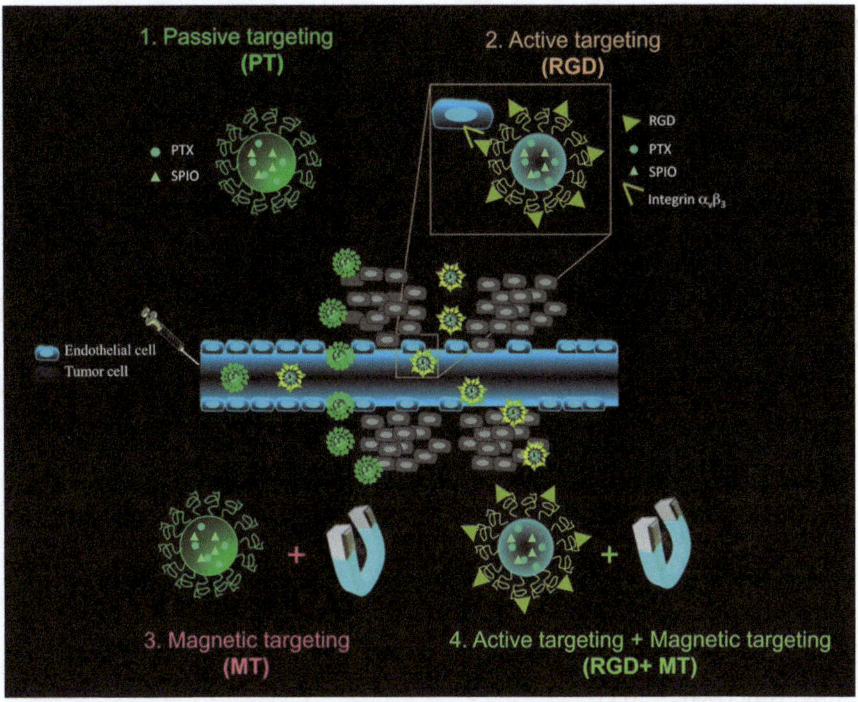

Figure 9.7 Schematic representation of the different tumor-targeting strategies compared in the present study. (1) Passive targeting *via* the EPR effect (PT); (2) Active targeting of αvβ3 integrin *via* RGD grafting (RGD); (3) Magnetic targeting *via* a magnet of 1.1 T placed on the tumor (MT) and (4) The combination of magnetic targeting and active targeting of αvβ3 integrin (RGD + MT). Reprinted from Journal of Controlled Release, 194, N. Schleich, C. Po, D. Jacobs, B. Ucakar, B. Gallez, F. Danhier, V. Préat, Comparison of active, passive and magnetic targeting to tumors of multifunctional paclitaxel/SPIO-loaded nanoparticles for tumor imaging and therapy, 82–91, Copyright (2014) with permission from Elsevier.[99]

(B) After extravasation has occurred, the presence of a magnetic field keeps the magnetic nanoparticles in the tumor's interstitium. Preventing the escape of the drug against naturally occurring osmotic pressure gradients[101] by means of magnetic attraction is the key to magnetic targeting. (C) Enhanced uptake of the nanoparticles by tumor cells from the interstitium occurs due to increased nanoparticle concentration. (D) If active targeting ligands are present, enhanced uptake by tumor cells from the interstitium is further increased.[99]

In Figures 9.7–9.9, the recent research by Schleich *et al.*[99] is summarized, in which they compared the efficacy of the following *in vivo* delivery strategies of the anticancer drug paclitaxel (PTX) to BALB/c (immunocompetent white) mice bearing murine colon carcinoma (CT 26). PTX, SPION (superparamagnetic iron oxides, and the RGD targeting sequence[102,103] were co-loaded into PLGA nanoparticles (d = 245 ± 10 nm, Zeta ζ potential > −0.42 mV). Four *in vivo* delivery strategies were compared: (1) passive targeting *via* the EPR effect, (2) active targeting of αvβ3 integrin *via* RGD grafting,[103] (3) magnetic targeting *via* a magnet placed on the tumor and (4) the combination of magnetic targeting and active targeting of αvβ3 integrin. The approaches are shown in Figure 9.7.

CT26 colon carcinoma cells were inoculated subcutaneously in the right flank or in the right leg (to avoid respiratory artifacts in MRI studies) of BALB/c mice (5×10^4 cells per mouse). The mice were divided into 5 groups:

Group 1: control group (injected with PBS for *in vivo* antitumor efficacy study and histological study and SPIO-loaded nanoparticles for *ex vivo* biodistribution and MRI),

Group 2: mice treated by SPIO/PTX-NP (passive targeting, PT),

Group 3: mice treated by SPIO/PTX-NP grafted with RGD peptide (active targeting or RGD),

Group 4: mice treated with SPIO/PTX-NP magnetic guided by placing a 1.1 T neodyme-iron-bore external magnet on the surface of the tumor for 1 h or 4 h (magnetic targeting or MT),

Group 5: mice treated with SPIO/PTX-NP using both magnetic and active targeting (RGD + MT).

MR imaging, shown in Figure 9.8, clearly indicates that magnetic targeting is superior to active targeting employing the targeting sequence RGD (arginine-glycine-aspartic acid). However, the combination of magnetic and active targeting clearly works best. Schleich *et al.* could clearly demonstrate the synergy of combined magnetic and αvβ3 integrin targeting.[99] The iron accumulations in tumor tissue, as determined by Electron Paramagnetic Resonance (EPR) spectroscopy, were practically indistinguishable after 1 h and 4 h. Active targeting (mice treated with RGD) led to a 2.5-fold accumulation increase in tumor tissue, compared to the group treated with PT (passive targeting). This accumulation rose to 5-fold when magnetic targeting was applied. The highest iron concentration was found in mice treated with a combination of magnetic and active targeting (8-fold increase, compared to passive targeting).

It is the conviction of the authors that these results have provided the cancer research community with a powerful paradigm for future research.

The data summarized in Figure 9.9 convincingly show how powerful this new paradigm is. In Figure 9.9a, the tumor growth curves of CT26-tumor bearing mice treated with co-loaded SPIO/PTX nanoparticles are shown, in comparison with all control experiments. The combination of magnetic and active targeting caused the most significant suppression of tumor growth, followed by magnetic and active targeting alone, which were indistinguishable. Figure 9.9b and c show the results of a survival study. Again, magnetic and active targeting showed significant synergy.

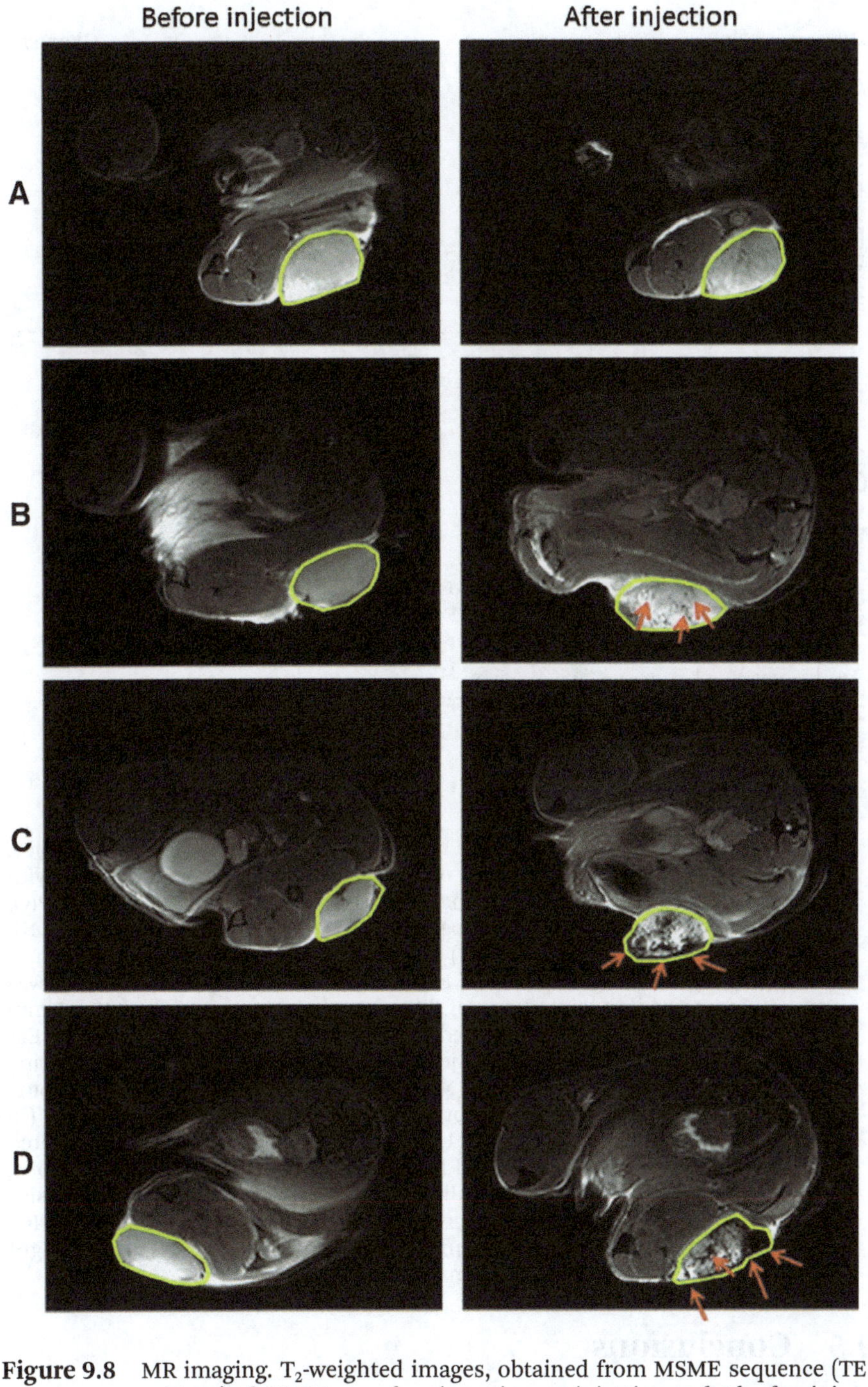

Figure 9.8 MR imaging. T_2-weighted images, obtained from MSME sequence (TE = 10 ms) of CT26-tumor bearing mice pre-injection and 4 h after injection. The position of the dark region in tumor is indicated by red arrows (n = 5). A: PT, B: RGD, C: MT, D: RGD + MT. Reprinted from Journal of Controlled Release, 194, N. Schleich, C. Po, D. Jacobs, B. Ucakar, B. Gallez, F. Danhier, V. Préat, Comparison of active, passive and magnetic targeting to tumors of multifunctional paclitaxel/SPIO-loaded nanoparticles for tumor imaging and therapy, 82–91, Copyright (2014) with permission from Elsevier.[99]

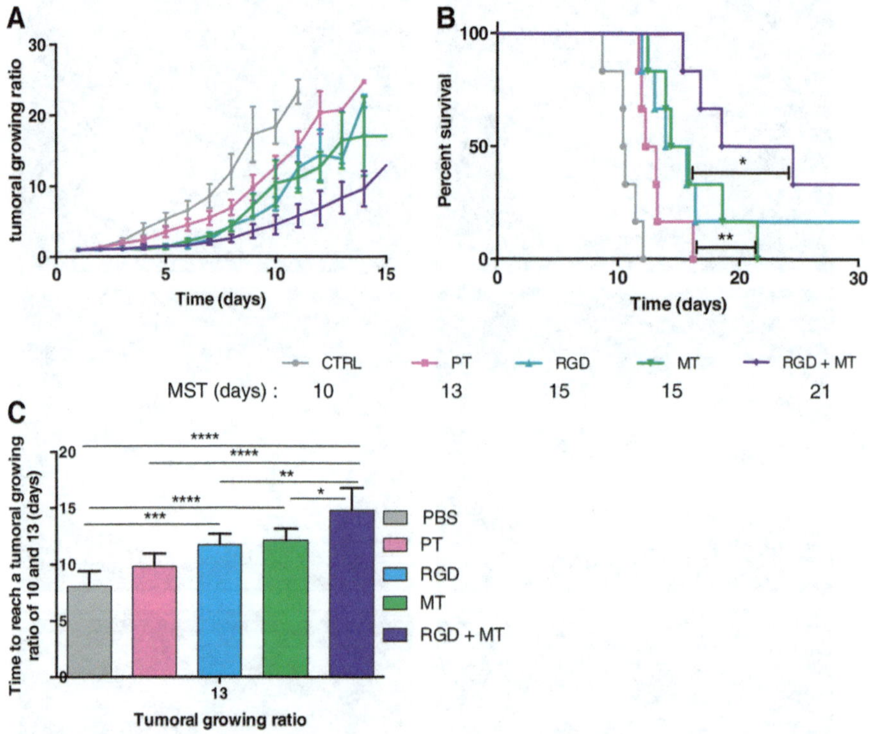

Figure 9.9 Tumoral growing ratio and survival rates of CT26-tumor bearing mice. (A) Tumor growth curves of CT26-tumor bearing mice treated with co-loaded SPIO/PTX nanoparticles (PT) (PTX dose = 5 mg kg^{-1}), SPIO/PTX-NP grafted with RGD-peptide (RGD), SPIO/PTX-NP followed by 4 h of magnetic targeting (MT) and SPIO/PTX-NP RGD followed by 4 h of magnetic targeting (RGD + MT) or untreated (PBS). The treatment was injected into the tail vein of the mice when tumors reached 27 ± 5 mm ($n = 6$). Results are presented as the average tumor growing ratio ± SEM *vs.* time after treatment injection. (B) Survival rates of tumor-bearing mice. Mean survival times (MST) in days were calculated for each group. (C) Comparison of time for tumor to reach a growing ratio of 13 (in days). *p b 0.05, **p b 0.01, ***p b 0.001, ****p b 0.0001. Reprinted from Journal of Controlled Release, 194, N. Schleich, C. Po, D. Jacobs, B. Ucakar, B. Gallez, F. Danhier, V. Préat, Comparison of active, passive and magnetic targeting to tumors of multifunctional paclitaxel/SPIO-loaded nanoparticles for tumor imaging and therapy, 82–91, Copyright (2014) with permission from Elsevier.[99]

9.5 Conclusions

The research discussed in this chapter provides a powerful new paradigm for maximizing the transport of magnetic nanoformulations into solid tumors, metastases and sentinel lymph nodes. A novel paradigm is required because recently serious doubts have been cast over the question of whether the extent of the EPR effect, which was originally established in rodent models, is sufficient for treating solid tumors in humans.[56] A combination of active and magnetic targeting strategies is discussed, together with PET, SPECT, and MRI imaging methods, which permit the quantitative assessment of drug transport into tumor tissue, as well as anatomical information to accurately monitor the biodistribution and target site accumulation of the nanocarriers.

References

1. R. Duncan and S. C. W. Richardson, *Mol. Pharmaceutics*, 2012, **9**, 2380–2402.
2. A. Fernandez-Fernandez, R. Manchanda and A. J. McGoron, *Appl. Biochem. Biotechnol.*, 2011, **165**, 1628–1651.
3. X. Ma, Y. Zhao and X.-J. Liang, *Acc. Chem. Res.*, 2011, **44**, 1114–1122.
4. V. Lopez-Davila, A. M. Seifalian and M. Loizidou, *Curr. Opin. Pharmacol.*, 2012, **12**, 414–419.
5. L. Miao, C. M. Lin and L. Huang, *J. Controlled Release*, 2015, **219**, 192–204.
6. B. Chaurasiya, A. Mahanty, D. Roy, Y. Shen, J. Tu and C. Sun, *Curr. Drug Metab.*, 2016, **17**, 783–798.
7. S. E. L. Detiger, R. M. Holewijn, R. J. W. Hoogendoorn, R. B. J. van, M. N. Helder, F. H. Berger, J. P. A. Kuijer and T. H. Smit, *Eur. Spine J.*, 2015, **24**, 1935–1943.
8. V. Gras, F. Grinberg and N. J. Shah, *Magn. Reson. Med.*, 2016, DOI: 10.1002/mrm.26353.
9. H.-L. Liu, M.-Y. Hua, H.-W. Yang, C.-Y. Huang, P.-C. Chu, J.-S. Wu, I. C. Tseng, J.-J. Wang, T.-C. Yen, P.-Y. Chen and K.-C. Wei, *Proc. Natl. Acad. Sci. U. S. A.*, 2010, **107**, 15205–15210.
10. O. Petracic, *Superlattices Microstruct.*, 2010, **47**, 569–578.
11. R. L. Hardman, F. El-Merhi, A. J. Jung, S. Ware, I. M. Thompson, H. T. Friel and Q. Peng, *J. Magn. Reson. Imaging*, 2011, **33**, 902–907.
12. J. S. Carter, J. S. Koopmeiners, J. E. Kuehn-Hajder, G. J. Metzger, N. Lakkadi, L. S. Downs Jr and P. J. Bolan, *J. Magn. Reson. Imaging*, 2013, **38**, 1501–1509.
13. K. Hueper, S. Rong, M. Gutberlet, D. Hartung, M. Mengel, X. Lu, H. Haller, F. Wacker, M. Meier and F. Gueler, *Invest. Radiol.*, 2013, **48**, 834–842.
14. S. P. Newman and I. R. Wilding, *Pharm. Sci. Technol. Today*, 1999, **2**, 181–189.
15. J. M. Rozental, *Neurol. Clin.*, 1991, **9**, 287–305.
16. M. Elsabahy and K. L. Wooley, *Chem. Soc. Rev.*, 2012, **41**, 2545–2561.
17. Y. Liu, H. Miyoshi and M. Nakamura, *Int. J. Cancer*, 2007, **120**, 2527–2537.
18. C. Sun, J. S. H. Lee and M. Zhang, *Adv. Drug Delivery Rev.*, 2008, **60**, 1252–1265.
19. M. R. Prausnitz and R. Langer, *Nat. Biotechnol.*, 2008, **26**, 1261–1268.
20. S. Chanda, M. Bagga and R. K. Tiwari, *Int. Res. J. Pharm.*, 2011, **2**, 72–78.

21. M. B. Brown, M. J. Traynor, G. P. Martin and F. K. Akomeah, *Methods Mol. Biol.*, 2008, **437**, 119–139.
22. C. H. Purdon, C. G. Azzi, J. Zhang, E. W. Smith and H. I. Maibach, *Crit. Rev. Ther. Drug Carrier Syst.*, 2004, **21**, 97–132.
23. L. Fang, L. Wang, Y. Yao, J. Zhang, X. Wu, X. Li, H. Wang, X. Zhang, X. Gong and J. Chang, *Nanomedicine (N. Y., NY, U. S.)*, 2017, **13**, 153–171.
24. D. Docter, D. Westmeier, M. Markiewicz, S. Stolte, S. K. Knauer and R. H. Stauber, *Chem. Soc. Rev.*, 2015, **44**, 6094–6121.
25. H. He, A. David, B. Chertok, A. Cole, K. Lee, J. Zhang, J. Wang, Y. Huang and V. C. Yang, *Pharm. Res.*, 2013, **30**, 2445–2458.
26. M. Kijanka, B. Dorresteijn, S. Oliveira and P. M. P. van Bergen en Henegouwen, *Nanomedicine (Lond.)*, 2015, **10**, 161–174.
27. S. Hallaj-Nezhadi, C. R. Dass and F. Lotfipour, *Future Oncol.*, 2013, **9**, 59–68.
28. S. Kunjachan, J. Ehling, G. Storm, F. Kiessling and T. Lammers, *Chem. Rev. (Washington, DC, U. S.)*, 2015, **115**, 10907–10937.
29. R. K. Jain and T. Stylianopoulos, *Nat. Rev. Clin. Oncol.*, 2010, **7**, 653–664.
30. S. Tang, C. Peng, J. Xu, B. Du, Q. Wang, R. D. Vinluan III, M. Yu, M. J. Kim and J. Zheng, *Angew. Chem., Int. Ed.*, 2016, **55**, 16039–16043.
31. R. Toy, E. Hayden, C. Shoup, H. Baskaran and E. Karathanasis, *Nanotechnology*, 2011, **22**, 115101.
32. P. Jurney, R. Agarwal, V. Singh, K. Roy, S. V. Sreenivasan and L. Shi, *J. Nanotechnol. Eng. Med.*, 2013, **4**, 031002.
33. S. Kunjachan, F. Gremse, B. Theek, P. Koczera, R. Pola, M. Pechar, T. Etrych, K. Ulbrich, G. Storm, F. Kiessling and T. Lammers, *ACS Nano*, 2013, **7**, 252–262.
34. R. S. Rachakatla, S. Balivada, G.-M. Seo, C. B. Myers, H. Wang, T. N. Samarakoon, R. Dani, M. Pyle, F. O. Kroh, B. Walker, X. Leaym, O. B. Koper, V. Chikan, S. H. Bossmann, M. Tamura and D. L. Troyer, *ACS Nano*, 2010, **4**, 7093–7104.
35. S. Balivada, R. S. Rachakatla, H. Wang, T. N. Samarakoon, R. K. Dani, M. Pyle, F. O. Kroh, B. Walker, X. Leaym, O. B. Koper, M. Tamura, V. Chikan, S. H. Bossmann and D. L. Troyer, *BMC Cancer*, 2010, **10**, 119.
36. T. Farrah, E. W. Deutsch, G. S. Omenn, D. S. Campbell, Z. Sun, J. A. Bletz, P. Mallick, J. E. Katz, J. Malmstroem, R. Ossola, J. D. Watts, B. Lin, H. Zhang, R. L. Moritz and R. Aebersold, *Mol. Cell. Proteomics*, 2011, **10**, M110.006353.
37. L. Vroman, *Nature (London, U. K.)*, 1962, **196**, 476–477.
38. S. Palchetti, D. Pozzi, M. Mahmoudi and G. Caracciolo, *J. Mater. Chem. B*, 2016, **4**, 4376–4381.
39. S. M. Moghimi and Z. S. Farhangrazi, *Nanomedicine (N. Y., NY, U. S.)*, 2013, **9**, 458–460.
40. M. Lundqvist, J. Stigler, G. Elia, I. Lynch, T. Cedervall and K. A. Dawson, *Proc. Natl. Acad. Sci. U. S. A.*, 2008, **105**, 14265–14270.

41. N. T. Huynh, E. Roger, N. Lautram, J.-P. Benoit and C. Passirani, *Nanomedicine (Lond.)*, 2010, **5**, 1415–1433.
42. H. Ding, V. Sagar, M. Agudelo, S. Pilakka-Kanthikeel, V. S. R. Atluri, A. Raymond, S. Thangavel and M. P. Nair, *Nanotechnology*, 2014, **25**, 055101.
43. M. P. Monopoli, C. Aaberg, A. Salvati and K. A. Dawson, *Nat. Nanotechnol.*, 2012, **7**, 779–786.
44. C. D. Walkey and W. C. W. Chan, *Chem. Soc. Rev.*, 2012, **41**, 2780–2799.
45. D. Docter, S. Strieth, D. Westmeier, O. Hayden, M. Gao, S. K. Knauer and R. H. Stauber, *Nanomedicine (Lond.)*, 2015, **10**, 503–519.
46. M. A. Dobrovolskaia and S. E. McNeil, *Nat. Nanotechnol.*, 2007, **2**, 469–478.
47. T.-G. Iversen, T. Skotland and K. Sandvig, *Nano Today*, 2011, **6**, 176–185.
48. I. Canton and G. Battaglia, *Chem. Soc. Rev.*, 2012, **41**, 2718–2739.
49. G. Sahay, D. Y. Alakhova and A. V. Kabanov, *J. Controlled Release*, 2010, **145**, 182–195.
50. L. N. Patel, J. L. Zaro and W.-C. Shen, *Pharm. Res.*, 2007, **24**, 1977–1992.
51. M. C. Morris, L. Chaloin, F. Heitz and G. Divita, *Cell-Penetrating Peptides*, 2002, pp. 93–113.
52. G. Von Heijne, *Protein Targeting, Transp. Translocat.*, 2002, 35–46.
53. N. Oh and J.-H. Park, *Int. J. Nanomed.*, 2014, **9**, 51–63, 13 pp.
54. L. W. Dobrucki, D. Pan and A. M. Smith, *Curr. Drug Targets*, 2015, **16**, 560–570.
55. Y. Nakamura, A. Mochida, P. L. Choyke and H. Kobayashi, *Bioconjugate Chem.*, 2016, **27**, 2225–2238.
56. F. Danhier, *J. Controlled Release*, 2016, **244**, 108–121.
57. F. S. Nouri, D. Banerjee and A. Hatefi, *Stem Cell Rev. Rep.*, 2015, **11**, 688–698.
58. M. T. Basel, T. B. Shrestha, S. H. Bossmann and D. L. Troyer, *Ther. Delivery*, 2014, **5**, 555–567.
59. D. Troyer, *J. Cancer Sci. Ther.*, 2013, **5**, 142–143.
60. Y. Sakurai, K. Kajimoto, H. Hatakeyama and H. Harashima, *Expert Opin. Drug Delivery*, 2015, **12**, 41–52.
61. T. Carter, P. Mulholland and K. Chester, *Immunotherapy*, 2016, **8**, 941–958.
62. V. H. Shargh, H. Hondermarck and M. Liang, *Nanomedicine (Lond.)*, 2016, **11**, 63–79.
63. A. Jain and K. Cheng, *J. Controlled Release*, 2017, **245**, 27–40.
64. M. Toporkiewicz, J. Meissner, L. Matusewicz, A. Czogalla and A. F. Sikorski, *Int. J. Nanomed.*, 2015, **10**, 1399–1414.
65. J. R. Beech, S. J. Shin, J. A. Smith and K. A. Kelly, *Curr. Pharm. Des.*, 2013, **19**, 6560–6574.
66. C. P. Cerrato, K. Kunnapuu and U. Langel, *Expert Opin. Drug Delivery*, 2017, **14**, 245–255.
67. J. S. Prakash and K. Rajamanickam, *Biomedicines*, 2015, **3**, 248–269.

68. D. Xiang, S. Shigdar, G. Qiao, S.-F. Zhou, Y. Li, M. Q. Wei, L. Qiao, H. A. Shamaileh, Y. Zhu, C. Zheng, C. Pu and W. Duan, *Curr. Gene Ther.*, 2015, **15**, 109–119.
69. B.-P. Jiang, X. Tan, X.-C. Shen, W.-Q. Lei, W.-Q. Liang, S.-C. Ji and H. Liang, *ACS Macro Lett.*, 2016, **5**, 450–454.
70. K. P. S. Kumar and D. Bhowmik, *Elixir Int. J.*, 2015, 33713–33717.
71. P. Padmanabhan, A. Kumar, S. Kumar, R. K. Chaudhary and B. Gulyas, *Acta Biomater.*, 2016, **41**, 1–16.
72. R. Bar-Shalom, A. Y. Valdivia and M. D. Blaufox, *Semin. Nucl. Med.*, 2000, **30**, 150–185.
73. E. D. Pressly, R. Rossin, A. Hagooly, K.-I. Fukukawa, B. W. Messmore, M. J. Welch, K. L. Wooley, M. S. Lamm, R. A. Hule, D. J. Pochan and C. J. Hawker, *Biomacromolecules*, 2007, **8**, 3126–3134.
74. N. K. Devaraj, E. J. Keliher, G. M. Thurber, M. Nahrendorf and R. Weissleder, *Bioconjugate Chem.*, 2009, **20**, 397–401.
75. M. M. Herth, M. Barz, M. Jahn, R. Zentel and F. Roesch, *Bioorg. Med. Chem. Lett.*, 2010, **20**, 5454–5458.
76. F. Roesch, *Curr. Radiopharm.*, 2012, **5**, 187–201.
77. C. Muller, M. Bunka, S. Haller, U. Koester, V. Groehn, P. Bernhardt, N. van der Meulen, A. Tuerler and R. Schibli, *J. Nucl. Med.*, 2014, **55**, 1658–1664.
78. R. Chakravarty, S. Goel, H. F. Valdovinos, R. Hernandez, H. Hong, R. J. Nickles and W. Cai, *Bioconjugate Chem.*, 2014, **25**, 2197–2204.
79. S. S. Gambhir, *Nat. Rev. Cancer*, 2002, **2**, 683–693.
80. K. C. P. Li, S. D. Pandit, S. Guccione and M. D. Bednarski, *Biomed. Microdevices*, 2004, **6**, 113–116.
81. K. Stockhofe, J. M. Postema, H. Schieferstein and T. L. Ross, *Pharmaceuticals*, 2014, **7**, 392–418, 327.
82. L. Karmani, D. Labar, V. Valembois, V. Bouchat, P. G. Nagaswaran, A. Bol, J. Gillart, P. Leveque, C. Bouzin, D. Bonifazi, C. Michiels, O. Feron, V. Gregoire, S. Lucas, T. VanderBorght and B. Gallez, *Contrast Media Mol. Imaging*, 2013, **8**, 402–408.
83. M. J. W. D. Vosjan, J. Vercammen, J. A. Kolkman, M. Stigter-van Walsum, H. Revets and G. A. M. S. van Dongen, *Mol. Cancer Ther.*, 2012, **11**, 1017–1025.
84. E. J. Keliher, J. Yoo, M. Nahrendorf, J. S. Lewis, B. Marinelli, A. Newton, M. J. Pittet and R. Weissleder, *Bioconjugate Chem.*, 2011, **22**, 2383–2389.
85. R. Cohen, D. J. Vugts, M. Stigter-van Walsum, G. W. M. Visser and G. A. M. S. van Dongen, *Nat. Protoc.*, 2013, **8**, 1010–1018, 1019 pp.
86. D. A. Heuveling, A. van Schie, D. J. Vugts, N. H. Hendrikse, M. Yaqub, O. S. Hoekstra, K. H. Karagozoglu, C. R. Leemans, G. A. M. S. van Dongen and R. de Bree, *J. Nucl. Med.*, 2013, **54**, 585–589.
87. L. R. Perk, G. W. M. Visser, M. J. W. D. Vosjan, M. Stigter-van Walsum, B. M. Tijink, C. R. Leemans and G. A. M. S. van Dongen, *J. Nucl. Med.*, 2005, **46**, 1898–1906.

88. X. Huang, F. Zhang, S. Lee, M. Swierczewska, D. O. Kiesewetter, L. Lang, G. Zhang, L. Zhu, H. Gao, H. S. Choi, G. Niu and X. Chen, *Biomaterials*, 2012, **33**, 4370–4378.
89. J. Cheon and J.-H. Lee, *Acc. Chem. Res.*, 2008, **41**, 1630–1640.
90. M. Liong, M. Fernandez-Suarez, D. Issadore, C. Min, C. Tassa, T. Reiner, S. M. Fortune, M. Toner, H. Lee and R. Weissleder, *Bioconjugate Chem.*, 2011, **22**, 2390–2394.
91. C. Tassa, S. Y. Shaw and R. Weissleder, *Acc. Chem. Res.*, 2011, **44**, 842–852.
92. S. H. Koenig and K. E. Kellar, *Magn. Reson. Med.*, 1995, **34**, 227–233.
93. A. Roch, Y. Gossuin, R. N. Muller and P. Gillis, *J. Magn. Magn. Mater.*, 2005, **293**, 532–539.
94. W. J. Rieter, J. S. Kim, K. M. L. Taylor, H. An, W. Lin, T. Tarrant and W. Lin, *Angew. Chem., Int. Ed. Engl.*, 2007, **46**, 3680–3682.
95. Y. Cui, M. Zhang, F. Zeng, H. Jin, Q. Xu and Y. Huang, *ACS Appl. Mater. Interfaces*, 2016, **8**, 32159–32169.
96. W.-H. Chen, G.-F. Luo, Q. Lei, F.-Y. Cao, J.-X. Fan, W.-X. Qiu, H.-Z. Jia, S. Hong, F. Fang, X. Zeng, R.-X. Zhuo and X.-Z. Zhang, *Biomaterials*, 2016, **76**, 87–101.
97. V. V. Mody, A. Cox, S. Shah, A. Singh, W. Bevins and H. Parihar, *Appl. Nanosci.*, 2014, **4**, 385–392.
98. R. A. Revia and M. Zhang, *Mater. Today (Kidlington)*, 2016, **19**, 157–168.
99. N. Schleich, C. Po, D. Jacobs, B. Ucakar, B. Gallez, F. Danhier and V. Preat, *J. Controlled Release*, 2014, **194**, 82–91.
100. Z. Li, C. Wang, L. Cheng, H. Gong, S. Yin, Q. Gong, Y. Li and Z. Liu, *Biomaterials*, 2013, **34**, 9160–9170.
101. C. Voutouri and T. Stylianopoulos, *J. Biomech.*, 2014, **47**, 3441–3447.
102. F. Danhier, B. Vroman, N. Lecouturier, N. Crokart, V. Pourcelle, H. Freichels, C. Jerome, J. Marchand-Brynaert, O. Feron and V. Preat, *J. Controlled Release*, 2009, **140**, 166–173.
103. E. A. Murphy, B. K. Majeti, L. A. Barnes, M. Makale, S. M. Weis, K. Lutu-Fuga, W. Wrasidlo and D. A. Cheresh, *Proc. Natl. Acad. Sci. U. S. A.*, 2008, **105**, 9343–9348.

CHAPTER 10

Cell-Based Magnetic Nanomaterials for Tracking and Therapy

MATTHEW T. BASEL*[a], TEJ B. SHRESTHA[a] AND DERYL L. TROYER[a]

[a]Department of Anatomy & Physiology, Coles Hall, Kansas State University, Manhattan, KS 66506, USA
*E-mail: mattbasel@hotmail.com

10.1 Introduction

Investigation into cell-based cytotherapies has been an active area of research due to the unique properties that cells bring to potential therapies including the ability to migrate and home to specific sites and the potential to act therapeutically at the site of injury. Much *in vitro* and preclinical work has been done that has shown significant promise in this area, but to translate cytotherapies from the bench to preclinical and even clinical studies requires an understanding of where the cells go in the body and how they act while they are there. Ideally, this would be answered by real-time tracking of the cells *in vivo* after application to measure migration, duration and final disposition of these cells.

Although there are many potential possibilities for tracking cells *in vivo*, including live-imaging fluorescence or positron emission tomography (PET), one particularly promising method for tracking the migration and

Smart Materials No. 26
Magnetic Nanomaterials: Applications in Catalysis and Life Sciences
Edited by Stefan H. Bossmann and Hongwang Wang

Published by the Royal Society of Chemistry, www.rsc.org

disposition of therapeutic cells in the body is to utilize magnetic nanoparticles as contrast agents for magnetic resonance imaging (MRI). The magnetic nanoparticles can be attached or loaded into the cell and the resulting T2 relaxation signal can be monitored in real-time *in vivo* using MRI. MR imaging has several benefits over other potential tracking methods. Post-mortem fluorescent microscopy has been well-utilized to track the fate of cells *in vivo*, but this method can only determine endpoint locations of cells and no real-time information. Live-imaging fluorescence can address this to some degree and provide some real-time tracking information, but live-imaging fluorescence suffers from lack of spatial resolution as well as low signal-to-noise ratios making tracking of individual cells very difficult. PET imaging can increase the signal-to-noise ratio and the spatial resolution but comes at the cost of introducing ionizing radiation into the system. MRI is superior to fluorescence imaging, optical imaging, CT scan and PET because it is non-ionizing, and has high spatial resolution, high signal-to-noise ratio, and intrinsic tissue contrast. MRI is also non-invasive and has the potential to even track single nanoparticles or single cells in real-time.

10.2 Nanoparticles

Magnetic nanoparticles are good T2 contrast agents. Magnetic nanoparticles (MNPs) are generally made by oxidation of iron metal, but are also made by alloys of Ni, Co, Zn with iron oxide. The most commonly used magnetic nanoparticle, superparamagnetic iron oxide nanoparticle (SPION), has a purported molecular formula of $Fe_2^{3+}O_3Fe^{2+}O$. A single nanoparticle has several thousands of crystal units within the formula. Each crystal has a single magnetic domain that has a greater magnetic dipole than the net dipole of all unpaired electrons of atoms of the crystal. Thermal motion is sufficient to freely rotate the domain resulting in zero net magnetic dipole. When a magnetic field is applied to the nanoparticles, all of the domains are aligned in the direction of the external magnetic field creating a large magnetic momentum, but this magnetic momentum is completely lost when the external magnetic field is removed. Therefore, superparamagnetic nanoparticles do not have magnetic dipoles and will not aggregate on their own or in an induced magnetic field, making them useful as contrast agents in magnetic resonance imaging (MRI).[1] SPION can be used as contrast agents in both longitudinal relaxation (R1) and transverse relaxation (R2), but are generally more useful for transverse relaxation. Transverse relaxation (R2) is defined as (R2 = 1/T2) where T2 is the time delay at which 63% of transverse magnetization is lost after a 90° RF pulse is applied during the MRI. Freely rotating protons in pure water move freely and have short translation correlation time and therefore long T2, but this effect is the opposite of protons in large molecules such as fats or proteins. SPIONs greatly reduce the transverse relaxation time because these particles induce a large anisotropic magnetic susceptibility due to unpaired electrons in the iron, and create a local magnetic field gradient (also known as inhomogeneity). The magnetic field gradient dephases the phase coherence decay of the transverse magnetization (magnetization along the transverse axis).[2] Observed transverse relaxation, that is, dephasing of transverse magnetization due to spin–spin relaxation T2 and magnetic inhomogeneity, is called T*.

Several types of magnetic nanoparticles are used as contrast agents for tracking cells *in vivo*, including superparamagnetic iron oxide nanoparticles (SPION, 50–500 nm) and ultra-small superparamagnetic nanoparticles (USPION, 4–50 nm) which create positive contrast in T1 weighted MRI but negative contrast in T2.[2] NPs with a larger core size are useful as negative contrast agents for the T2 weighted imaging sequence, while smaller core size nanoparticles are better for enhancing the positive contrast in T1 weighted imaging sequences.[3]

10.3 Disease Models

Cytotherapy has become a significant area of research in several different diseases due to the wide-ranging possibilities associated with cytotherapy. Similarly, multiple different disease models have used magnetic nanoparticles to track cytotherapeutic cells to the disease site.

10.3.1 Brain and Neurodiseases/Disorders

Brain injuries, disorders and diseases have been notoriously hard to treat for a variety of reasons including the blood–brain barrier and lack of effective treatment options. Cytotherapy, therefore, has been a promising option in multiple disease states. Neural stem cells, with their potential ability to differentiate into either new neurons or glial cells, have potential for treating a wide variety of diseases or injuries in the brain and their ability to traffic and stay at the site of injury or disease can be tracked using magnetic nanoparticles.[4–6] Stroke, or ischemic brain injury, could be a potential target for neural stem cells; these have been labeled with magnetic nanoparticles and their ability to home to stroke locations has been followed (see Figure 10.1).[7–9] Traumatic brain injury also could potentially benefit from neural stem cell treatment, and neural stem cells have been tracked to sites of traumatic brain injury.[10,11] Similarly, neural stem cells have been tested for their ability to home to important sites in neurodegenerative diseases.[9,11]

Mesenchymal stem cells, with their ability to suppress inflammation and secrete neural growth factors, also have potential for use in a variety of brain injuries. Mesenchymal stem cells loaded with magnetic nanoparticles have been tracked to stroke,[12,13] cerebral palsy,[14] olfactory bulb lesions,[15] neurodegenerative sites,[16] traumatic brain injury,[17] and inflammatory lesions in the brain.[18] Other cells have been tracked to significant sites in the brain as well, including astrocytes,[19] dental pulp cells,[20] endothelial precursor cells,[21] and pericytes.[22]

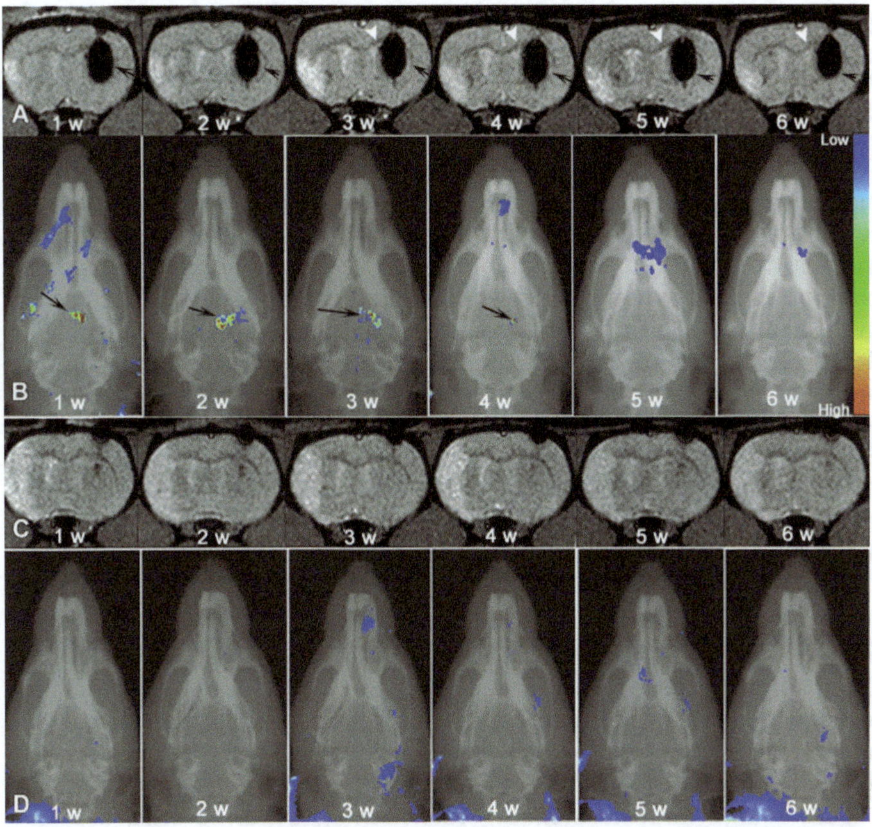

Figure 10.1 *In vivo* MR imaging and fluorescence imaging of the grafted cells. Longitudinal coronal T2* weighted images (A, C) and fluorescence images overlaid on radiograph (B, D) show a persistent hypointense area (within 6-week follow-up) or focal gradually declining fluorescence signal (within 4-week follow-up) representing labeled NSCs (arrows) in the left striatum of animals grafted with cationic polymersome-labeled cells (A, B). At 3 weeks after transplantation, a spotty hypointense area was found in the ipsilateral corpus callosum on T2* weighted images in animals grafted with cationic polymersome-labeled cells (arrowheads). There was no developmental hypointense signal or fluorescence signal observed in animals grafted with neutral polymersome-labeled cells (C, D). Reprinted from Biomaterials Volume 35, 11, Xuehua Wen *et al. in vivo* monitoring of neural stem cells after transplantation in acute cerebral infarction with dual-modal MR imaging and optical imaging. Pages 4627–4635, Copyright 2014 with permission from Elsevier.

Similar to tracking cells to the brain, several cytotherapeutic cells have been loaded with magnetic nanoparticles and tracked to the peripheral nerve. For example, adipose stem cells have been tracked in their ability to home and close wounds in the sciatic nerve.[23] Mesenchymal stem cells have been tracked to wounds in the sciatic nerve[24] and in the inner ear.[25]

10.3.2 Cardiac Disorders

Myocardial infarction is another area where cytotherapy may hold significant treatment promise. Cells that can differentiate into cardiomyocytes and/or endothelial cells could help repair damaged tissue while anti-inflammatory cells could reduce ongoing damage. Mesenchymal stem cells may be able to differentiate into both cardiomyocytes and endothelial cells[26] and have been loaded with magnetic nanoparticles for imaging in the heart (Figure 10.2).[27] Embryonic stem cells hold similar pluripotent potential and have been loaded with magnetic nanoparticles for imaging in the heart.[28,29] Endothelial progenitor cells may be able to treat vascular defects and have been tracked to the heart as well.[30] Similarly, cardiac progenitor cells may be able to regenerate cardiac tissue damaged after myocardial infarction and have been tracked using magnetic nanoparticles in the heart.[31]

Cells have been tracked to non-cardiac vascular injuries as well. For example, adipose-derived stem cells have been tracked to carotid artery injuries to determine their potential to repair the damaged tissue.[32]

10.3.3 Cancer

Cytotherapy has received significant interest in cancer due to the cells' ability to actively carry and/or produce anti-cancer products into the tumor. Magnetic nanoparticle-loaded cytotherapeutic cells have been directly tracked to gliomas,[16] hepatocellular carcinomas,[33] pancreatic adenocarcinomas (Figure 10.3),[34] and melanomas.[35] Alternatively, dendritic cells, used for immunotherapy, have been loaded with magnetic nanoparticles and tracked to draining lymph nodes (Figure 10.4)[36] in lymphomas[37] and pancreatic adenocarcinomas.[38]

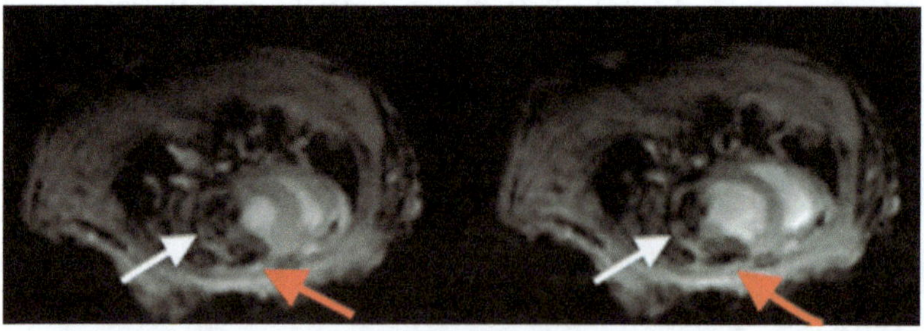

Figure 10.2 Cardiac MRI of Balb/c mouse in the BMCS-treated group. Short-axis images with GRE cine sequence. Signals of the anterior wall (red arrow) and lateral wall (white arrow) of the left ventricle decreased significantly, corresponding with the injection sites. Reprinted from Cytotherapy Volume 17, Issue 10, Jian Cao *et al.* Dual-modular molecular imaging to trace transplanted bone mesenchymal stromal cells in an acute myocardial infarction model. Pages 1365–1373, Copyright 2015 with permission from Elsevier.

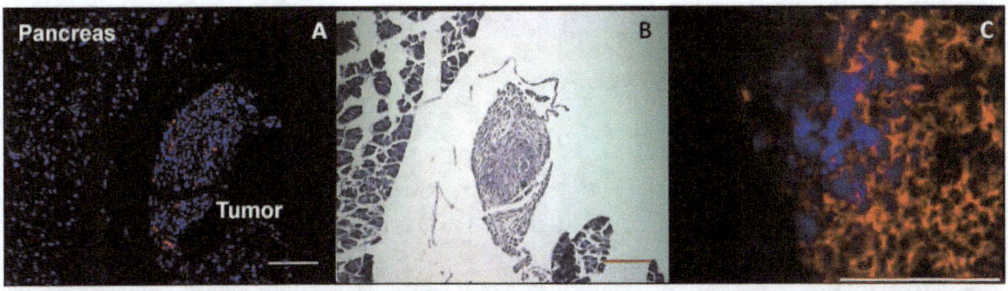

Figure 10.3 Monocyte/macrophage-like cells (Mo/Ma) infiltrate Pan02 tumors. (**A** and **B**) Mo/Ma loaded with PKH26 were injected into mice bearing intraperitoneal Pan02 tumors. 6 days later mice were euthanized and tumors were harvested. (**A**) Hoechst nuclear counterstained section shows Mo/Ma labeled with PKH26 in tumor. (**B**) Hematoxylin and eosin staining of serial sections shows irregular morphology demonstrating that the targeted area is a tumor. Scale bars = 100 μm. (**C**) Mo/Ma labeled with Hoechst before injection were injected into mice bearing Pan02-expressing firefly luciferase tumors. 5 days later mice were euthanized and tumors were harvested. Sections were stained with rabbit α-firefly luciferase and DYLIGHT® 650-goat α-rabbit (Abcam, Cambridge, MA) (sections were not counterstained with Hoechst). Immunohistochemistry verifies that the Mo/Ma infiltrate pancreatic tumors. Reprinted from International Journal of Nanomedicine, Volume 7, Matthew Basel, *et al.* Cell-delivered magnetic nanoparticles caused hyperthermia-mediated increased survival in a murine pancreatic cancer model. Pages 297–306, Copyright 2012 with permission from Dove Medical Press Ltd.

10.3.4 Other Diseases

Magnetic nanoparticles have been used to track cells to a wide variety of other tissues as well. Mesenchymal stem cells may have regenerative abilities in cartilage[39] and have been imaged under the knee cap[40] and in rheumatoid arthritis models.[41] Treating kidney damage can pose significant challenges that pluripotent stem cells may be able to address and mesenchymal stem cells have been tracked to the kidney[42,43] and specifically to damaged areas of the kidney.[44] Similarly, magnetic nanoparticle loaded mesenchymal stem cells can be tracked to the liver[43] to sites of ischemia[45] or fibrosis.[46]

10.4 Considerations

To use magnetic nanoparticles for MRI tracking, at least four areas need to be addressed: uptake and loading of nanoparticles into cells; viability and division of cells after loading; ability of cells to differentiate after loading; ability of the cells to migrate after loading.

10.4.1 Uptake and Loading

Loading magnetic nanoparticles into the desired cells is obviously the first hurdle that must be addressed to utilize magnetic nanoparticle-MRI *in vivo* tracking. Fortunately, many cells used in cytotherapy show some level of

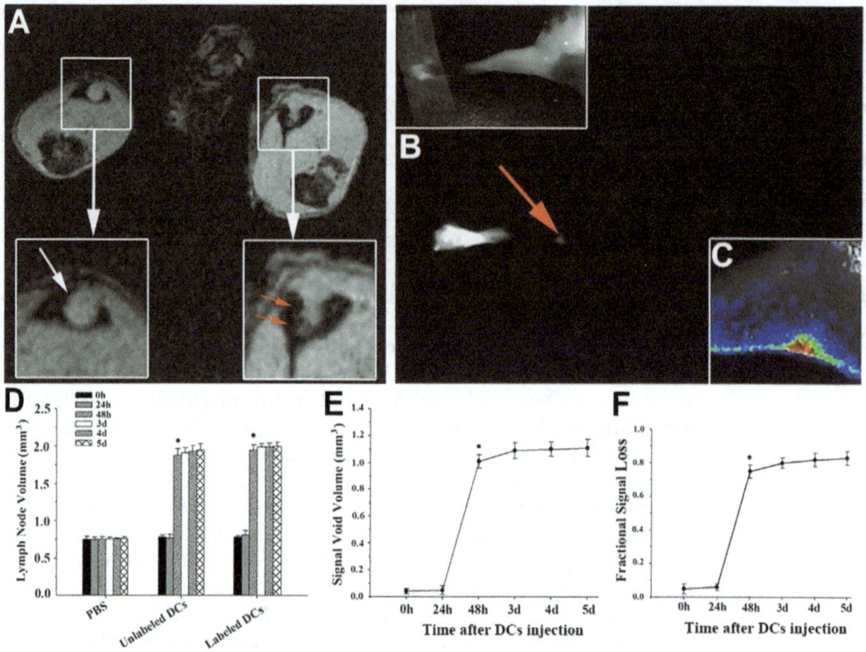

Figure 10.4 SPIO-NIR797-labeled DCs homing to the draining LNs *in vivo* assay. (A) Labeled DCs homing to the right popliteal LN by MRI. Red arrow indicates the location of signal loss within the popliteal LN. White arrow shows no signal changes within the control side. (B) Labeled DCs homing to the popliteal LN by NIRF imaging. Red arrow indicates NIRF in the popliteal LN. (C) Enlarged pseudo-color image of labeled DCs within the popliteal LN. (D) LN volume was measured from images acquired before injection (0 h) and 24 h, 48 h, 3 days, 4 days and 5 days after injection of PBS, unlabeled DCs or labeled DCs. (E) Signal void volume and (F) fractional signal loss were measured from images acquired before injection (0 h) and 24 h, 48 h, 3 days, 4 days and 5 days after injection of labeled DCs. $*P < 0.05$. Reprinted from Cytotherapy Volume 16, Issue 5, Yu-Chen Chen *et al.* Magnetic resonance and near-infrared imaging using a novel dual-modality nano-probe for dendritic cell tracking *in vivo*. Pages 699–710, Copyright 2014 with permission from Elsevier.

phagocytosis[10,14,15,21,32,40,47] and actively internalize the magnetic nanoparticles simply by incubating the cells with the nanoparticles at the correct dose. While this may work well for some cell types, especially naturally phagocytic cells like monocytes or macrophages, some cell types do not load effectively with detectable doses of nanoparticles through simple incubation.[17] To overcome this limitation, a wide range of methods for increasing internalization have been tested. An early method for increasing internalization that is still used is co-incubating the nanoparticles with a transfection reagent, similar to DNA delivery into cells. Commercially available transfection kits can be used with the nanoparticles[28] or simple polymers such as poly-L-lysine are often used.[20,32] The transfection reagents act primarily to increase positive charge, facilitating productive interaction with the membranes of the cells and are particularly effective at increasing internalization of magnetic nanoparticles. But there are potential drawbacks as well, particularly worries about toxicity of the transfection reagents on the loaded cells,[39] and they therefore should not be used unless needed.

A closely related mechanism for increasing internalization is stably wrapping polymers around the magnetic nanoparticles, turning them into core/shell magnetic nanoparticle/polymersomes. The polymersomes act as stably incorporated transfection reagents and are often made of similar materials. For example, steric-acid-grafted polyethylimine was used to wrap superparamagnetic iron oxide nanoparticles to create a polyethylimine polymersome with a magnetic nanoparticle core that efficiently labeled embryonic stem cells.[48] A similar, though more complicated, block copolymer containing L-aspartic acid and polyethylimine was used to create magnetic nanoparticle cored polymersomes that efficiently labeled mesenchymal stem cells[45] and also could act as a transfection reagent for DNA. Other similar polymers can be used to create transfectable polymersomes including a polyetherimide/poly-D-L-lactide diblock copolymer[7] or aminopolyvinyl alcohol.[40] These transfection reagents incorporated into the nanoparticles as polymersomes appear to be less toxic than classic transfection reagents while preserving the increase in internalization. Similar to incorporating transfection reagents into the nanoparticle complex, internalization can be increased by incorporating cell-penetrating peptides. For example, incorporating the TAT peptide allowed efficient internalization of iron oxide nanoparticles into epidermal neural crest stem cells.[18]

A novel method for internalizing magnetic nanoparticles into target cells is the use of a virus to act as an infectious agent delivering magnetic nanoparticles. For example, magnetic nanoparticles have been wrapped in hemagglutinating virus of Japan envelopes to internalize them into neural stem cells.[6] Compared to a common transfection reagent (LIPOFECTAMINE 2000), the virus-wrapped nanoparticles had up to twenty times more magnetic nanoparticle internalization while remaining non-toxic to the neural stem cells.

A simple method for increasing internalization is to change the ligands stabilizing the nanoparticle. For example, magnetic nanoparticles stabilized with citric acid were internalized into mesenchymal stem cells at much higher concentrations than commercially available nanoparticles without reducing cell proliferation or differentiation potential.[49] Similar results have been obtained with meso-2,3-dimercaptosuccinic acid stabilized magnetic nanoparticles,[50] although why these ligands increased internalization is not necessarily clear.

Finally, nanoparticles can have specific targeting moieties incorporated into them to increase internalization. For example, hyaluronic acid polymers interact with CD44, and wrapping magnetic nanoparticles in hyaluronic acid based polymers can increase internalization of the nanoparticles.[17]

10.4.2 Viability and Proliferation

After loading, the next consideration in using magnetic nanoparticles to track cells is viability and proliferation after loading. As foreign substances, magnetic nanoparticles have the potential to be toxic or cytostatic, disrupting the ability to utilize the cells after loading. Generally, iron oxide nanoparticles have been shown to be non-toxic and do not significantly reduce viability or proliferation in loaded cells.[51] Furthermore, loading cells with magnetic nanoparticles does not significantly change morphology or metabolism.[51] Given the number of studies performed on a variety of different iron oxide nanoparticles, this would suggest that these nanoparticles are highly biocompatible and pose few toxic threats to the cell. This may be publication bias, though, as nanoparticles that show substantial loading toxicity likely do not continue as research subjects to the point of publication. Verifying the lack of toxicity of nanoparticles on the target cell continues to be an important step in validating new nanoparticles and needs to be performed for any new nanoparticle preparation.

Because of ease of use, nearly all toxicity studies have been performed *in vitro* on cultured cells. While this data is important, especially for labeling cells *ex vivo* for tracking *in vivo*, this may or may not strictly correlate to *in vivo* viability and proliferation. Whether nanoparticle loading affects viability or proliferation *in vivo* seems to be a relatively unexplored area in the literature, but it is conceivable that loaded nanoparticles could cause either positive or negative changes in cell viability *in vivo* that are not seen *ex vivo*.

10.4.3 Differentiation

Given that many active areas of cytotherapy research involve investigating the potential of stem or stem-like cells to traffic to a diseased or damaged tissue and differentiate into a useful cell type, the ability of loaded cells to differentiate is an important consideration. Differentiation ability is measured by culturing stem cells in specific defined differentiation media that facilitate the differentiation of stem cells into different lineage and then observing the differentiation either by morphology and protein products or genetically through PCR. Differentiation ability has been measured in magnetic nanoparticle loaded mesenchymal stem cells,[39,45,49] adipose-derived stem cells,[50] neural stem/progenitor cells[5,6,9,11] and embryonic stem cells.[48] Generally magnetic nanoparticles have been demonstrated to not significantly change the ability of stem cells to differentiate into expected lineages.

Differentiation potential has been particularly studied in bone-marrow-derived mesenchymal stem cells and neural stem/progenitor cells. In mesenchymal stem cells differentiation ability is tested in three different lineages: adipogenic; osteogenic; and chondrogenic. Differentiation is generally verified by staining for a lineage specific product: for adipocytes, lipid inclusions are stained with Oil Red O; for osteoblasts, increased phosphate levels are demonstrated through staining alkaline phosphatase or calcium deposits are visualized with Alizarin Red; and for chondrocytes, proteoglycans are stained with Alcian Blue. Differentiation can also be verified using PCR to detect gene products, for example detecting expression of SOX9, ACAN and COL2A1 in chondrocytes. Numerous studies have demonstrated that loading mesenchymal stem cells with magnetic nanoparticles does not significantly affect osteogenic or adipogenic differentiation.[45,49] Chondrogenic differentiation, on the other hand, has been shown to be impaired by loading with magnetic nanoparticles in a dose-dependent manner,[49,52] as measured by Alcian Blue staining. There is disagreement about this in the literature though, with many publications demonstrating the ability of mesenchymal stem cells loaded with magnetic nanoparticles to differentiate into chondrocytes as measured by Alcian Blue staining[53,54] or expression of chondrogenic genes.[39] Since the exact make-up of the nanoparticles and the surface coating were different in each example, it is unclear if there is a lack of consensus on this point or if different nanoparticles and surface coatings affect differentiation in variable ways. The disparity in results, though, emphasizes the need to demonstrate that differentiation is not compromised after loading with magnetic nanoparticles. In a similar way, some publications have claimed that osteogenic differentiation is also compromised in magnetic nanoparticle loaded mesenchymal cells.[52]

Neural stem/progenitor cell differentiation ability is usually tracked to two lineages: neuronal and glial. These lineages can be demonstrated by antibody staining for lineage specific proteins: often GFAP for glial cells and β-III tubulin for neurons.[5,9,11] Neuronal differentiation can also be measured by measuring the growth of neural processes (axons and dendrites) over time after differentiation.[6] Results have been more consistent in neural stem cells demonstrating that loaded neural stem cells retain their stem-like character, as demonstrated by expression of SOX2,[5] and their ability to differentiate into both neurons and glial cells.[5,6,9,11]

10.4.4 Homing and Migration

Measuring the ability of cells to home and migrate to the site of injury or disease is the purpose of loading cells with magnetic nanoparticles, so it is important to demonstrate that the nanoparticles do not change or hinder the cells' ability to migrate. Migration of cells can be measured *in vitro* or *in vivo* in several different ways. Common *in vitro* methods include transwell migration assays toward or away from a gradient or scratch assays to assess the ability of cells to close wound gaps. *In vivo* methods usually involve endpoint comparisons of loaded and unloaded cells at a target site. As with differentiation, there have been many different reported results on how magnetic nanoparticles affect homing and migration.

Many studies have shown no statistically significant difference in migration ability after loading cells with magnetic nanoparticles. For example, neural stem cells loaded with FERIDEX I.V. (Berlex Laboratories) showed no reduction in migration through the rostal migratory stream in the neonatal brain as compared to unloaded controls.[9] Similarly, loading mesenchymal stem cells with superparamagnetic iron oxide nanoparticles caused minimal difference in migration in an *in vitro* scratch assay as compared to unloaded controls.[55,56]

In contrast, several studies have reported decreased migratory ability after loading with magnetic nanoparticles. For example, loading macrophages with superparamagnetic iron oxide nanoparticles with various coatings reduced macrophage migration *in vivo* toward a CX3CL1 gradient and also reduced migration toward a FBS gradient.[57] Similarly, monocyte-derived macrophages showed dose-dependent reduced migration *in vivo* toward LPS-induced inflammation in the brain after loading with polymer-coated iron oxide nanoparticles.[58] In direct contrast to other reports, mesenchymal stem cells showed significantly impaired migration toward FBS or CCL25 in *in vivo* transwell assays after loading with several different iron oxide nanoparticles.[49]

It is unclear if more investigation needs to be done to reconcile these results or if these results mirror the unique differences of every cell–nanoparticle pair. Suggesting the latter, some reports report no significant change in migration in some cell–nanoparticle pairs, while noting significantly reduced migration in others. For example, one publication reported that endothelial cells loaded with several different iron oxide nanoparticles showed no difference in migration in an *in vitro* scratch assay compared to unloaded controls. Monocytes, on the other hand, showed dose-dependent reduction in migration after loading with lauric acid-coated iron oxide nanoparticles but no significant decrease in migration with dextran coated iron oxide nanoparticles.[59] There have been some reports that suggest that magnetic nanoparticles can increase homing and migration ability. For example, loading mesenchymal stem cells with hyaluronic acid-coated magnetic nanoparticles caused increased CXCR4 expression and a concurrent increase in migration toward SDF-1α as compared to unloaded control.[17] These results underline the importance of characterizing each cell–nanoparticle pair for migration and homing ability.

Interestingly, some studies have even used magnetic nanoparticles as a method for increasing migration to a specific site by utilizing strategically placed magnets. The magnetic nanoparticles should react to the magnetic field and exert force on the loaded cell. This force may cause the cell to migrate toward the magnet. For example, loading dendritic cells with magnetic nanoparticles caused increased migration *in vitro* in a transwell assay and *in vivo* toward lymph nodes in a correctly placed magnetic field.[60] Similarly, loading neural stem cells with magnetic nanoparticles increased their migration toward a magnet both *in vitro* and *in vivo* in the brain, increasing cell migration nearly threefold as compared to without the magnet.[8,10]

10.5 Using Magnetic Nanoparticles for Cytotherapy

Beyond merely using nanoparticles to track cells that are therapeutically valuable, an emerging field in magnetic nanoparticle research is using magnetic nanoparticles to confer an additive or synergistic treatment effect to cytotherapy cells. Numerous stem and defensive cells have been shown to traffic to tumors efficiently, mediated by chemokine gradients that direct cellular traffic toward them. Given this targeting ability, loading such cells *ex vivo* with magnetic nanomaterials should allow more selective accumulation of such particles in tumors. This has numerous advantages: 1. It avoids removal by the reticulo-endothelial system; 2. It shields the nanomaterials from the immune system; 3. It enhances accumulation in the interior of the tumor since the delivery cells can actively migrate against the high interstitial pressures with tumors; 4. It retains the nanomaterials in the tumor.

Core/shell/iron oxide magnetic nanoparticles (MNPs) have been incubated *ex vivo* with mouse 17.2 neural progenitor cells (NPCs), with subsequent uptake and minimal toxicity to the cells. The MNP loaded NPCs were administered *via* the lateral tail vein to mice bearing subcutaneous mouse melanomas. Mice were exposed to an alternating magnetic field (AMF) 4, 5 and 6 days after NPC administration with subsequent reduction in tumor volume compared to controls.[35]

In a similar approach, mouse Raw 264.7 monocyte/macrophage (Mo/Ma) cells were loaded with MNPs and administered intraperitoneally to mice with disseminated mouse pancreatic tumors (Figure 10.5). MNP loaded cells were given 3 times at 4 day intervals, and mice were exposed to an alternating magnetic field 3 days after each transplant. It was shown that MNP-loaded MNPs specifically migrated to tumor tissue, and a significant survival advantage was shown in the treated mice (Figure 10.6).[34]

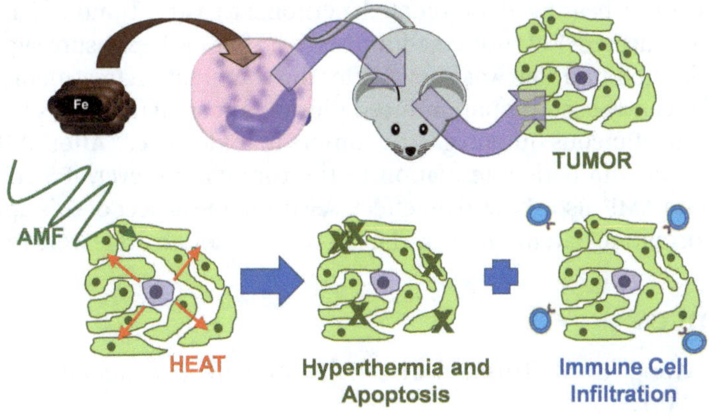

Figure 10.5 Schematic of cell-loaded magnetic nanoparticles used for local hyperthermia treatment. Reprinted from International Journal of Nanomedicine, Volume 7, Matthew Basel, *et al.* Cell-delivered magnetic nanoparticles caused hyperthermia-mediated increased survival in a murine pancreatic cancer model. Pages 297–306, Copyright 2012 with permission from Dove Medical Press Ltd.

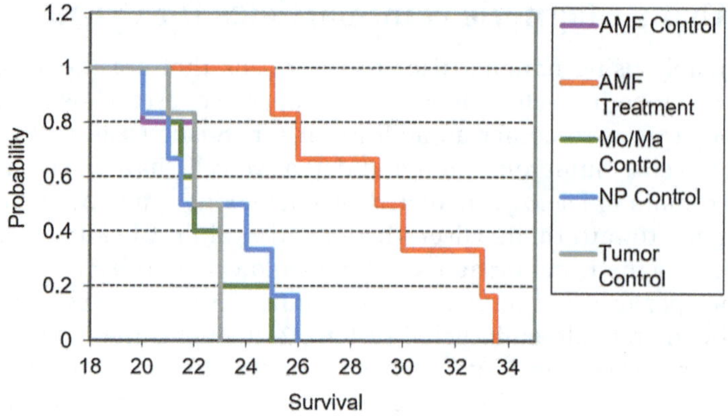

Figure 10.6 Treatment with cell-delivered nanoparticles and local alternating magnetic fields significantly increased survival in a disseminated pancreatic cancer model.

In a subsequent report, Ruan *et al.* showed that iron oxide nanoparticles loaded *ex vivo* into mesenchymal stem cells (MSCs) that were administered intravenously followed by exposure to alternating magnetic fields could almost completely inhibit growth of subcutaneous gastric tumors.[61]

In another approach, magnetic nanoparticles have been injected intraperitoneally to allow peritoneal macrophages to phagocytoze them. Toraya-Brown and colleagues harvested these peritoneal macrophages loaded with the MNPs by peritoneal lavage, and these cells were transplanted intraperitoneally into other mice bearing peritoneal tumors followed by AMF and tumor attenuation.[64] Ikehara and colleagues employed oligomannose-coated liposomes (OMLs) encasing MNPs for intraperitoneal uptake by macrophages which then carried the payload to mice bearing disseminated peritoneal gastric tumors.[62] In some cases, OMLs encased 5-FU was co-administered; after AMF exposure significant reduction in tumor burden was shown after the combination treatment.

Li *et al.* recently used human-induced pluripotent stem (iPS) cells to deliver MNPs to subcutaneous human gastric tumors in nude mice.[63] After AMF exposure, there was dramatic attenuation of the tumors; however, iPS cells with MNPs but no AMF also showed an effect. A worrisome aspect of this approach could be formation of teratomas by the iPS cells, although none were reported.

References

1. Y. X. Wang, S. M. Hussain and G. P. Krestin, *Eur. Radiol.*, 2001, **11**, 2319–2331.
2. L. M. De Leon-Rodriguez, A. F. Martins, M. C. Pinho, N. M. Rofsky and A. D. Sherry, *J. Magn. Reson. Imaging*, 2015, **42**, 545–565.
3. H. Ittrich, K. Peldschus, N. Raabe, M. Kaul and G. Adam, *Rofo*, 2013, **185**, 1149–1166.

4. S. Sart, F. C. Bejarano, M. A. Baird, Y. Yan, J. T. Rosenberg, T. Ma, S. C. Grant and Y. Li, *Cytotherapy*, 2015, **17**, 98–111.
5. W. B. Shen, C. Plachez, A. Chan, D. Yarnell, A. C. Puche, P. S. Fishman and P. Yarowsky, *Int. J. Nanomed.*, 2013, **8**, 4593–4600.
6. S. Miyoshi, J. A. Flexman, D. J. Cross, K. R. Maravilla, Y. Kim, Y. Anzai, J. Oshima and S. Minoshima, *Mol. Imaging Biol.*, 2005, **7**, 286–295.
7. X. Wen, Y. Wang, F. Zhang, X. Zhang, L. Lu, X. Shuai and J. Shen, *Biomaterials*, 2014, **35**, 4627–4635.
8. M. Song, Y. J. Kim, Y. H. Kim, J. Roh, S. U. Kim and B. W. Yoon, *Hum. Gene Ther.*, 2010, **21**, 603–610.
9. F. Sicard, M. Ehrhart-Bornstein, D. Corbeil, S. Sperber, A. W. Krug, C. G. Ziegler, V. Rettori, S. M. McCann and S. R. Bornstein, *Proc. Natl. Acad. Sci. U. S. A.*, 2007, **104**, 2007–2012.
10. W. B. Shen, C. Plachez, O. Tsymbalyuk, N. Tsymbalyuk, S. Xu, A. M. Smith, S. L. Michel, D. Yarnell, R. Mullins, R. P. Gullapalli, A. Puche, J. M. Simard, P. S. Fishman and P. Yarowsky, *Cell Transplant.*, 2016, **25**, 1085–1099.
11. E. Y. Egawa, N. Kitamura, R. Nakai, Y. Arima and H. Iwata, *Biomaterials*, 2015, **54**, 158–167.
12. R. M. Guo, N. Cao, F. Zhang, Y. R. Wang, X. H. Wen, J. Shen and X. T. Shuai, *Eur. Radiol.*, 2012, **22**, 2328–2337.
13. J. T. Rosenberg, K. L. Sellgren, A. Sachi-Kocher, F. Calixto Bejarano, M. A. Baird, M. W. Davidson, T. Ma and S. C. Grant, *Cytotherapy*, 2013, **15**, 307–322.
14. A. Chen, B. Siow, A. M. Blamire, M. Lako and G. J. Clowry, *Stem Cell Res.*, 2010, **5**, 255–266.
15. G. J. Delcroix, M. Jacquart, L. Lemaire, L. Sindji, F. Franconi, J. J. Le Jeune and C. N. Montero-Menei, *Brain Res.*, 2009, **1255**, 18–31.
16. M. S. Thu, J. Najbauer, S. E. Kendall, I. Harutyunyan, N. Sangalang, M. Gutova, M. Z. Metz, E. Garcia, R. T. Frank, S. U. Kim, R. A. Moats and K. S. Aboody, *PLoS One*, 2009, **4**, e7218.
17. X. Huang, F. Zhang, Y. Wang, X. Sun, K. Y. Choi, D. Liu, J. S. Choi, T. H. Shin, J. Cheon, G. Niu and X. Chen, *ACS Nano*, 2014, **8**, 4403–4414.
18. J. S. Jackson, J. P. Golding, C. Chapon, W. A. Jones and K. K. Bhakoo, *Stem Cell Res. Ther.*, 2010, **1**, 17.
19. G. Elvira, I. Garcia, J. Gallo, M. Benito, P. Montesinos, E. Holgado-Martin, A. Ayuso-Sacido, S. Penades, M. Desco, A. Silva and J. A. Garcia-Sanz, *Stem Cell Res.*, 2015, **14**, 114–129.
20. T. Struys, A. Ketkar-Atre, P. Gervois, C. Leten, P. Hilkens, W. Martens, A. Bronckaers, T. Dresselaers, C. Politis, I. Lambrichts and U. Himmelreich, *Cell Transplant.*, 2013, **22**, 1813–1829.
21. E. Carenza, V. Barcelo, A. Morancho, L. Levander, C. Boada, A. Laromaine, A. Roig, J. Montaner and A. Rosell, *Nanomedicine*, 2014, **10**, 225–234.
22. C. H. Liu, J. Q. Ren, Z. You, J. Yang, C. M. Liu, R. Uppal and P. K. Liu, *FASEB J.*, 2012, **26**, 1652–1662.

23. K. Li, J. Qin, X. Wang, Y. Xu, Z. Shen, X. Lu and G. Zhang, *Cytotherapy*, 2013, **15**, 1275–1285.
24. Y. Watada, D. Yamashita, M. Toyoda, K. Tsuchiya, N. Hida, A. Tanimoto, K. Ogawa, S. Kanzaki and A. Umezawa, *Neurosci. Res.*, 2015, **95**, 21–26.
25. T. C. Tseng and S. H. Hsu, *Biomaterials*, 2014, **35**, 2630–2641.
26. Z. Wen, S. Zheng, C. Zhou, J. Wang and T. Wang, *J. Cell. Mol. Med.*, 2011, **15**, 1032–1043.
27. J. Cao, X. Li, N. Chang, Y. Wang, J. Lei, D. Zhao, K. Gao and Z. Jin, *Cytotherapy*, 2015, **17**, 1365–1373.
28. J. P. Vallee, M. Hauwel, M. Lepetit-Coiffe, W. Bei, K. Montet-Abou, P. Meda, S. Gardier, P. Zammaretti, T. P. Kraehenbuehl, F. Herrmann, J. A. Hubbell and M. E. Jaconi, *Stem Cells Transl. Med.*, 2012, **1**, 248–260.
29. N. Himes, J. Y. Min, R. Lee, C. Brown, J. Shea, X. Huang, Y. F. Xiao, J. P. Morgan, D. Burstein and P. Oettgen, *Magn. Reson. Med.*, 2004, **52**, 1214–1219.
30. A. Chaudeurge, C. Wilhelm, A. Chen-Tournoux, P. Farahmand, V. Bellamy, G. Autret, C. Menager, A. Hagege, J. Larghero, F. Gazeau, O. Clement and P. Menasche, *Cell Transplant.*, 2012, **21**, 679–691.
31. E. Kustermann, W. Roell, M. Breitbach, S. Wecker, D. Wiedermann, C. Buehrle, A. Welz, J. Hescheler, B. K. Fleischmann and M. Hoehn, *NMR Biomed.*, 2005, **18**, 362–370.
32. J. B. Qin, K. A. Li, X. X. Li, Q. S. Xie, J. Y. Lin, K. C. Ye, M. E. Jiang, G. X. Zhang and X. W. Lu, *Int. J. Nanomed.*, 2012, **7**, 5191–5203.
33. J. Zhao, J. Vykoukal, M. Abdelsalam, A. Recio-Boiles, Q. Huang, Y. Qiao, B. Singhana, M. Wallace, R. Avritscher and M. P. Melancon, *Nanotechnology*, 2014, **25**, 405101.
34. M. T. Basel, S. Balivada, H. Wang, T. B. Shrestha, G. M. Seo, M. Pyle, G. Abayaweera, R. Dani, O. B. Koper, M. Tamura, V. Chikan, S. H. Bossmann and D. L. Troyer, *Int. J. Nanomed.*, 2012, **7**, 297–306.
35. R. S. Rachakatla, S. Balivada, G. M. Seo, C. B. Myers, H. Wang, T. N. Samarakoon, R. Dani, M. Pyle, F. O. Kroh, B. Walker, X. Leaym, O. B. Koper, V. Chikan, S. H. Bossmann, M. Tamura and D. L. Troyer, *ACS Nano*, 2010, **4**, 7093–7104.
36. Y. C. Chen, S. Wen, S. A. Shang, Y. Cui, B. Luo and G. J. Teng, *Cytotherapy*, 2014, **16**, 699–710.
37. P. M. Ferguson, A. Slocombe, R. D. Tilley and I. F. Hermans, *PLoS One*, 2013, **8**, e65318.
38. Z. Zhang, W. Li, D. Procissi, K. Li, A. Y. Sheu, A. C. Gordon, Y. Guo, K. Khazaie, Y. Huan, G. Han and A. C. Larson, *Radiology*, 2015, **274**, 192–200.
39. S. Saha, X. B. Yang, S. Tanner, S. Curran, D. Wood and J. Kirkham, *J. Tissue Eng. Regener. Med.*, 2013, **7**, 461–469.
40. F. Schulze, A. Dienelt, S. Geissler, P. Zaslansky, J. Schoon, K. Henzler, P. Guttmann, A. Gramoun, L. A. Crowe, L. Maurizi, J. P. Vallee, H. Hofmann, G. N. Duda and A. Ode, *Small*, 2014, **10**, 4340–4351.

41. H. Markides, O. Kehoe, R. H. Morris and A. J. El Haj, *Stem Cell Res. Ther.*, 2013, **4**, 126.
42. J. H. Lee, M. J. Jung, Y. H. Hwang, Y. J. Lee, S. Lee, D. Y. Lee and H. Shin, *Biomaterials*, 2012, **33**, 4861–4871.
43. C. Bos, Y. Delmas, A. Desmouliere, A. Solanilla, O. Hauger, C. Grosset, I. Dubus, Z. Ivanovic, J. Rosenbaum, P. Charbord, C. Combe, J. W. Bulte, C. T. Moonen, J. Ripoche and N. Grenier, *Radiology*, 2004, **233**, 781–789.
44. R. Moghadasali, M. Azarnia, M. Hajinasrollah, H. Arghani, S. M. Nassiri, M. Molazem, A. Vosough, S. Mohitmafi, M. Najarasl, Z. Ajdari, R. S. Yazdi, M. Bagheri, H. Ghanaati, B. Rafiei, Y. Gheisari, H. Baharvand and N. Aghdami, *Cytotherapy*, 2014, **16**, 734–749.
45. C. Wu, J. Li, P. Pang, J. Liu, K. Zhu, D. Li, D. Cheng, J. Chen, X. Shuai and H. Shan, *Biomaterials*, 2014, **35**, 8249–8260.
46. Y. Zhong, Z. Tang, R. Xu, N. Lin, M. Deng, H. Fang, J. Lin, K. Zhu, Y. Liu and Z. Kang, *Cytotherapy*, 2013, **15**, 1266–1274.
47. H. Wang, T. B. Shrestha, M. T. Basel, R. K. Dani, G. M. Seo, S. Balivada, M. M. Pyle, H. Prock, O. B. Koper, P. S. Thapa, D. Moore, P. Li, V. Chikan, D. L. Troyer and S. H. Bossmann, *Beilstein J. Nanotechnol.*, 2012, **3**, 444–455.
48. J. Liu, L. Wang, J. Cao, Y. Huang, Y. Lin, X. Wu, Z. Wang, F. Zhang, X. Xu and G. Liu, *Nanoscale*, 2014, **6**, 9025–9033.
49. K. Andreas, R. Georgieva, M. Ladwig, S. Mueller, M. Notter, M. Sittinger and J. Ringe, *Biomaterials*, 2012, **33**, 4515–4525.
50. J. Qin, K. Li, C. Peng, X. Li, J. Lin, K. Ye, X. Yang, Q. Xie, Z. Shen, Y. Jin, M. Jiang, G. Zhang and X. Lu, *Biomaterials*, 2013, **34**, 4914–4925.
51. Y. Qi, G. Feng, Z. Huang and W. Yan, *Mol. Biol. Rep.*, 2013, **40**, 2733–2740.
52. Y. K. Chang, Y. P. Liu, J. H. Ho, S. C. Hsu and O. K. Lee, *J. Orthop. Res.*, 2012, **30**, 1499–1506.
53. J. Salamon, D. Wicklein, M. Didie, C. Lange, U. Schumacher, G. Adam and K. Peldschus, *Rofo*, 2014, **186**, 367–376.
54. C. Y. Yang, J. K. Hsiao, M. F. Tai, S. T. Chen, H. Y. Cheng, J. L. Wang and H. M. Liu, *Mol. Imaging Biol.*, 2011, **13**, 443–451.
55. G. Muhammad, A. Jablonska, L. Rose, P. Walczak and M. Janowski, *Acta Neurobiol. Exp. (Wars)*, 2015, **75**, 144–159.
56. A. Scharf, S. P. Holmes, M. Thoresen, J. Mumaw, A. Stumpf and J. Peroni, *Stem Cells Int.*, 2016, **2016**, 8610964.
57. J. M. Rojas, L. Sanz-Ortega, V. Mulens-Arias, L. Gutierrez, S. Perez-Yague and D. F. Barber, *Nanomedicine*, 2016, **12**, 1127–1138.
58. H. I. Tong, W. Kang, Y. Shi, G. Zhou and Y. Lu, *Int. J. Pharm.*, 2016, **505**, 271–282.
59. J. Matuszak, P. Dorfler, J. Zaloga, H. Unterweger, S. Lyer, B. Dietel, C. Alexiou and I. Cicha, *Clin. Hemorheol. Microcirc.*, 2015, **61**, 259–277.
60. H. Jin, Y. Qian, Y. Dai, S. Qiao, C. Huang, L. Lu, Q. Luo, J. Chen and Z. Zhang, *Theranostics*, 2016, **6**, 2000–2014.

61. J. Ruan, J. Ji, H. Song, Q. Qian, K. Wang, C. Wang and D. Cui, *Nanoscale Res. Lett.*, 2012, **7**, 309.
62. Y. Ikehara, T. Niwa, L. Biao, S. K. Ikehara, N. Ohashi, T. Kobayashi, Y. Shimizu, N. Kojima and H. Nakanishi, *Cancer Res.*, 2006, **66**, 8740–8748.
63. C. Li, J. Ruan, M. Yang, F. Pan, G. Gao, S. Qu, Y. L. Shen, Y. J. Dang, K. Wang, W. L. Jin and D. X. Cui, *Cancer Biol. Med.*, 2015, **12**, 163–174.
64. S. Toraya-Brown, *Integr. Biol.*, 2013, **5**, 159–171.

Subject Index